# ANIMAL CELL TECHNOLOGY: CHALLENGES FOR THE 21ST CENTURY

JAACT/ESACT'98 Organizing Committee

| | | |
|---|---|---|
| Meeting Chairman: | R. Sasaki | (Japan) |
| Meeting Secretaries: | K. Ikura | (Japan) |
| | M. Nagao | (Japan) |
| | S. Masuda | (Japan) |

Organizing Boards:

| | | |
|---|---|---|
| T. Amachi (Japan) | A. Ametani (Japan) | K. Aramaki (Japan) |
| M. Carrondo (Portugal) | S. Dosako (Japan) | A. Enomoto (Japan) |
| T. Fujimori (Japan) | Y. Fuke (Japan) | T. Furudera (Japan) |
| T. Fushiki (Japan) | B. Griffiths (UK) | S. Hashizume (Japan) |
| H. Hoshi (Japan) | S. Iijima (Japan) | K. Ikura (Japan) |
| K. Inouye (Japan) | K. Issiki (Japan) | Y. Ikeda (Japan) |
| Y. Ito (Japan) | S. Kaminogawa (Japan) | T. Kawada (Japan) |
| K. Kawamura (Japan) | Y. Kitagawa (Japan) | J. Lehmann (Germany) |
| C. MacDonald (UK) | S. Masuda (Japan) | T. Matsuda (Japan) |
| R. Matsuno (Japan) | O. Merten (France) | K. Nagai (Japan) |
| K. Nagaike (Japan) | M. Nagao (Japan) | H. Ohigashi (Japan) |
| R. Sasaki (Japan) | T. Sato (Japan) | H. Sawada (Japan) |
| K. Shin (Japan) | K. Shinohara (Japan) | S. Shirahata (Japan) |
| Y. Shirai (Japan) | T. Suzuki (Japan) | E. Suzuki (Japan) |
| M. Tanokura (Japan) | K. Ueda (Japan) | J. Wérenne (Belgium) |
| K. Yagasaki (Japan) | M. Yoshikawa (Japan) | |

# Animal Cell Technology: Challenges for the 21st Century

Proceedings of the joint international meeting of the Japanese Association for Animal Cell Technology (JAACT) and the European Society for Animal Cell Technology (ESACT) 1998, Kyoto, Japan

*edited by*

Kouji Ikura

Masaya Nagao

Seiji Masuda

and

Ryuzo Sasaki

KLUWER ACADEMIC PUBLISHERS
DORDRECHT / BOSTON / LONDON

A C.I.P. Catalogue record for this book is available from the Library of Congress.

ISBN 0-7923-5805-8

---

Published by Kluwer Academic Publishers,
P.O. Box 17, 3300 AH Dordrecht, The Netherlands.

Sold and distributed in North, Central and South America
by Kluwer Academic Publishers,
101 Philip Drive, Norwell, MA 02061, U.S.A.

In all other countries, sold and distributed
by Kluwer Academic Publishers,
P.O. Box 322, 3300 AH Dordrecht, The Netherlands.

*Printed on acid-free paper*



Printed in the Netherlands.

# Contents

V. Glycoengineering



VI. Immunologicals, Monoclonal Antibodies and Vaccines



VII. Transplantation and Artificial Organs



VIII. Gene Therapy

IX. Transgenic Animals



X. Safety and Regulations



XI. Cell Regulatory Factors and Signal Transduction

XII. Functional Substances in Food and Natural Sources



XIII. Animal Cells for *in vitro* Assay

## Preface

The second joint meeting of the Japanese Association for Animal Cell Technology (JAACT) and the European Society for Animal Cell Technology (ESACT) was held in Kyoto, Japan, July 26-30, 1998. This meeting focused on important and exciting recent developments in the field of animal cell technology from basic and applied aspects.

The first joint meeting was held at Veldhoven, The Netherlands, in 1994. Thanks to ESACT, the first joint meeting was quite successful and participants from all over the world had the most exciting and enjoyable discussions. The Proceedings of this first meeting were published (Animal Cell Technology edited by E. C. Beuvery, J. B. Griffiths and W. P. Zeijlemaker, 1995, Kluwer Academic Publishers). As we expected, the 1998 second joint meeting was even more successful than the first meeting, because of a number of revolutionary new discoveries in life sciences were made and many of these were presented in the meeting.

This volume of Proceedings contains the reflection of this vast amount of new information. There is not doubt that animal cell technology plays far more important roles in the improvement of the quality of life in the 21st century and therefore this volume should greatly contribute to find the important, urgent and rewarding targets to be challenged.

The Kyoto joint meeting was divided into thirteen sessions:

I. Asian Biotechnology in Future
II. Cell Culture and Engineering
III. Production of Biologicals
IV. Functional Cell Lines
V. Glycoengineering
VI. Immunologicals, Monoclonal Antibodies and Vaccines
VII. Transplantation and Artificial Organs
VIII. Gene Therapy
IX. Transgenic Animals
X. Safety and Regulations
XI. Cell Regulatory Factors and Signal Transduction
XII. Functional Substances in Food and Natural Sources
XIII. Animal Cells for *in vitro* Assay

All presentation (oral and poster) have been brought together according to these themes.

Finally we dedicate these Proceedings to the late Professor Hiroki Murakami. He founded JAACT, acted the first president of JAACT and promoted the joint meeting with great enthusiasm.

The Editors

# THE ROLE OF HONG KONG SAR IN THE DEVELOPMENT OF BIOTECHNOLOGY INDUSTRY IN CHINA

Albert Yen Chang

**Hong Kong Institute of Biotechnology, Ltd., CUHK, Shatin, N.T. Hong Kong SAR**

The biotechnology industry is one of the fastest growing industries in China and, with a market size of 1.2 billion consumers, the potential for its growth is enormous. With a vast network of state-supported academic and research institutions and a large number of scientists trained in topnotch universities worldwide, China also has a superior R&D infrastructure to support the discovery of biotechnology products. Nevertheless, for China's biotechnology industry to be globally competitive, it needs to upgrade its product development processes for the purpose of enabling its products to obtain registration approvals worldwide and be more cost-effective. Specifically, issues such as quality assurance, compliance to regulatory requirements, cost of goods and effective management must be addressed and resolved. In this regard, the Hong Kong Special Administrative Region (HKSAR) is poised to play a significant role in the development of biotechnology industry in China.

An effective work force is mandatory for any industrial sector to gain a competitive edge in the global marketplace. HKSAR has much to offer in this respect, including 7 tertiary institutions, 2 technical colleges, 7 technical institutes and 24 industry training centres, all publicly funded, to educate as well as to offer on-the-job training for its 6.3 million residents to produce a bilingual work force with a global mindset. On 1965, the government established the University Grants Committee (UGC) for the development and funding of higher education and to administer public grants to these schools. A few years later, the government set up a Research Grant Council within the UGC to advise it and monitor the use of public research grants in these institutions. In 1997-98, the RGC disbursed HK$423 million in grants for academic research, a portion of which supported upstream biotechnology projects[1].

In addition to the 7 tertiary institutions that carry out upstream research, HKSAR also has a downstream development laboratory, Hong Kong Institute of Biotechnology, Ltd. (HKIB), which was established in 1989 with a one-time donation of HK$170 million from the Hong Kong Jockey Club Charities Trust with a mission to provide the catalyst and essential infrastructure for the emergence of a successful biotechnology industry in Hong Kong. Currently, HKIB has the following programmes.

### *Human Therapeutics*

At the present, human therapeutics remain the most profitable products in the global biotechnology industry. In 1994, HKIB established a multi-purpose Bioprocessing Unit to develop manufacturing processes of biopharmaceuticals and it was upgraded to a GMP Manufacturing Technology Centre (GMP-MTC). GMP-MTC is fully equipped to develop and produce clinical-grade biomaterials under Good Manufacturing Practice (GMP) guidelines for Phase I and II clinical trials of innovative products in human subjects.

HKIB's GMP-MTC has a molecular biology laboratory to carry out its own bioengineering projects as well as to provide contract gene cloning services to global clients. The Centre's process development laboratory has the capability to develop manufacturing processes using microbial, plant, insect and mammalian cell cultures. Its pilot plant has a fermentation and bioprocessing capacity of 100 to 350 litres. The Centre's Class 10,000 clean room accommodates its purification suite to process clinical grade protein or DNA products under GMP conditions. Its Quality Control (QC) Laboratories and Quality Assurance (QA) Office ensure that its own manufacturing operations meet the World Health Organization's (WHO) and Food and Drug Administration's (FDA) GMP guidelines.

Since 1994, HKIB has actively searched for worthwhile projects to utilize its pilot plant. Malaria remains one of the world's major health problems and up to 500 million people may be infected with the parasite, resulting in approximately two million deaths annually. There is an urgent need to develop an effective vaccine to prevent the spread of malaria. Therefore, in 1996, HKIB applied for and was subsequently awarded, via a competitive process, an R&D Partnership Grant in malaria vaccine development by the UNDP/World Bank/WHO Special Programme for Research Training in Tropical Diseases. The R&D Partnership Grant was used to train research personnel in HKIB's GMP-MTC to scale up production of a transmission-blocking malaria vaccine candidate developed by Dr. David Kaslow, head of the Malaria Vaccine Section of the National Institutes of Allergy and Infectious Diseases (NIAID) at the National Institute of Health (NIH) in the U.S. The vaccine candidate is based on Pfs25, the predominant surface protein of Plasmodium falciparum zygotes. The antigen TBV25H (Transmission-Blocking Vaccine based on Pfs25 with a Histidine tag) is expressed in yeast and its purification process uses a nickel-NTA agarose column.[2] HKIB signed a Cooperative Research and Development Agreement (CRADA) with NIAID, NIH in January 1998.

HKIB filed a Drug Master File (Type I) on its Phase I microbial fermentation facility with the FDA in May 1998. The Facility was also inspected by the Therapeutic Goods Administration (TGA) of the Australian Government and a GMP Certification on the Facility was issued by the TGA in October 1998. NIAID and HKIB will jointly file Investigational New Drug (IND) applications with the FDA and HKSAR's Department of Health to carry out Phase I clinical trials of the TBV25H vaccine in Hong Kong and the U.S. simultaneously. HKIB will manufacture the bulk clinical trial materials for the Phase I human studies.

HKIB also plans to use its GMP facility to develop vaccines and biopharmaceuticals for other life-threatening diseases such as AIDS, schistosomiasis, Dengue fever, etc. A collaboration between HKIB and the Institute for International Vaccine Development in the U.S. to develop HIV vaccines is currently under discussion. HKIB is also actively seeking opportunities to offer its facilities and services in the development of human therapeutics to global pharamaceutical companies in the expansion of their businesses in China and Southeast Asia.

***Plant Micropropagation Programme***

As agriculture remains one of the top-priority industries in China, it is imperative for the HKSAR to develop and maintain a judicious amount of agriculture R&D, although the HKSAR *per se* lacks sufficient arable land to sustain an agriculture industry. Therefore, HKIB established a plant micropropagation programme with the following objectives:

- To develop mass propagation protocols of high value-added ornamental and medicinal plants.

- To develop plant cell culture technologies in a bioreactor culture system on a two-to-20-litre scale.
- To investigate morphogenesis events involved in the development of propagation protocols in order to better understand micropropagation processes and to optimise conditions.

HKIB has successfully produced cell culture-based micropropagation protocols for the following ornamental herbaceous plants: *Ananas*, *Anoectochilus formosanus*, *Cordyline terminalis*, *Drosera burmanii*, *Gladiolus spp.*, *Jovibarba*, *Kalanchoe blossfeldiana*, *Lilium spp.*, *Oxalis triangularis*, *Platycerium bifurcatum*, *Saintpaulia ionantha*, *Scindapsus aureus* and *Spathoglottis plicata*. In collaboration with a group of local businessmen, desk-top ornaments containing these live herbaceous plants are being produced for the mass consumer market. Under the persuasion of its business partners, HKIB has also embarked on the development of the micropropagation processes of the following woody plants: *Bauhinia purpurea*, *Armeniaca*, *Cerasus spp.*, *Camellia* and *Coffea arabica* for the same purpose. In addition, HKIB will transfer the developed protocols and processes to companies that are interested in large-scale production of these plants in the open field.

HKIB is also developing manufacturing processes for ginseng cells in liquid cultures. A fast-growing strain has been isolated and successfully cultured in two-litre bioreactors. Its ginsenosides content has also been verified. The process will be transferred to food-and-beverage companies to provide the active ingredients in ginseng-based drinks and foods.

***The Incubation Programme***

The development of biotechnology products is time-consuming and capital-intensive. For HKSAR to develop a biotechnology industry, the region must provide low-cost laboratory facilities for entrepreneurs to develop their technologies into products. It is for this purpose that HKIB furnished a 16,000-square-foot, second-floor space in its five-story R&D building into an Incubator Facility. The Facility provides furnished laboratories, offices, shared common-use equipment, technical and management services, for companies to start their R&D activities in Hong Kong SAR. Such facilities and services enable companies to minimise up-front investment during the initial product- or service-development stages. Companies can rent space at a flat monthly charge or sign up for a back-loading scheme that allows HKIB to share profits with the entrepreneurs. There are eight labs, each of which can accommodate four to six persons. Presently, the Facility has the following local start-up biotechnology companies:

- HKIB/Techpool Research Laboratory Ltd.

  This is a joint venture company between HKIB and Hong Kong Techpool Ltd. with the business goal of adding value to human urine-derived biopharmaceuticals manufactured by Techpool in Guangdong Province.

- Leadergene Ltd.

  The Company was founded to utilize a unique expression vector in bacteria that facilitates the secretion of cloned gene products. Its initial products focus on generic biopharmaceuticals.

- Cell Therapy Technologies Centre Ltd.

  The Company focuses on utilizing cell therapy technologies such as autologous cord blood banking, etc., in medical applications.

*Consultation Services*

On August 30, 1995, the Pharmacy and Poisons Board of the Hong Kong Government Department of Health issued a schedule for implementing GMP guidelines in Hong Kong's pharmaceutical industry. The implementation schedule was divided into three phases. The target completion dates for Phase 1, 2 and 3 were December 31, 1996, 1998 and 2000, respectively. HKIB has a unit to provide consultation on GMP implementation to the local pharmaceutical industry. It assisted two local companies in implementing Phase 1 GMP guidelines for its manufacturing operation in 1996. Another local company which had been selling its products in Australia sought help from the Unit in responding to deficiencies cited by the Australian Government's regulatory agency, TGA, after an inspection in 1996. The Unit successfully provided consultation services for the Company to upgrade its manufacturing operation to meet the requirements. Furthermore, the Unit has supplied a wide range of consultation services to a number of local companies in their attempts to meet Phase 1 GMP requirements.

The success of bringing a biotechnology innovation to the marketplace also requires effective and efficient management of the R&D process which is extremely complex for biotechnology products. HKSAR has an abundance of management talents with global experience. Being one of the major financial centres in the world, HKSAR is also well-positioned to offer financial services to China's biotechnology industry in search for venture capitals.

In conclusion, Hong Kong SAR's future role in the development of China's biotechnology and pharmaceutical industries will focus on product development, i.e. manufacturing processes, formulation, supplying clinical trial materials under GMP standards, clinical trial monitoring and data management under GCP, and product marketing. With its strengths in technology supply, human resources, Hong Kong SAR is poised to play a significant role in the development of China's biotechnology and pharmaceutical industries.

1. Hong Kong - *A New Era*. Published by Hong Kong Government Information Services Department, 1998.

2. Kaslow, D.C. and Shilock J. "Production, Purification and Immunogenicity of a Malaria Transmission-blocking Vaccine Candidate:TBV25H Expressed in Yeast and Purified Using Nickel-NTA Agarose." Bio/Technology Vol. 12, 494-499 (1994).

***Keywords***: Hong Kong Institute of Biotechnology, Ltd.; human therapeutics; vaccines; GMP pilot plant; FDA; TGA; plant micropropagation, incubation programme, consultation services, Hong Kong Science Park.

# INCREASE IN INTRACELLULAR CALCIUM INFLUX ENHANCES MONOCLONAL ANTIBODY PRODUCTIVITY

Sun Ho Park, Jae Sung Park, and Min Hee Nam[1]
*School of Chemical Engineering and Materials Engineering,*
*Keimyung University, Taegu 704-701, Korea*
*[1]National Yeongnam Agricultural Experiment Station,*
*RDA, Milyang 627-130, Korea*
*Telephone: 82-53-5805457; Fax: 82-53-633-4929*
*E.mail)park@kmucc.keimyung.ac.kr*

## 1. Introduction

The endoplasmic reticulumn (ER) is the site where not only the secretory and lysosomal proteins, resident luminal ER proteins, Golgi and lysosomal membrane proteins are synthesized, but most extensive folding and oligomeric assembly take place (Gething and Sambrook, 1992; Hurtley and Helenius, 1989). The ER is also a major intracellular reservoir of $Ca^{2+}$ and serves as a major intracellular site of $Ca^{2+}$ sequestration and is thought to function prominently in maintenance of $Ca^{2+}$ homeostasis. The sequestered $Ca^{2+}$ may function to maintain various proteins processing events within the ER. Although the exact concentration of $Ca^{2+}$ in the ER is not known with certainty, measurements using electron probe X-ray microanalysis indicates that in different cells it can be between 1.8 and 5.4 mM. Meanwhile, the concentration of free(unbound) $Ca^{2+}$ can be two or three orders of magnitude greater than the 1 $\mu$M or lower concentration in the cytosol (Galina *et al.*, 1993).

$Ca^{2+}$ has been demonstrated to be required for optimal processing of glycoproteins by the ER. $Ca^{2+}$ stored by the organelle is subject to rapid release by hormones that generate inositol trisphosphate, by unsaturated fatty acids by $Ca^{2+}$ ionophore such as ionomycin (Albert and Tashjian, 1984) and by agents such as thapsigargin (Thastup, 1990) that blocks active transport of $Ca^{2+}$ into the ER. But $Ca^{2+}$ can be rapidly stored by ryanodine (Pfeffer and Rothman, 1987) that enhances on $Ca^{2+}$ of ER.

In this work, hybridoma cells (5F12AD3) were treated with A23187, ryanodine, and thapsigargin, respectively. Intracellular $Ca^{2+}$ levels of the treated cells were qualitatively observed by using confocal microscope and the changes in hybridoma cell growth and monoclonal antibody production were examined.

## 2. Experiments

### 1. Measurement of intracellular $Ca^{2+}$

The hybridoma cell line, 5F12AD3 was used. Intracellular $Ca^{2+}$ concentrations of hybridoma cells were measured by using confocal microscope (ZEISS) at 488nm(*Williams et al.*, 1990). Intracellular $Ca^{2+}$ indicator used was Fura-2/AM (sigma chemical company, st. Louis, U.S.A). The $5x10^5$ cells were attached at 37℃ on surface of slide on petri dish containing 1mL medium. After 2hr incubation, the medium was discarded and the fresh medium of 10-12mL was fed on petri dish. After 2 day incubation, the medium was discarded and the cells on slide was treated with 7$\mu$M A23187, 10$\mu$M ryanodine, and 0.5$\mu$M thapsigargin. After treatment, cells were washed with medium and equilibrated for 1 hr. Then the cells were washed twice with S-CES buffer and loading 2$\mu$M Fura-2/AM with S-CES containing 1mM $CaCl_2$ and incubated for 30 minutes at 37℃. The cells were rewashed with S-CES buffer. Intracellular $Ca^{2+}$ of cells were measured by using confocal microscope.

### 2. Measurements of cell growth and MAb concentration

Hybridoma cells (5F12AD3) were grown at 10mL working volume in 25 $cm^2$ tissue culture flask (T-flask). Cells were cultured at IMDM containing 1.49mM $CaCl_2$. After cells were cultured for 3 days, the cells were treated at various concentrations of A23187, ryanodine, and thapsigargin at about $1x10^6$ cells/mL and incubated at 37℃ for 30 minutes. The treated cells were recultivated in spent IMDM. The cells were inoculated in IMDM containing 10%(v/v) horse serum at cell density of $5x10^4$ cells/mL. Total and viable cells were counted with a haemocytometer using the tryphan blue method.

Quantitative MAb concentrations were measured by using ELISA methods at 405nm and the specific MAb production rate can be estimated for each sampling time $\Delta t(t^{-1}, t)$ as follows:

$$q_p = (\frac{\Delta P}{\Delta t})\frac{1}{\bar{x}_v}$$

where $\bar{x}_v$ is the average value of the two concentration sample.

P is the MAb concentration at a given time.

## 3. Results and discussion

Figure 1 shows the intracellular $Ca^{2+}$ of hybridoma cells treated with A23187, ryanodine and thapsigargin. $Ca^{2+}$ ionophore, A23187, treatment caused $Ca^{2+}$ to flood into the cytoplasm both from the extracellular medium and from the ER. A specific inhibitor of the ER $Ca^{2+}$ ATPase(SERCA pump), thapsigargin, rapidly caused depletion of ER stores of calcium. Meanwhile, $Ca^{2+}$ release channel blocker, ryanodine, caused increase of $Ca^{2+}$ influx to ER. It should be noted that the $Ca^{2+}$ was more localized within the ER in case of 10 $\mu$M ryanodine treatment(Fig. 1 C).

Figure 2 shows the typical profiles of hybridoma cell growth and MAb production when the cells were treated with $Ca^{2+}$-mobilizing agents. The treatments, in all cases, caused inhibition of cell growth even though the 5 $\mu$M ryanodine treatment enhanced maximum cell concentration up to $2.6 \times 10^6$ cells/mL(data not shown). The ryanodine treated cells also showed maximum MAb concentration compared to control and other treatments, indicating that the increase in $Ca^{2+}$ influx to ER can enhance the hybridoma cell growth and MAb production. Interestingly, as shown in Figure 3, A23187 and ryanodine treatments caused remarkable increases in specific MAb productivities. This is mainly due to the increase in intracellular calcium concentration as shown in Figure 1. Since A23187 depletes $Ca^{2+}$ from the ER, it is possible that unfolded or unassembled heavy and light chains of immunoglobulin could be rapidly secreted. In contrast, a rise in ER $Ca^{2+}$ by ryanodine treatment seems to be very effective for the proper folding and assembly of secretory proteins, and also slight increase in cytosolic $Ca^{2+}$ would be helpful for stimulating the exocytosic by enhancing the fusing of regulated secretory vesicles within the plasma membrane(Lodish and Kong, 1990).

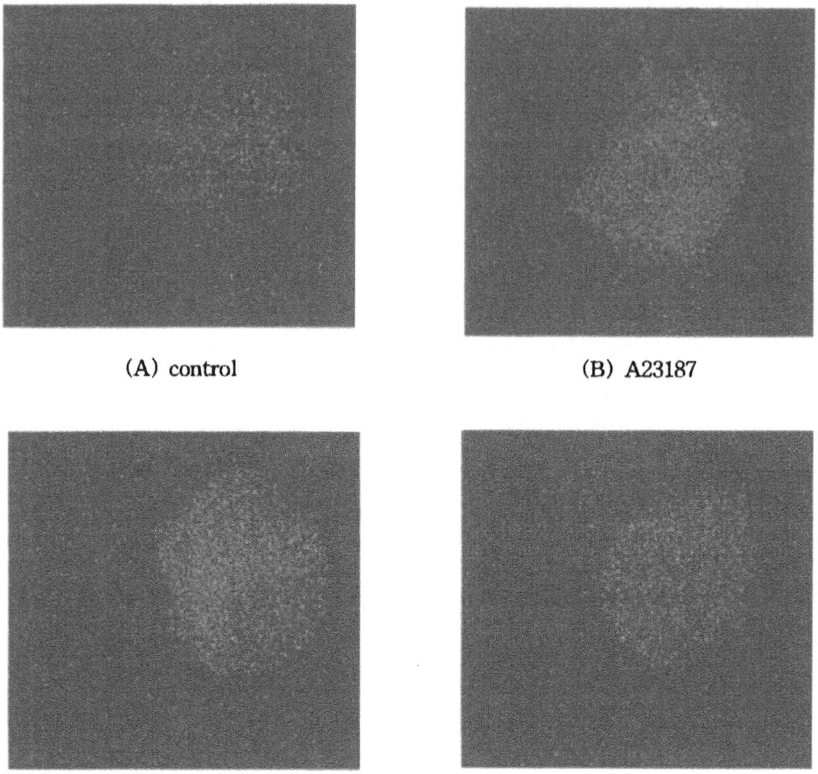

(A) control (B) A23187

(C) Ryanodine (D) Thapsigargin

*Figure 1.* Confocal microscopic images of Fura-2/AM loaded cells. (A) control. Treatment with (B) 7 $\mu$M A23187, (C) 10 $\mu$M ryanodine, and (D) 0.5 $\mu$M thapsigargin.

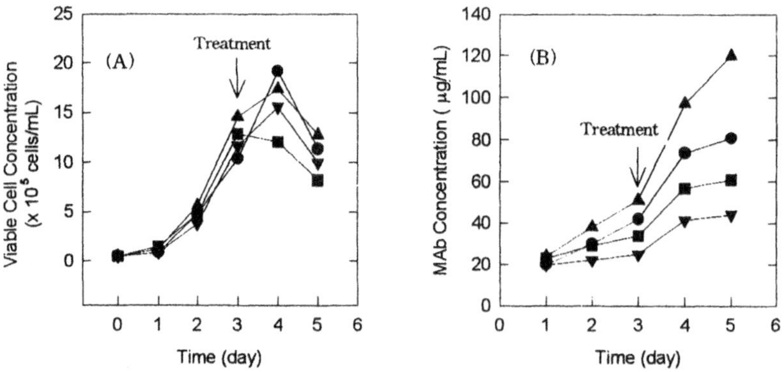


*Figure 2.* Effect of intracellular on cell growth(A) and MAb concentration(B); (●) control, (■) 7 $\mu$M A23187, (▲) 10 $\mu$M ryanodine, (▼) 0.5 $\mu$M thapsigargin.

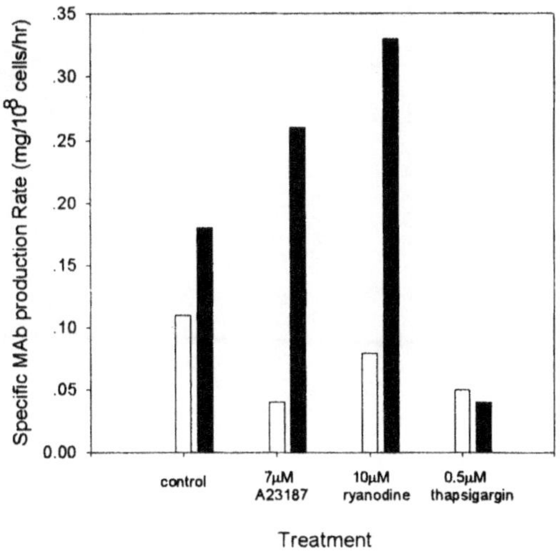


*Figure 3.* Enhancement of specific MAb production rate by intracellular $Ca^{2+}$ manipulation; (□) just before treatment, (■) 1day after treatment.

## 4. Acknowledgements

This research was supported by a grant from the Korea Science and Engineering Foundation(951-1104-002-2).

## 5. References

Albert, P.R. and Tashjian, A. H. Jr. (1984), Relationship of thyrotropin-releasing hormone-induced spike and plateau phases in cytosolic free $Ca^{2+}$ concentrations to hormone secretion. Selective blockade using ionomycine and nifedipine, *J. Biol. Chem.*, **259**, 15350-15363.

Gething, M. and Sambrook, J. (1992), Protein folding in the cell, *Nature* **355**, 33-45.

Galina, K., Margaret, A. B. and Chares, O. B. (1993), Role of endoplasmic reticular calcium in oligosaccharide processing of α1-Antitrypsin, *J. Bio. Chem.* **268(3)**, 2001-2008.

Hurtley, S. M. and Helenius, A. (1989), Protein oligomerization in the ER, *Ann. Res. Cell Bio.*. **5**, 277-307.

Lodish, H. F. and Kong, N. (1990), Perturbation of cellular calcium blocks exit of secretory proteins from the endoplasmic reticulum, *J. Biol. Chem.*, **265(19)**, 10893-10899.

Pfeffer, S. R. and Rothman, J. E. (1987), Biosynthetic protein transport and sorting by the endoplasmic and golgi, *Ann. Rev. Biochem.*, **56**, 829-852.

Thastup, O. (1990), Role of $Ca^{2+}$-ATPases in requlation of cellular $Ca^{2+}$ signaling, as studied with the selective microsomal $Ca^{2+}$-ATPase inhibitor, thapsigargin, *Agents Actions*, **29**, 8-15.

# ENERGY METABOLISM AND ANTIBODY PRODUCTION BASED ON FLUX ANALYSIS IN HYBRIDOMA

**Takeshi Omasa, Tomoya Iemura, Keisuke Furuichi, Yoshio Katakura, Michimasa Kishimoto and Ken-ichi Suga**
***Department of Biotechnology,***
***Graduate School of Engineering, Osaka University***
***2-1 Yamada-oka, Suita, 565-0871 JAPAN***
**Telephone: +81-6-6879-7437, Facsimile: +81-6-6879-7439**
**e-mail: omasa@bce.bio.eng.osaka-u.ac.jp**

**ABSTRACT**: The specific antibody production rate is closely related to the specific ATP production rate. In order to increase the specific antibody production rate by increasing the specific ATP production rate, the effects of oxygen concentration reinforcement (4.7, 6.9 and 10.3 $mg\ell^{-1}$) or metabolic intermediate addition (pyruvate, malate and citrate) on ATP production were investigated under glutamine-limited continuous cultivation with a constant specific growth rate. As a result, the specific ATP production rate decreased under high oxygen concentration cultivation. However, the metabolic fluxes related to ATP production were increased and higher specific ATP and antibody production rates were achieved by addition of metabolic intermediates.

## 1. Introduction

Generally, mammalian cells utilize glucose and glutamine as primary energy sources. The ATP produced is used for cell growth and antibody production. In a previous study, we found that an increase in the specific ATP production rate caused specific antibody production rates to increase [7]. Antibody production was enhanced by increasing the fluxes of the metabolic pathway related to ATP production. In this study, we try to increase the specific ATP production rate by increasing the concentration of dissolved oxygen or by addition of intermediates to the medium.

## 2. Materials and Methods

The cell line employed in the experiments was the mouse-mouse hybridoma 3A21, which produces an anti-RNase A monoclonal antibody (IgG) [1]. The serum-free medium RDF-ITES with bovine serum albumin (BSA) [2-5] was used for cultivation. In the glutamine-limited continuous cultivation, the feed glutamine concentration was 0.16 $g\ell^{-1}$. The cell, glucose, lactate, glutamine, ammonia, and antibody concentrations were analyzed as described previously [2-5]. The specific oxygen consumption rate was also calculated as described previously [3]. The cell volume was measured by a Coulter counter (Coulter model ZM). The average dry cell weight was estimated from the average cell volume, because the cell density was constant under various specific growth rates. Amino acid concentrations were analyzed using an amino acid analysis system (Waters, Pico Tag system). The kinetic parameters (specific growth rate; glucose, glutamine and amino acid consumption rates; and specific antibody, lactate, and ammonia production rates) were calculated according to the procedure previously outlined [2-5] The metabolic fluxes were calculated based on the pathway proposed by Zupke and Stephanopoulos. [6,7].

## 3. Results and Discussion

### 3.1 EFFECT OF DISSOLVED OXYGEN CONCENTRATION ON THE SPECIFIC ATP AND ANTIBODY PRODUCTION RATES

Yano and Nishizawa reported that at higher dissolved oxygen concentrations than the air-saturated value of 6.86 $mg\ell^{-1}$, the specific antibody production rate increased [8]. We carried out continuous cultivations at dissolved oxygen concentrations of 4.7, 6.9, and 10.3 $mg\ell^{-1}$. At concentrations greater than 10.3 $mg\ell^{-1}$, we could not obtain steady-state conditions.

The flux pathways and analysis of fluxes are shown in Table 1 and Figure 1, respectively. The values of $r_1$ to $r_4$, which are related to glycolysis, for the 10.3 $mg\ell^{-1}$ cultivation were greater than those for the 4.7 and 6.9 $mg\ell^{-1}$ cultivations. However, the fluxes related to the TCA cycle were lower in the 10.3 $mg\ell^{-1}$ cultivation than in the 4.7 and 6.9 $mg\ell^{-1}$ cultivations, as was the specific ATP production rate, flux $r_{15}$. We consider that this decrease in ATP production and increase in glycolysis, were due to decreased ATP production in the TCA cycle. Consequently, the specific antibody production rate also decreased with the decrease in the specific ATP production rate (Figure 2). Therefore, the increase in DO concentration was not effective for increase of ATP production.

Table. 1 Metabolic pathways for calculation of the metabolic fluxes. $r_{15}$ represents the flux of ATP consumption for cell growth and antibody production.

| |
|---|
| $r_1: Gluc + 2ATP \rightarrow 2GTP + 2ADP$ |
| $r_2: GAP + ADP + NAD \rightarrow PEP + ATP + NADH$ |
| $r_3: PEP + ADP \rightarrow Pyr + ATP$ |
| $r_4: Pyr + NADH \rightarrow Lac + NAD$ |
| $r_5: OAA + Pyr + NAD \rightarrow Cit + NADH + CO_2$ |
| $r_6: Cit + NAD \rightarrow \alpha KG + NADH + CO_2$ |
| $r_7: Cit + ATP \rightarrow OAA + ADP$ |
| $r_8: \begin{cases} \alpha KG + FAD + NAD + GDP \rightarrow Mal + FADH_2 + NADH + CO_2 + GTP \\ GTP + ADP \rightarrow ATP + GDP \end{cases}$ |
| $r_9: Mal + NAD \rightarrow OAA + NADH + 2CO_2$ |
| $r_{10}: Pyr + NADPH + CO_2 \rightarrow Mal + NADP$ |
| $r_{11}: OAA + ATP \rightarrow PEP + ADP + CO_2$ |
| $r_{12}: Gln \rightarrow Glu + NH_3$ |
| $r_{13}: Glu + Pyr \rightarrow \alpha KG + Ala$ |
| $r_{14}: \begin{cases} 2NADH + 6ADP + O_2 \rightarrow 2NAD + 6ATP \\ FADH_2 + 2ADP + \frac{1}{2}O_2 \rightarrow FAD + 2ATP + H_2O \end{cases}$ |
| $r_{15}: ATP \rightarrow ADP$ |

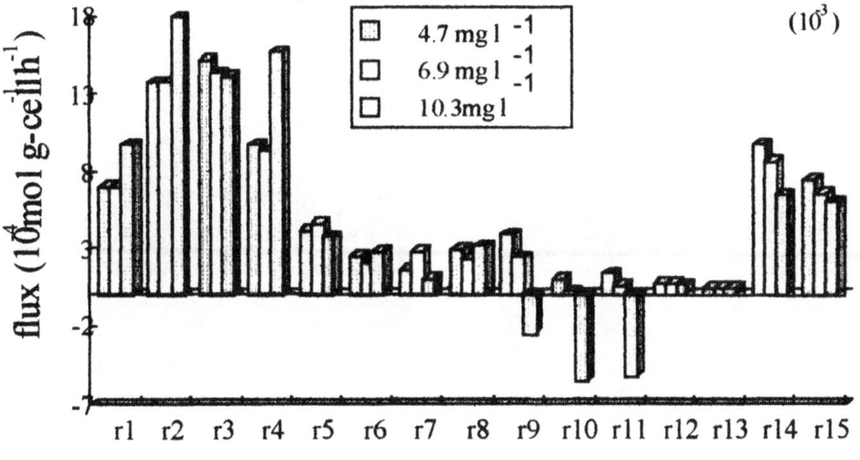


*Figure 1* Effect of DO concentration on metabolic fluxes

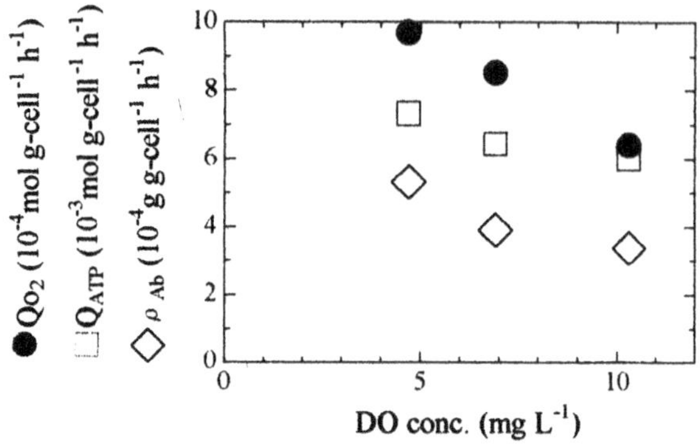


*Figure 2* Effect of DO concentration on specific oxygen consumption ($Q_{O2}$), ATP ($Q_{ATP}$) and antibody production ($\rho_{Ab}$) rates

## 3.2 EFFECT OF ADDITION OF INTERMEDIATES ON THE SPECIFIC ATP AND ANTIBODY PRODUCTION RATE

In order to increase the specific ATP production rate, the effects of metabolic intermediate addition (pyruvate, malate and citrate) on ATP production were investigated under glutamine-limited continuous cultivations with constant specific growth rate ($\mu$). The results are shown in Figure 3.

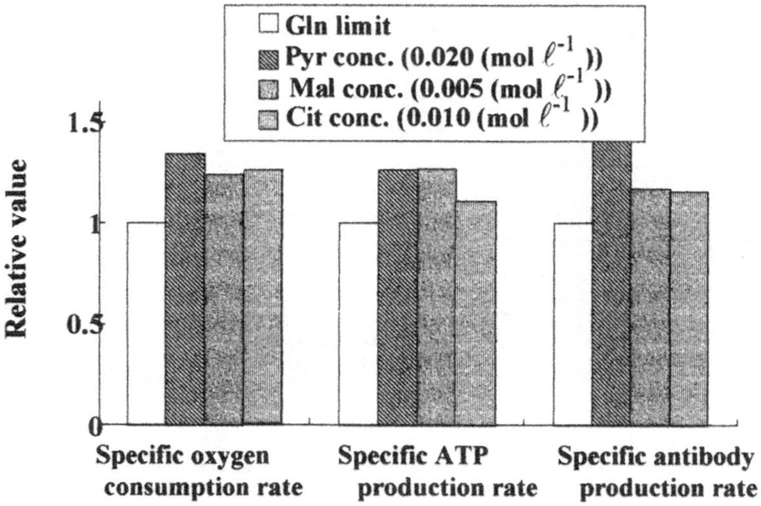


*Figure 3* Comparison between specific oxygen consumption, ATP production and antibody production rates ($\mu$: 0.0312~0.0360 ($h^{-1}$))

These specific rates were relative values based normal glutamine-limited continuous cultivation. By the addition of these intermediates, the specific production rates could be increased. In the case of the addition L-malate and citrate, intermediates directly incorporated into only the TCA cycle, TCA cycle fluxes were changed markedly (data not shown). However, in the case of the addition of pyruvate, not only TCA cycle flux, but also the bypass pathway flux from TCA to glycolysis are increased. Consequently, pyruvate was the most effective intermediate for enhancement of the specific antibody production rate.

## 4. References

1. Katakura,Y., Kobayashi, E., Kurokawa, Y., Omasa, T., Fujiyama, K., and Suga, K.: Cloning of cDNA and characterization of anti-RNase A monoclonal antibody 3A21, *J. Ferment. Bioeng.*, **82**, (1996) 312-314.
2. Omasa, T., Higashiyama, K., Shioya, S., and Suga, K.: Effect of lactate concentration on hybridoma culture in lactate-controlled fed-batch operation' *Biotechnol. Bioeng.*, **39**, (1992) 556-564
3. Omasa, T., Ishimoto, M., Higashiyama, K., Shioya S. and Suga, K.: The enhancement of specific antibody production rate in glucose- and glutamine-controlled fed-batch culture, *Cytotechnology*, **8**,. (1992) 75-84
4. Kobayashi M., Kato, S., Omasa, T., Shioya, S.and Suga K.: Enhancement effects of BSA and linoleic acid on hybridoma cell growth and antibody production, *Cytotechnology*, **15**, (1994) 51-56
5. Omasa, T., Kobayashi, M., Nishikawa, T., Shioya, S., Suga, K., Uemura, S., Kitani, T., and Imamura Y.: Enhancement of antibody production by growth factor addition in perfusion and hollow-fiber culture systems, *Biotechnol. Bioeng.*, **48**, (1995) 673-680.
6. Zupke C. and Stephanopoulos, G.: Intracellular flux analysis in hybridomas using mass balances *in vitro* $^{13}$C NMR, *Biotechnol. Bioeng.*, **45**, (1995) 292-303.
7. Furuichi, K., Iemura, T., Omasa, T., Katakura Y. and Suga K.,: Analysis of antibody production based on energy metabolism in hybridoma, In: *Animal Cell Technology: Basic & Applied Aspects*, **9,** (1997) 261-265.
8. Yano, T. and Nishizawa, Y.: Effect of dissolved oxygen concentration on growth of mouse-mouse hybridoma, In: *Animal Cell Technology: Basic & Applied Aspects*, **5**, (1993) 509-514

# TRANSIENT TRANSFECTION INDUCES EARLY CYTOSOLIC CALCIUM SIGNALING IN CHO-K1 CELLS

A.K. PREUSS[1], H.M. PICK[1], F. WURM[2] & H. VOGEL[1]
[1]*Laboratory of Physical Chemistry of Polymers and Membranes, Department of Chemistry, Swiss Federal Institute of Technology, CH-1015 Lausanne, Switzerland*
[2]*Laboratory of Cellular Biotechnology, Department of Chemistry, Swiss Federal Institute of Technology, CH-1015 Lausanne, Switzerland*

## 1. Abstract

For the controlled production of recombinant proteins in mammalian cells by using transient transfection methods, it may be desirable not only to manipulate, but also to early diagnose success and extent of expression. Here, we applied laser scanning confocal microscopy to on-line monitor second messenger $Ca^{2+}$ signaling in Chinese hamster ovary (CHO-K1) cells during and after transfection. The calcium phosphate / DNA coprecipitation technique was our method of choice, because it is easily applicable for both mini and large scale transfection. In CHO cells we observed a strong $Ca^{2+}$ response already a few minutes after addition of the transfecting solution. Virtually all CHO cells exhibited asynchronous cytosolic $Ca^{2+}$ oscillations 3 hours after transfection onset. Somewhat surprisingly, we observed that the glycerol shock commonly used to enhance the productivity in CHO host cells at the same time "soothed" their $Ca^{2+}$ signaling, reducing the oscillations. Time lapse studies revealed shock induced morphological changes in the endoplasmic reticulum of CHO cells. Our finding points to intracellular $Ca^{2+}$ signaling as a possible early indicator of the transient transfection fate of CHO-K1 cells.

## 2. Introduction

Chinese hamster ovary (CHO-K1) cells are commonly used for heterologous gene expression (Geisse et al, 1996). On the basis of an improved version of the calcium phosphate transfection technique (Jordan et al., 1996), we generally obtain a transfection rate of 50% of the cells, using the green fluorescent protein (GFP) as a reporter gene. With other adherently grown cells, for example human embryonic kidney (HEK) 293 cells, we even reach up to 90 % transfectants. Recent results indicate that the calcium phosphate approach can also be applied for cell suspension based systems (Jordan et al., 1998). For controling the transfection rate in CHO-K1 cells, we feel that a product independent early diagnosis within hours would well meet the flexible and very fast approach of the calcium phosphate based transient transfection method and its upscale application for suspension systems.

We report, for the first time, the use of laser scanning confocal microscopy for on-line monitoring of the intracellular $Ca^{2+}$ fluctuations during the transfection process in CHO-

K1 cells. We show that the glycerol shock treatment usually required for the enhancement of transfection efficiency (Staedel et al., 1994) strongly affects their $Ca^{2+}$ signaling. Moreover, these signal changes are accompanied by substantial morphological changes of the major intracellular $Ca^{2+}$ store.

This report about monitoring intracellular $Ca^{2+}$ signaling in transfected CHO cells may be the first step towards a *product-independent* diagnostic tool of the transfection rate for biotechnological processing.

## 3. Materials and Methods

### 3.1 CELL CULTURE

Our studies were performed on adherent Chinese hamster ovary (CHO-K1) and human embryonic kidney (HEK 293)cell lines. The cultures were kept in DMEM/F12 medium supplemented with 2.2 % fetal calf serum in an incubator with humidified air (5% $CO_2$, 37°C). Best growth conditions were maintained by splitting the cultures 2-3 times per week at ratios between 1:4 to 1:6.

### 3.2 PLASMID CONSTRUCTS AND TRANSIENT TRANSFECTION

Cells were transfected with calcium phosphate-precipitated DNA according to Jordan et al. (1996). Twenty hours before transfection, exponentially growing cells were seeded ($2\text{-}4 \times 10^5$ cells/ml) into multiwell plates or Labtek chambered cover glass (Nunc, Denmark). As reporter for gene expression, we used the green fluorescent protein encoding vector (Clontech, USA).

### 3.3 ON-LINE MONITORING OF INTRACELLULAR SIGNALING

Fluorescence and transmission time lapse studies were performed with a laser scanning confocal microscope (Zeiss LSM 510). Intracellular $Ca^{2+}$ signaling was recorded using 488 nm as excitation wavelength. The laser power was kept below 0.3mW. The frame size was 512x512 pixels.

Imaging of cytosolic $Ca^{2+}$ changes was performed on cells incubated for 70 min with a 10 μM solution of the cell-permeant acetoxymethyl (AM) ester of $Ca^{2+}$ indicator stain fluo-3 (Molecular Probes, USA). After the loading procedure, cells were thoroughly washed in fresh medium and monitored at room temperature.

All fluorescence data were background corrected. A false color code was used for comparison of relative $Ca^{2+}$ changes. For visualization of the endoplasmic reticulum, cells were loaded with the indicator dye $DiOC_5$ (Molecular Probes, USA) according to the instructions of manufacturer. $DiOC_5$ is a cell permeant dye, which stains intracellular endoplasmic reticulum membranes emitting a fluorescein-like fluorescence.

## 4. Results and Discussion

Cytosolic $Ca^{2+}$ responses were recorded in sixteen fluo-3 loaded CHO cells both before and after administration of the transfecting reaction mixture (Fig 1).

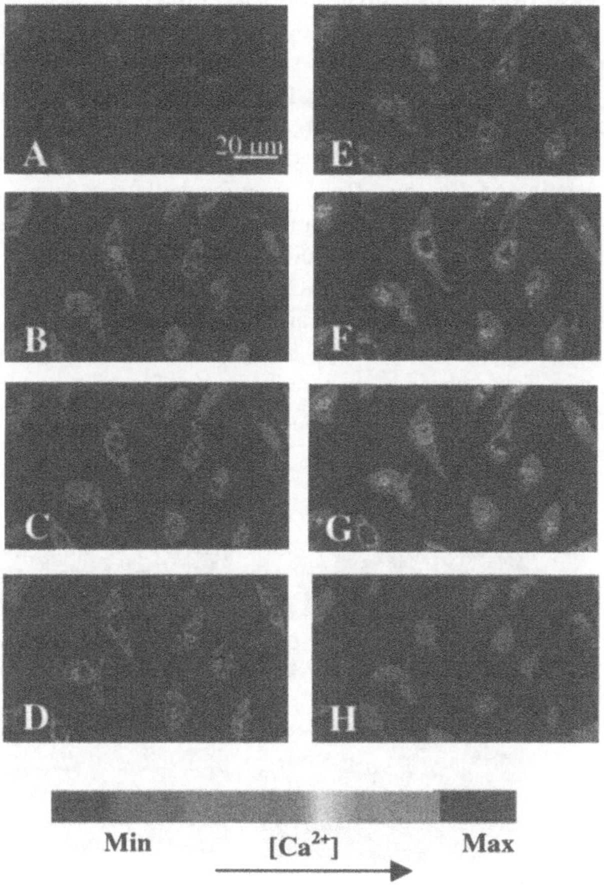


*Figure 1.* False color micrographs indicating the relative cytosolic $Ca^{2+}$changes in 16 CHO-K1 cells before (A), during exposure to the calcium phosphate / DNA transfection mixture (B - G), and 5 minutes after glycerol shock (H). Details see text.

Addition of the transfection mixture elicited a marked and immediate cellular response, which could not be mimicked by addition of an equivolume of cell medium. The initial $Ca^{2+}$ increase was followed by declining $Ca^{2+}$ levels with isolated fluctuations (240s after transfection onset, Fig 1B). After three hours, the cells exhibited a high $Ca^{2+}$ concentration, accompanied by marked, more or less sinoidal oscillations.

The periods are in the range of around 20 s (Fig. 1D-G, images taken with 9 min interval). Although the use of a calcium phosphate based transfection method seems to imply that the observed signal increases might stem from ion influx into the cells, such sinoidal cytosolic $Ca^{2+}$ oscillations have been known to result, at least in part, from repetitive release from intracellular stores (Kraus et al., 1996, Thomas et al., 1996; Scheenen et al., 1996). A temporary, or longer lasting, decline of store luminal $Ca^{2+}$ concentration, has been found to have significant impact on cellular protein synthesis (Greber and Gerace, 1995; Rosen et al., 1995; Wei and Hendershot, 1996; Bading et al., 1997; Dolmetsch et al., 1998; Lee et al., 1998; Obermoeller et al., 1998). The question whether or not the transfection-induced $Ca^{2+}$ oscillations is related to internal store release is investigated by the effect of store-depleting drugs on the transfection efficiency in CHO cells. Preliminary studies on HEK cells showed significantly lower signal increases and oscillations (manuscript in preparation).

The strong down regulation of the $Ca^{2+}$ response after the glycerol shock (10% in DMEM medium for 2.5 min, and 4 h after transfection) was accompanied by marked morphological changes in both shape and position of the $DiO_5$ stained endoplasmic reticulum (Fig 2). These changes appeared to persist for the time of observation (1 hour) and after reperfusion with glycerol free cell medium (Fig 2 D, 30 min after end of shock).

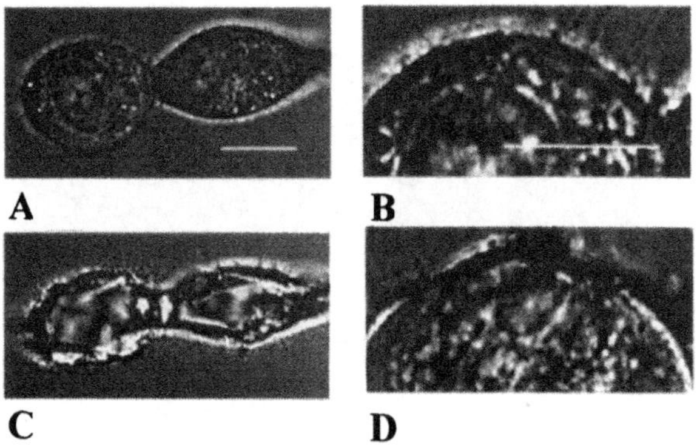


*Figure 2.* CHO-K1 cells before (A, B), during (C) and 1 min after (D) glycerol shock. Size bars are 10 μm.

### 5. Conclusion

Our work points to a possible key function of intracellular $Ca^{2+}$ signaling for heterologous, recombinant protein production. This function appears to complement its known regulatory role for cellular protein synthesis. The on-line observation of $Ca^{2+}$ signaling in CHO cells may have several advantages. Firstly, the development of product-independent diagnostic tools for the early assessment of the transfection rate in small cell samples. Secondly, improving the transfection efficiency of the calcium phosphate method by manipulating the store luminal $Ca^{2+}$ content. Thirdly, optimizing the transfection enhancing property of the glycerol shock treatment.

**Acknowledgement:**

This research was partly financed by an EPFL grant for interdepartmental research.

### 6. References

Bading H, Hardingham GE, Johnson CM, Chawla S (1997) Gene regulation by nuclear and cytoplasmic calcium signals. *Biochem Biophys Res Commun* 236: 541-543

Berridge MJ (1997) Elementary and global aspects of calcium signaling. *J Exp Biol* 200: 315-319

Dolmetsch RE, Xu K, Lewis RS (1998) Calcium oscillations increase the efficiency and specificity of gene expression. *Nature* 392: 933-936

Geisse S, Gram H, Kleuser B, Kocher HP (1996) Eukaryotic expression systems: a comparison. *Protein Expr Purif* 8: 271-282

Greber UF, Gerace L (1995) Depletion of calcium from the lumen of endoplasmic reticulum reversibly inhibits passive diffusion and signal-mediated transport into the nucleus. *J Cell Biol* 128: 5-14

Jordan M, Schallhorn A, Wurm FM (1996) Transfecting mammalian cells: optimization of critical parameters affecting calcium-phosphate precipitate formation. *Nucleic Acids Res* 24:596-601

Jordan M, Köhne C, Wurm FM (1998) Calcium-phosphate mediated DNA transfer into HEK-293 cells in suspension: control of physicochemical parameters allows transfection in stirred media. Transfection and protein expression in mammalian cells *Cytotechnology* 26: 39-47

Kraus M, Wolf B, Wolf B (1996) Crosstalk between cellular morphology and calcium oscillation patterns. Insights from a stochastic computer model. *Cell Calcium* 19: 461-472

Lee MA, Dunn RC, Clapham DE, Stehno-Bittel L (1998) Calcium regulation of nuclear pore permeability. *Cell Calcium* 23: 91-101

Preuss A, Pick H, Connor JA, Vogel H (1998) Transient transfection induces different calcium signaling in CHO K1 versus HEK 293 cells. *Manuscript in preparation*

Obermoeller LM, Chen Z, Schwartz AL, Bu G (1998) $Ca^{2+}$ and receptor-associated protein are independently required for proper folding and disulfide bond formation of the low density lipoprotein receptor-related protein. *J Biol Chem* 273: 22374-22381

Rosen LB, Ginty DD, Greenberg ME (1995) Calcium regulation of gene expression. *Adv Second Messenger Phosphoprotein Res* 30: 225-253

Scheenen WJ, Jenks BG, van Dinter RJ, Roubos EW (1996) Spatial and temporal aspects of $Ca^{2+}$ oscillations in Xenopus laevis melanotrope cells. *Cell Calcium* 19: 219-227

Staedel C, Remy JS, Hua Z, Broker TR, Chow LT, Behr JP (1994) High-efficiency transfection of primary human keratinocytes with positively charged lipopolyamine DNA complexes. *J Invest Dermatol* 102: 768-772

Thomas AP, Bird GS, Hajnoczky G, Robb-Gaspers LD, Putney JW Jr (1996) Spatial and temporal aspects of cellular calcium signaling. *FASEB J* 10:1505-1517

Wei J, Hendershot LM (1996) Protein folding and assembly in the endoplasmic reticulum. *EXS* 77: 41-55

**Keywords**

CHO-K1 HEK 293 transient transfection calcium phosphate

cytosolic calcium signaling laser scanning confocal microscopy

# AN ON-LINE FEEDING STRATEGY FOR FED-BATCH AND DIALYSIS CULTURES OF HYBRIDOMA CELLS

J. O. SCHWABE, R. PÖRTNER, H. MÄRKL
*Technische Universität Hamburg-Harburg, Bioprozeß- und Bioverfahrenstechnik, D-21071 Hamburg, Germany*

## 1. Abstract

An on-line feeding strategy based on oxygen uptake rate (OUR) was used to minimise the formation of inhibiting metabolites and to increase the yield of monoclonal antibodies in fed-batch cultures of hybridoma cells by a balanced supply of substrates. Concentrated feed was supplied according to the off-line estimated stoichiometric ratio between oxygen and glucose consumption (GC). The antibody concentration in fed-batch increased three-fold compared to the conventional batch culture by applying this strategy. Metabolites such as lactate and ammonia accumulated during fed-batch cultivation and it was not possible to avoid inhibition by ammonia. Fed-batch was efficiently improved by using the concept of dialysis and 'nutrient-split' feeding, where concentrated medium is fed directly to the cells and toxic metabolites are removed over a dialysis membrane into a buffer solution. This resulted in a ten-fold increase of the antibody concentration compared to the batch. Amino acid concentrations were analysed to identify limiting conditions during the cultivation and to analyse the performance of the nutrient supply in the fed-batch and dialysis fed-batch. The potential of the dialysis fed-batch with 'nutrient-split' feeding is discussed with respect to substrate loss, efficiency and process stability in comparison to the conventional fed-batch.

## 2. Cell Line and Culture Conditions

The hybridoma cell line IV F 19.23 was cultivated in a 1:1 mixture of Iscove's MDM and Ham's F12 supplemented with 2 mmol $l^{-1}$ glutamine, 2.5 g $l^{-1}$ $NaHCO_3$, 0.01% (w/v) PEG and 50 mmol $l^{-1}$ ethanolamine. The medium was completed with 1% (v/w) of a protein-free, iron-rich supplement [Franěk and Dolníková, 1991]. For the fed-batch phase and the 'nutrient-split' feeding a 10-fold concentrated, salt-free medium (BIOCONCEPT, D) containing amino acids, vitamins and 50 mmol $l^{-1}$ glucose and a concentrated glutamine solution (200 mmol $l^{-1}$, Life Technologies, D) were supplied separately.

## 3. On-Line Feeding Strategy

The feeding strategy used in fed-batch and dialysis fed-batch cultures first described by Zhou et al. (1995) is based on the stoichiometric ratio $\alpha_j$ of glucose consumption (equ. 1), determined by off-line analysis, and cumulative oxygen consumption *COC* (time intervall: $t_{j-1}, t_j$), calculated from on-line oxygen uptake rate *OUR*. The feeding rates (F) for glucose (Glc) and glutamine (Gln) containing feed is calculated according to the equations 2a/b:

$$\alpha_j = \frac{\int_{t_{j-1}}^{t_j} OUR \cdot V_i dt}{c_{Glc}(t_j) \cdot V_i(t_j) - c_{Glc}(t_{j-1}) \cdot V_i(t_{j-1}) - c_{Glc,F} \int_{t_{j-1}}^{t_j} F_{Glc}\, dt} \tag{1}$$

$$F_{Glc}(t_i) = \frac{\int_{t_{i-1}}^{t_i} OUR \cdot V_i dt}{\alpha_j \cdot c_{Glc,F}(t_{i-1} - t_i)} \qquad F_{Gln}(t_i) = F_{Glc}(t_i) \cdot \frac{c_{Glc,F}}{c_{Gln,F}} \cdot \beta_{Gln,Glc} \tag{2a/b}$$

The ratio of glucose to glutamine uptake $\beta_{Gln,Glc} = 0.25$ was obtained from previous suspension cultures (Pörtner and Schäfer, 1996).

## 4. Dialysis and 'Nutrient-Split'-Feeding

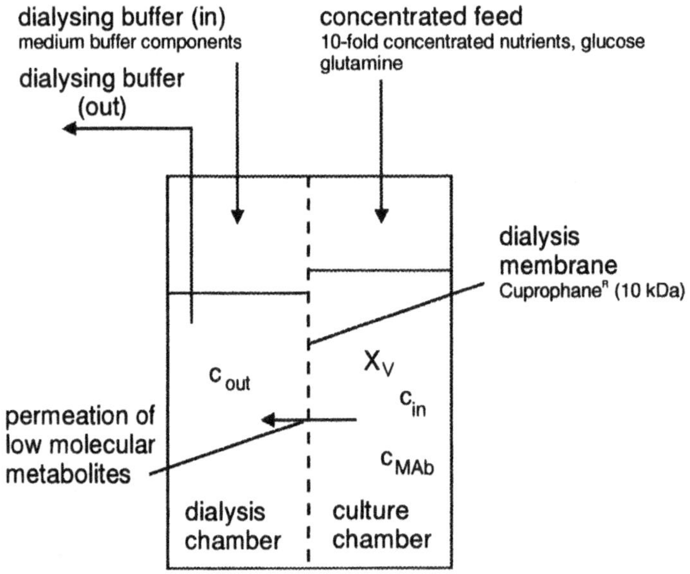


**Figure 1: Concept of dialysis and 'nutrient-split'-feeding.** Supply of concentrated nutrients to the culture chamber and removal of metabolites by dialysing with buffer solution in the dialysis chamber.

Figure 1 shows the concept of dialysis and 'nutrient-split'-feeding in a reactor which is separated into two chambers by a dialysis membrane (Cuprophane$^{R}$, 10 kDa), a culture chamber containing the cells and a dialysis chamber. During fed-batch phase concentrated nutrients are fed directly to the culture chamber. Low molecular metabolites permeate through the membrane according to the concentration gradient from culture to dialysis chamber. During dialysis phase a buffer is used to exchange the medium in the dialysis chamber to maintain the gradient in metabolite concentration and to remove inhibiting metabolites. This concept is realised in the Membrane Dialysis Reactor (Bioengineering, CH) shown in figure 2.

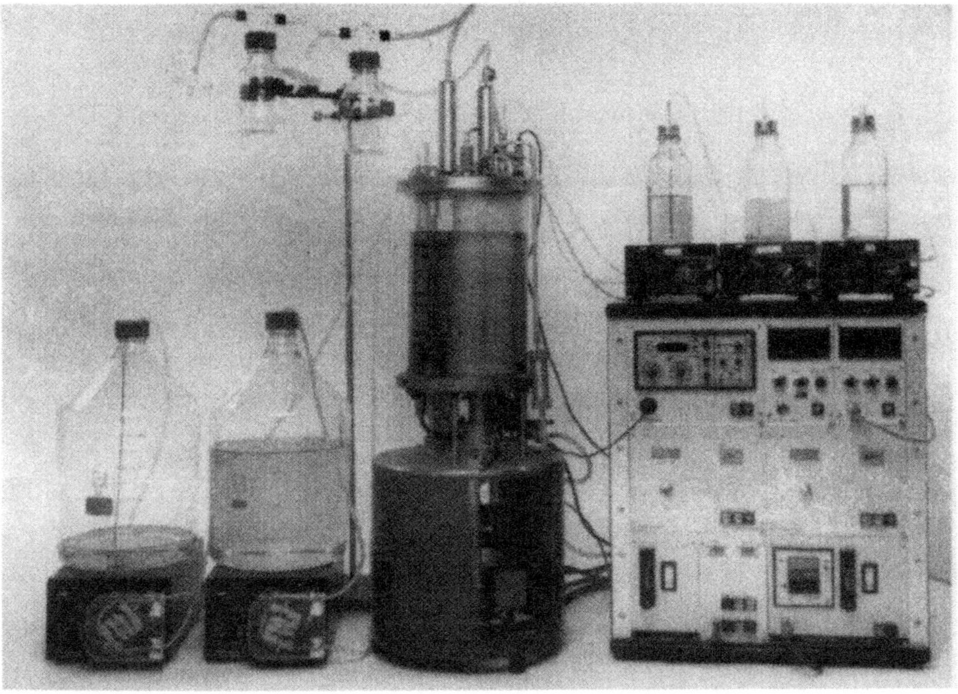

**Figure 2:** Membrane Dialysis Reactor

## 5. Results

The fed-batch with the on-line feeding strategy resulted in an increase of 1.5-fold in viable cell concentration $X_v$ (Fig. 3) and 3-fold in antibody concentration $c_{MAb}$ (Fig. 4b) compared to the conventional batch. The strategy was not able to avoid the accumulation of metabolites such as ammonia after 80 h of cultivation and culture growth was inhibited.

The dialysis fed-batch combines the on-line feeding strategy with the concept of 'nutrient-split' feeding to supply essential substrate directly to the culture and reduce the effect of inhibiting metabolites by dialysis. Viable cell concentration $X_v$ increased 5-fold (Fig. 3) and the monoclonal antibody concentration $c_{MAb}$ 10-fold respectively (Fig. 4b) compared to the conventional batch. Nutrient supply was sufficient to maintain a high cell concentration of $9 \cdot 10^6$ cell $ml^{-1}$ for about 60 h. Oxygen transfer was insufficient at high cell concentration, which caused problem with the on-line feeding algorithm based on OUR.

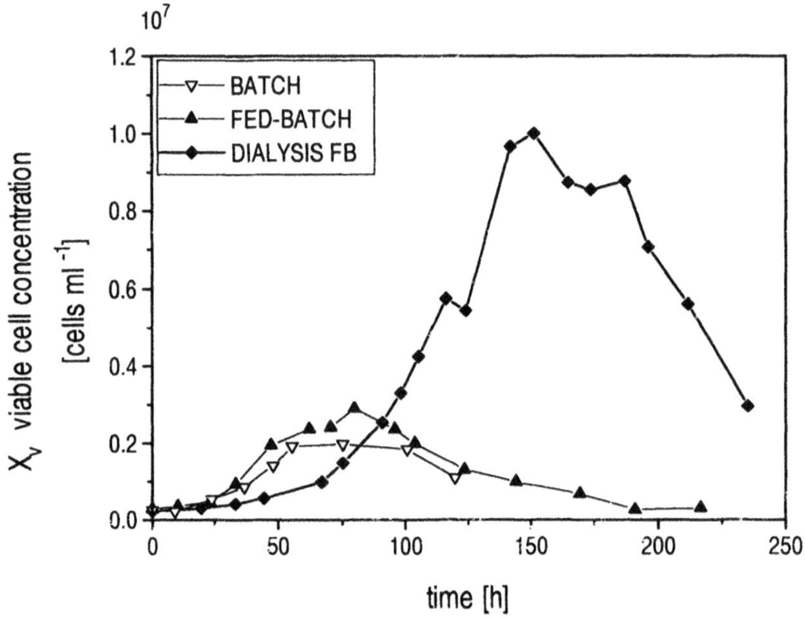


**Figure 3:** Comparison of viable cell concentration in batch, fed-batch and dialysis fed-batch

The concentration of ammonia was significantly lower in dialysis fed-batch and was effectively reduced by the exchange of dialysing buffer (Fig. 4a). The essential amino acids (determined by HPLC), which showed notable uptake in batch culture, were compared in fed-batch and dialysis fed-batch (Fig. 5). In fed-batch the significant amino acids accumulated in fed-batch phase (47 h - 217 h) whereas in dialysis fed-batch a stable level of amino acids was maintained during 'nutrient-split' feeding (dialysis: 124 h- 211 h, feeding: 141 h - 211 h). Media components were efficiently used due to the controlled feeding of concentrated nutrients. Tryptophane showed high relative uptake in both fed-batch and dialysis fed-batch.

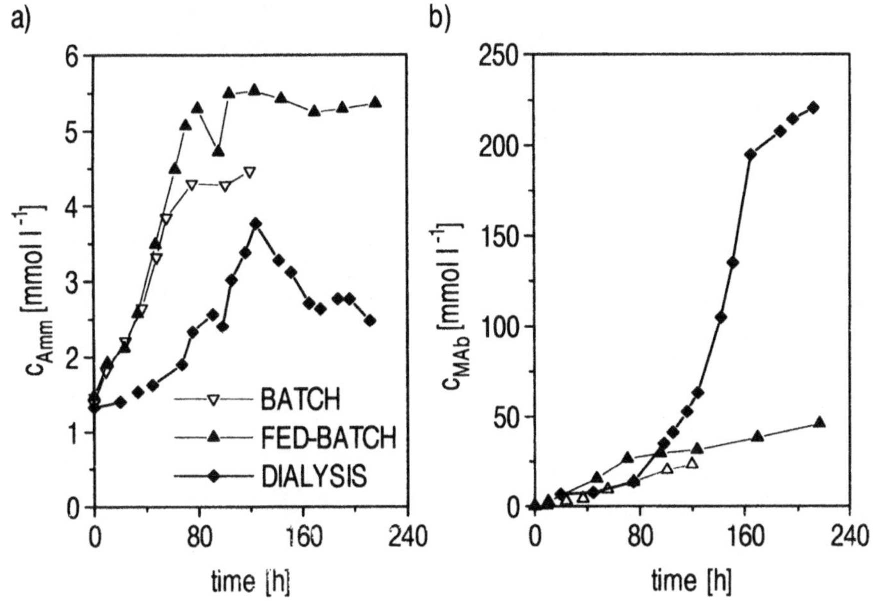


**Figure 4:** Comparison of a) ammonia concentration $c_{Amm}$ and b) antibody concentration $c_{MAb}$ in batch, fed-batch and dialysis fed-batch

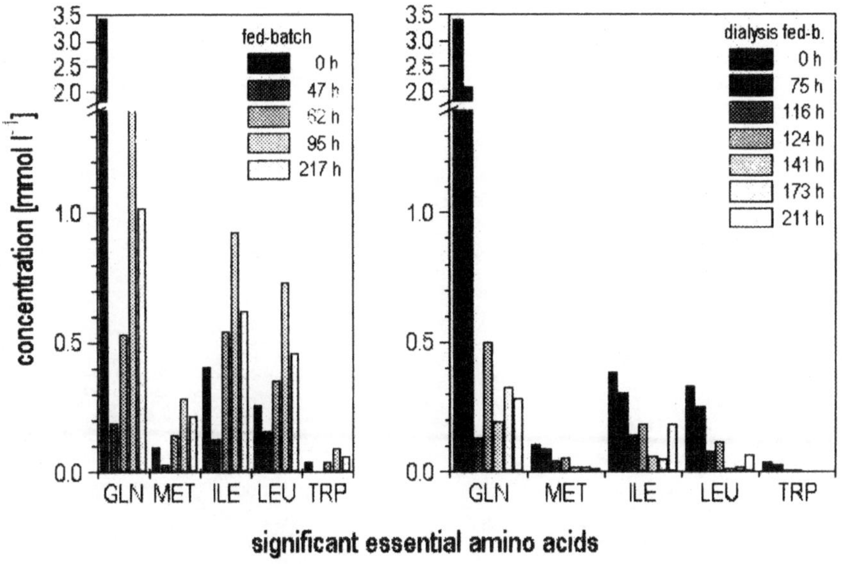


**Figure 5:** Significant essential amino acids compared in fed-batch and dialysis fed-batch at different cultivation stages

[1] Fran k, F., and Dolníková, J. 1991. Hybridoma growth and monoclonal antibody production in iron-rich protein-free medium: Effect of nutrient concentration. Cytotechnology **7**: 33-38

[2] Pörtner, R., Schäfer, Th.: Modelling hybridoma cell growth and metabolism - A comparision of selected models and data. J. Biotechnol. 49 (1996) 119-135

[3] Zhou, W., Rehm, J., Hu, W.-S.: High viable cell concentration fed-batch culture of hybridoma cells through on-line nutrient feeding. Biotechnol. Bioeng. 46 (1995) 579-587

# EFFECTS OF SERUM AND OTHER MEDIUM COMPONENTS ON SUSPENSION CULTURE OF CHO CELLS PRODUCING TISSUE PLASMINOGEN ACTIVATOR

**M. TAKAGI, H.C. HIA, J.H. JANG and T. YOSHIDA**

*International Center for Biotechnology, Osaka University, 2-1, Yamada-oka, Suita, Osaka 565, Japan*

Effects of the concentrations of serum and other medium components on cell metabolism in a suspension culture, in a spinner bottle, of CHO 1-15500 cells (ATCC CRL-9606) producing tissue plasminogen activator(tPA) were investigated in order to formulate an appropriate medium composition for perfusion culture. Ham's F-12 medium supplemented with 500 nM of methotrexate and 0-10% newborn-calf serum was employed. Cell and tPA concentrations were assayed using trypan blue and ELISA, respectively. Concentrations of glucose, lactate, glutamine and glutamic acid were determined enzymatically, and those of ammonia using a colorimetric method. The culture period could be separated into two phases: the cell growth phase, and the cell maintenance phase during which a high cell density was maintained. The specific growth rate during the cell growth phase increased (Fig. 1), and cell decreasing rate during the cell maintenance phase decreased (Fig. 2) with increase in the concentration of serum in the medium, while the specific glucose consumption rate and the conversion rate of glucose to lactate did not vary in both phases. The ratio of ammonia production to glutamine consumption during the cell growth phase increased as the concentration of serum in the medium increased although the specific glutamine consumption rate decreased. It is suggested that the metabolic flux to amino acid synthesis became dominant in the culture at low serum concentrations. Increase of specific tPA production rate was observed in both culture phases at reduced serum concentrations (Fig. 3). The effects of the concentrations of other medium components including glucose, glutamine, lactate and ammonia on metabolism and tPA production were also investigated. There was negative correlation between specific growth rate and specific tPA production rate and the effect of serum concentration was markedly larger than the effect of concentration of other 4 components (Fig. 4).

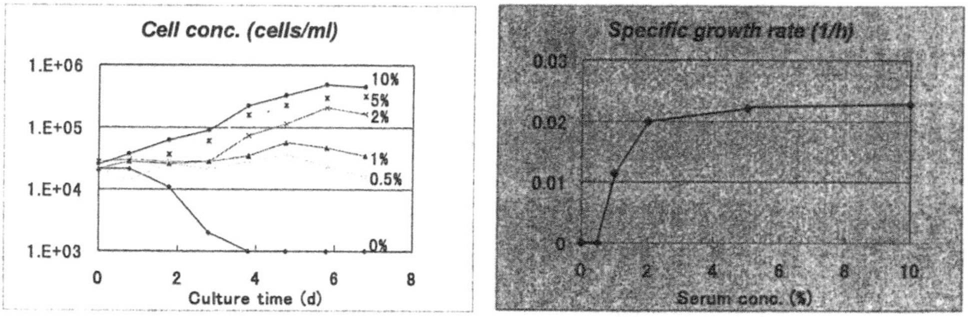


Fig. 1 Effect of serum concentration on growth of CHO cells during cell growth phase.

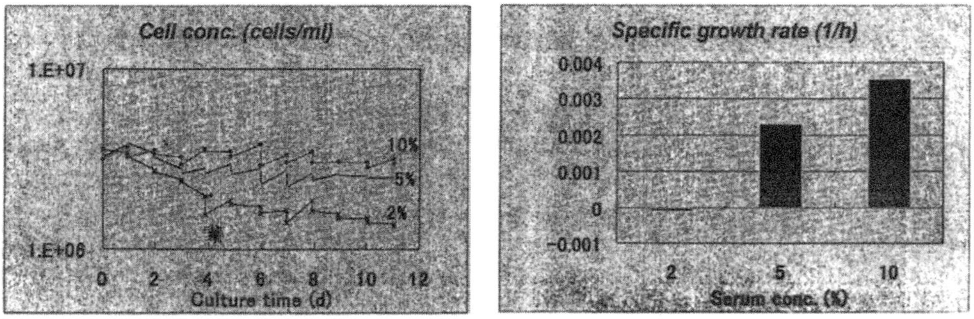


Fig. 2 Effect of serum concentration on concentration of CHO cells during cell maintenance phase.

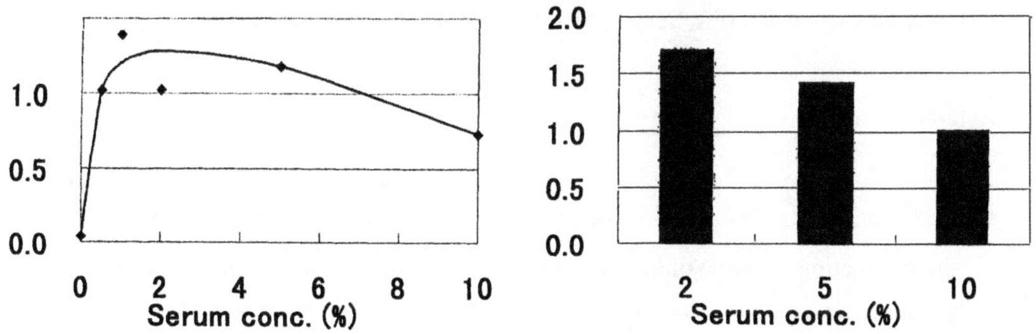


Fig. 3 Effect of serum concentration on specific tPA production rate. Left; cell growth phase, right; cell maintenance phase.

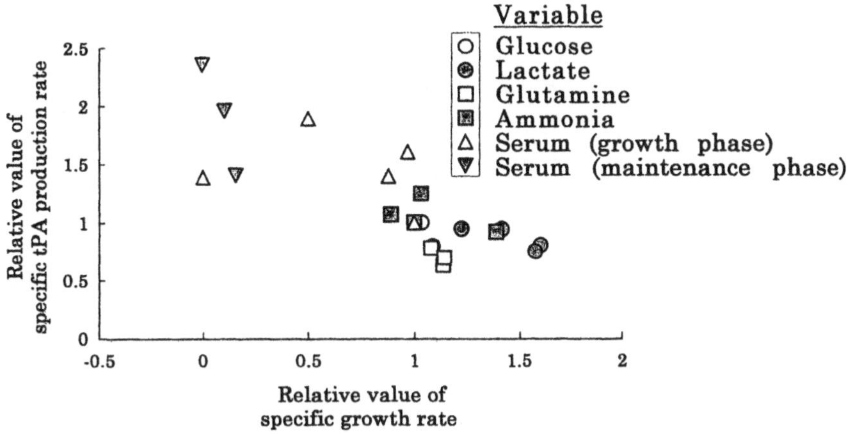


Fig. 4 Relation between specific rates of cell growth and tPA production. The relative value of specific rates against those for the culture using standard medium were employed in each culture experiments.

Key words: CHO cells, tPA, Serum, Medium components, Specific metabolic rates

References

M. Takagi and K. Ueda (1994) A low-serum medium with BSA and ferrous sulfate for the production of tissue plasminogen activator by human embryo lung cells. J. Ferment. Bioeng., 77, 394-399.

# EFFECTS OF OSMOTIC PRESSURE ON METABOLISM OF CHO CELLS PRODUCING TISSUE PLASMINOGEN ACTIVATOR IN ADHESION CULTURE COMPARED WITH THAT IN SUSPENSION CULTURE

**M. TAKAGI, H. HAYASHI and T. YOSHIDA**

*International Center for Biotechnology, Osaka University, 2-1, Yamada-oka, Suita, Osaka 565, Japan*

The effects of osmotic pressure, of between 300 and 500 mOsm/kg, on the metabolism of Chinese Hamster Ovary (CHO) cells producing tissue plasminogen activator (tPA) were compared in adhesion and suspension cultures. In both suspension and adhesion cultures, the specific rates of glucose consumption ($\nu_G$), lactate production ($q_L$) and tPA production ($q_{tPA}$) increased as osmotic pressure increased up to certain critical levels, above which all of the rates decreased (Fig. 1). However, specific growth rate ($\mu$) decreased with increase in osmotic pressure over the range examined, and this slope grew steeper in the range of osmotic pressures higher than the critical level. The decrease in $\mu$ was steeper in the adhesion culture than in the suspension culture. The critical osmotic pressure for adhesion culture (400 mOsm/kg) was lower than that for suspension culture (450 mOsm/kg), indicating that adhesion cultures were more sensitive to increase in osmotic pressure than suspension cultures. However, the specific rates obtained in adhesion cultures were generally 1.5- to 3-fold higher than those obtained in suspension cultures. Cell volume increased with increase in osmotic pressure in both suspension and adhesion cultures (Fig. 2) but no morphological change was noted in the suspension culture, in contrast, cell height decreased and cell adhesion area markedly increased with increase in osmotic pressure in the adhesion culture (Figs. 2, 3). This morphological change in adhesion cultures could be one explanation for the higher sensitivity of adherent cells to increase of osmotic pressure than suspended cells.

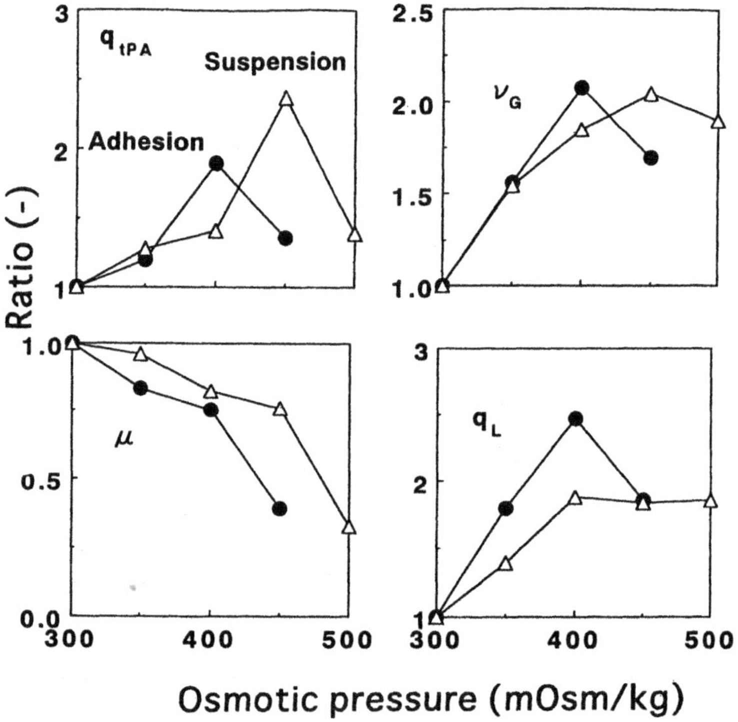


Fig. 1 Comparison of the effects of osmotic pressure on the specific rates between adhesion and suspension cultures of CHO cells.

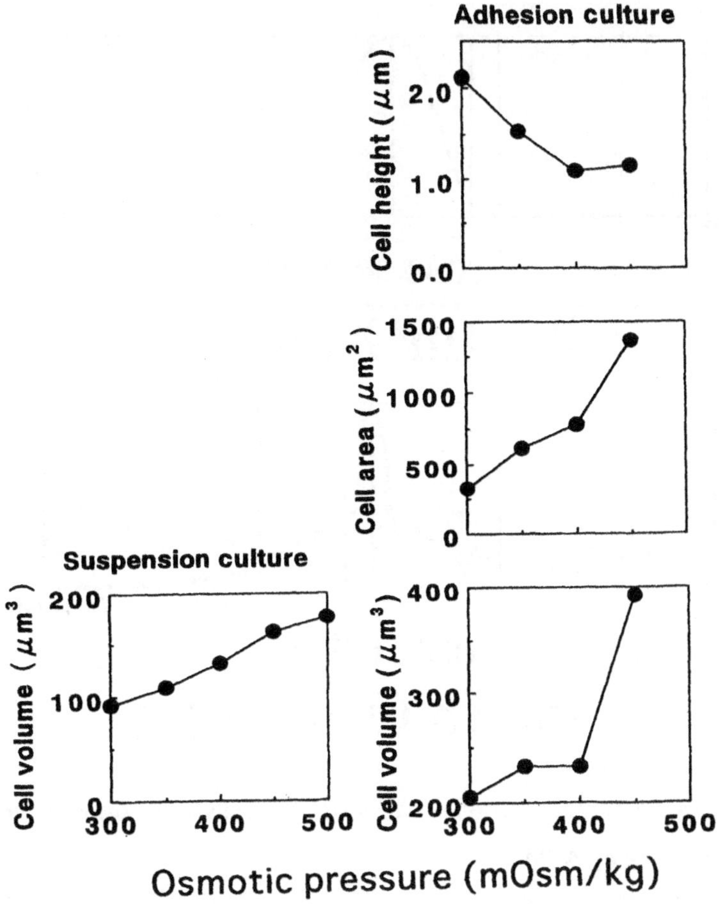


Fig. 2 Effect of osmotic pressure on the volume and morphology of CHO cells in suspension and adhesion cultures. Suspended and adhesive cells were analyzed by Coulter counter and atomic force microscope, respectively (AFM).

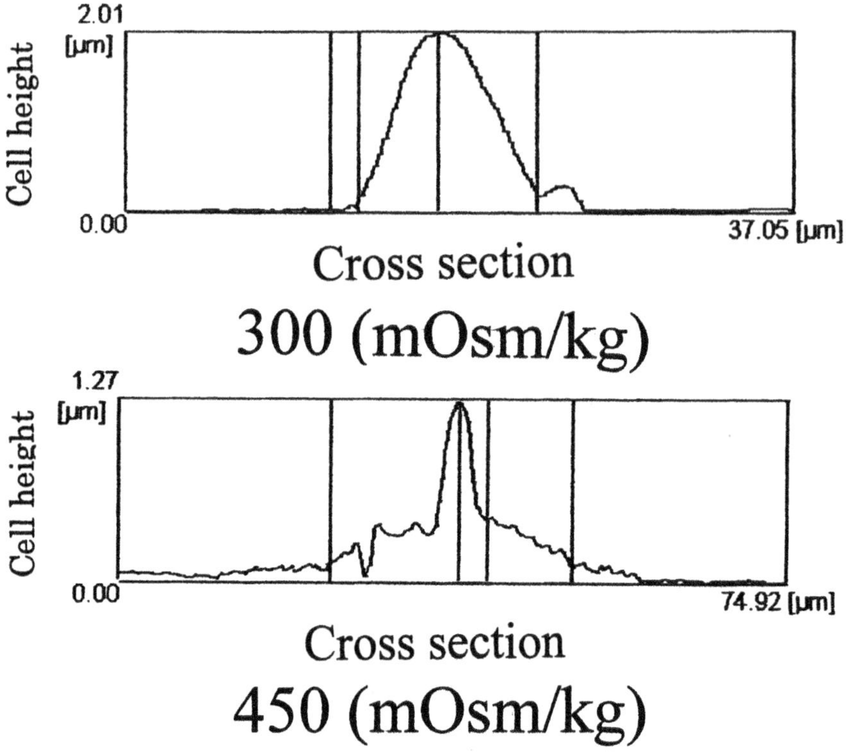


Fig. 3 Observation of cross section of adhesive CHO cells under osmotic pressures of 300 and 450 mOsm/kg by AFM.

Key words: Osmotic pressure, Cell morphology, Suspension culture, Adhesion culture, CHO cells

References

J. Lin, M. Takagi, I. Qu, P. Gao, and T. Yoshida (1998) Improvement of monoclonal antibody production by gradual increase in osmotic pressure.Cytotechnology, in press.

J. Lin, M. Takagi, Y. Qu, P. Gao and T. Yoshida (1998) Metabolic flux change in hybridoma cells under high osmotic pressure. Recent development in biochemical engineering.: Kagakukougakkai symposium series, volume 1, 177-181. Edited by the special study group for "Biochemical Engineering" under The Society of Chemical Engineers (S. Tone, K. Kawakami, M. Goto, H. Ijima), Japan (English).

J. Lin, M. Takagi, and T. Yoshida (1998) Enhancement of monoclonal antibody production under high osmotic pressure. 7th International Conference on Computer Application in Biotechnology (CAB7) 6-P4.

# CULTIVATION OF CHO CELLS ON NOVEL MICROCARRIERS

A.Y. HU, Z. ZHANG & M. AL-RUBEAI
*Centre for Bioprocess Engineering, School of Chemical Engineering, University of Birmingham, Edgbaston, Birmingham B15 2TT, UK*
*(http://www.bham.ac.uk/ChemEng/actg/actghome.htm)*

*Keywords*: Cell attachment, CHO cells, Porous microcarriers.

## 1. Abstract

CHO cells were successfully cultured on a novel type of microcarriers (ImmobaSil FS) in the bioreactors. Various factors influencing cell attachment, including serum concentration, cell to microcarrier ratio, agitation speed and agitation profile were investigated and the results were compared with those obtained using CultiSpher-G, and glass beads. It was found that the optimum serum concentration, cell-to- carrier ratio and agitation speed for the batch reactor with a working volume of 400 ml during the initial attachment period were 10% (v/v), $12x10^3$:1 and 60 rpm, respectively. Intermittent stirring/non-stirring regimen of 20/40 minutes resulted in an optimum rate of cell attachment (0.27 $h^{-1}$). However, intermittent and continuous stirring schemes had no significant effect on the total number of attached cells. ImmobaSil FS carriers performed better than CultiSpher-G and glass beads in term of the rate of cell attachment.

## 2. Introduction

Immobilisation of cells on porous microcarriers provides several advantages over free suspension culture for large-scale mammalian cell process including high cell density and concentration of secreted products (Fleischaker, 1987; Griffiths and Thornton, 1982; Griffiths, 1988). However, there are still problems associated with the various stages of cell culture, such as cell inoculation, attachment and growth (Clark *et al.*, 1980; Forestell *et al.*, 1992). The severity of these problems varies with different microcarriers, cell lines and culture systems used. Consequently, the conditions that affect cell attachment, distribution on the surface of microcarriers and growth rate need to be determined for a particular system. Although porous microcarriers provide a large surface area and protective environment for cells, the occurrence of oxygen diffusion resistance inside the beads often results in a non-uniform distribution of the cells immobilised within the matrix (Preissmann *et al.*, 1997). In this study, CHO cells producing β-galactosidase were cultured on a novel type of microcarriers (ImmobaSil FS) in the bioreactors. The ImmobaSil FS microcarriers, which are silicone-based material have a potential to provide a large surface area for cell growth and allow the diffusion of oxygen across the matrix structure of the carriers. Therefore, an investigation was made to determine the effect of inoculum concentration, media composition and stirring conditions on the level of cell attachment.

## 3. Materials and methods

Recombinant CHO cells producing β-galactosidase kindly provided by Dr. Steve Gorfien (Life Technologies Inc., USA) were used in these experiments. The cells were amplified with dihydrofolate reductase by adding 0.8 μM methotrexate (Sigma, UK) and were maintained in spinner flasks at 37°C and 5% carbon dioxide. Culture medium used was RPMI 1640 (Sigma, UK) supplemented with 5% foetal bovine serum (GibcoBRL, Scotland). ImmobaSil FS carriers were supplied by Ashby Scientific Ltd. (UK). ImmobaSil FS, CultiSpher-G (Percell, Sweden) and glass beads (Sigma, USA) were washed with phosphate-buffered saline solution and autoclaved. All glass vessels were siliconised before use. Cell attachment experiments were carried out in the 400 ml working volume SIXFORS bioreactors (Infors, UK). For the study of initial cell inoculum cell density ranged from $2.0 \times 10^5$ to $10.0 \times 10^5$ cells $ml^{-1}$, RPMI 1640 supplemented with 5% FBS and 60 rpm of agitation were used. For the study of the effect of serum on cell attachment serum-free medium CHO SFM-II (Sigma, UK), RPMI 1640 supplemented with 0% to 10% (v/v) serum, initial cell concentration of $8.0 \times 10^5$ cells $ml^{-1}$and 60 rpm of agitation were used. For the study of the effect of agitation initial cell concentration of $8.0 \times 10^5$ cells $ml^{-1}$, RPMI 1640 supplemented with 10% FBS and different agitation speeds ranged from 20 to 120 rpm were used. For the effect of stirring mode RPMI 1640 supplemented with 10% FBS, initial cell concentration of $8.0 \times 10^5$ cells $ml^{-1}$ and 60 rpm of agitation were used. The variation of the number of cells attaching on ImmobaSil FS microcarriers was quantified by the MTT method. 40 carriers were transferred into a 96-well plate. Each well contained a single microcarrier, in which 100 μl MTT solution (5mg $ml^{-1}$) was added. The optical intensity of each sample was measured by a spectrophotometer (Pharmacia, Sweden). The number of attached cells on microcarriers was measured by counting the number of cells which remained freely suspended in a microcarrier-free supernatant at various times up to 7 hours and at different conditions after inoculation. The method is based on the assumption that there was no significant cell growth and death during the experiment.

## 4. Results and discussion

Table 1 shows the effect of initial cell-to-carrier ratio on cell attachment for ImmobaSil FS microcarriers. The optimal ratio of cell-to-carrier for the cell attachment was $12 \times 10^3$:1, at which 64.3% of the cells had attached on the microcarriers at the incubation time of 7 hour. This result is also consistent with the previous work showing that a minimum number of cells per carrier was required for normal cell growth to occur (Hu *et al.*, 1985).

**TABLE 1.** The effect of inoculation level on CHO cell attachment for ImmobaSil FS microcarrier cultures (7 hours incubation).

| Inoculum (cell-to-bead) | Inoculum (cells $ml^{-1}$ medium) | Final density (cells $ml^{-1}$ carrier) |
|---|---|---|
| $15\times10^3$:1 | $10.0\times10^5$ | $96\times10^5$ |
| $12\times10^3$:1 | $8.0\times10^5$ | $104\times10^5$ |
| $6\times10^3$:1 | $4.0\times10^5$ | $66\times10^5$ |
| $3\times10^3$:1 | $2.0\times10^5$ | $31\times10^5$ |

The effect of serum concentration on the rate of cell attachment is shown in Table 2. Clearly, 10% FBS in RPMI 1640 produced the fastest cell attachment among the selected conditions, and only 30% of cells remained unattached at the time of 7 hours. The presence of serum might have promoted the interaction between cells and the substratum of the microcarriers. The attachment fraction in the serum-free medium (CHO SFM-II) supplemented with 10% serum was much lower than that in RPMI 1640 supplemented with 10% serum. This may be due to the presence of non-adhesive factors in the serum-free medium, which allow cells to be adapted to suspension cultures.

**TABLE 2.** Cell attachment on microcarriers at different serum concentrations.

| Medium composition | Inoculum (cells $ml^{-1}$ medium) | Unattached cell fraction |
|---|---|---|
| RMMI 1640 + 0% FBS | $8.0\times10^5$ | 0.54 |
| RMMI 1640 + 5% FBS | $8.0\times10^5$ | 0.40 |
| RMMI 1640 + 10% FBS | $8.0\times10^5$ | 0.30 |
| RMMI 1640 + 15% FBS | $8.0\times10^5$ | 0.35 |
| CHO SFM-II + 0% FBS | $8.0\times10^5$ | 0.63 |
| CHO SFM-II + 10% FBS | $8.0\times10^5$ | 0.51 |

Table 3 shows the effect of agitation speed on the rate of cell attachment. The lower the agitation speeds, the lower the concentration of the unattached cells at the incubation time of 7 hours. However, it was observed that the agitation speeds lower than 60 rpm were not efficient enough to keep all the microcarriers fully suspended. It is therefore considered that 60 rpm to be the optimal agitation speed which gives a reasonably high cell attachment rate and allows the carriers to be fully suspended. The lower rate of cell attachment generated by the higher agitation speeds might have resulted from the increased hydrodynamic forces acting on the cells and microcarriers against the van der Waals interactions (Gerson, 1981).

**TABLE 3.** Cell attachment on microcarriers at different agitation speeds.

| Agitation speed | Inoculum (cells $ml^{-1}$ medium) | Unattached cell fraction |
|---|---|---|
| 20 RPM | $8.0\times10^5$ | 0.13 |
| 40 RPM | $8.0\times10^5$ | 0.19 |
| 60 RPM | $8.0\times10^5$ | 0.31 |
| 80 RPM | $8.0\times10^5$ | 0.48 |
| 100 RPM | $8.0\times10^5$ | 0.63 |
| 120 RPM | $8.0\times10^5$ | 0.68 |

Table 4 shows the effect of stirring mode on cell attachment. The stirring scheme had no significant effect on the total number of attached cells at 7 hr of incubation.

Continuous stirring gave the lowest rate of cell attachment, but an intermittent stirring/non-stirring regimen of 20/40 minutes resulted in a maximum rate of cell attachment (results not shown).

**TABLE 4.** The effect of different stirring modes on cell attachment.

| Stirring mode | Inoculum (cells $ml^{-1}$ medium) | Unattached cell fraction |
|---|---|---|
| Continuous stirring | $8.0\times10^5$ | 0.320 |
| 10 min stirring /50 min non-stirring | $8.0\times10^5$ | 0.315 |
| 15 min stirring /15 min non-stirring | $8.0\times10^5$ | 0.314 |
| 20 min stirring /40 min non-stirring | $8.0\times10^5$ | 0.310 |

The variation in number of attached cells on single microcarrier is shown in Table 5. The MTT absorbance figures in the table represent the mean and standard error of the optical density of released formazan resulted from the reduction of MTT by viable cells obtained using the MTT method. The similarity between mean values indicates that the stirring modes had no significant effect on the number of attached cells. However, the large difference between the standard error values indicates that the distribution of cells on microcarriers was affected by the stirring modes.

**TABLE 5.** Variation in cell attachment between carriers using different stirring modes (6 hours incubation).

| Stirring mode | MTT absorbance |
|---|---|
| Continuos stirring | $0.0221 \pm 0.0062$ |
| 10 min stirring /50 min non-stirring | $0.0223 \pm 0.0140$ |
| 15 min stirring /15 min non-stirring | $0.0238 \pm 0.0076$ |
| 20 min stirring /40 min non-stirring | $0.0251 \pm 0.0115$ |

The comparison of the rate of cell attachment between the ImmobaSil FS carriers, CultiSpher-G (porous microcarriers) and glass beads (solid microcarriers) is shown in Table 6. The unattached cell fraction of the ImmobaSil FS was reduced to 0.32 at 7 hr of incubation in comparison to 0.61 and 0.53 for Glass beads and CultiSpher-G, respectively. This indicates that the ImmobaSil FS microcarriers had a higher loading capacity of cells compared to the others.

**TABLE 6.** Cell attachment on various microcarriers at an initial cell concentration of $8\times10^5$ cells $ml^{-1}$.

| Microcarrier type | Inoculum (cells $ml^{-1}$ medium) | Unattached cell fraction |
|---|---|---|
| CultiSpher-G microcarriers | $8.0\times10^5$ | 0.53 |
| ImmobaSil FS microcarriers | $8.0\times10^5$ | 0.32 |
| Glass beads | $8.0\times10^5$ | 0.61 |

## 5. Conclusions

In this study the optimal conditions for the anchorage-dependent CHO cells to attach on ImmobaSil FS microcarriers were identified. High inoculum cell-to-carrier ratio, high serum concentration, low agitation speed and intermittent stirring were the conditions required for optimal attachment rate. ImmobaSil FS microcarriers were found to be superior to other microcarriers in term of the number of attached cells.

## 6. Acknowledgement

The authors are grateful to Ashby Scientific Ltd. (http://www.ashby-scientific.co.uk) for providing the ImmobaSil FS porous carriers.

## 7. References

Clark, J.M., Hirtenstein, M.D. and Gebb, C. (1980) Critical parameter in the microcarrier culture of animal cells. *Developments Biologicals Standardization* **46**, 117-124.

Fleischaker, R. (1987) *Microcarrier cell culture.* in *Large scale cell culture technology* (Lydersen, B.K., ed.), pp. 59-79, Hanser, New York.

Forestell, S.P., Kalogerakis, N., Behie, L.A. and Gerson, D.F. (1992) Development of the optimal inoculation conditions for microcarrier cultures. *Biotechnology and Bioengineering* **39**, 305-313.

Gerson, D.F. (1981) *Methods in surface physics for immunology.* in *Immunological methods* (Lefkovits, I. & Pernis, B., eds.), pp. 105-138, Academic Press, New York.

Griffiths, J.B. (1988) *Overview of cell culture systems and their scale-up.* in *Animal cell biotechnology 3* (Spier, R.E. & Griffiths, J.B., eds.), pp. 179-220, Academic Press, London.

Griffiths, J.B. and Thornton, B. (1982) Use of microcarrier culture for the production of herpes simplex virus (type 2) in MRC-5 cells. *Journal of Chemical Technology and Biotechnology* **32**, 324-329.

Hu, W.S., Meier, J. and Wang, D.I. (1985) Mechanistic analysis of the inoculum requirement for the cultivation of mammalian cells on microcarriers. *Biotechnology and Bioengineering* **27**, 585-595.

Preissmann, Wiesmann, R., Buchholz, R., Werner, R.G. and Noé, W. (1997) Investigation on oxygen limitations adherent cells growing on macroporous microcarriers. *Cytotechnology* **24**, 121-134.

# BLEEDING STRATEGY FOR THE LONG-TERM PERFUSION CULTURE OF HYBRIDOMA

J.W. YUN[1], S.Y. LEE[1], B.W. CHOI[1], H.K. OH[1], J.S. LEE[1], B.H. CHUN[1,2], W.G. BANG[2], T.H. BYUN[1], AND S.Y. PARK[1]
[1]*Biological Production Unit, Central Research Center, Korea Green Cross Corporation, 227 Kugal-Ri, Kiheung-Eup, Yongin-Si, Kyunggi-Do, 449-900, Korea*
[2]*Department of Agricultural Chemistry, Korea University, Seoul 136-701, Korea*

We cultivated the hybridoma producing vWF antibody in a 10L bioreactor in order to investigate the characteristics of two different cultures, intermittent and continuous bleeding. To compare with bleeding culture non-bleeding culture was also performed. Non-bleeding culture achieved a cell density of 2.1 x $10^7$ cells/ml at $16^{th}$ day with 110 mg/l of vWF mAb titer. However, the culture was stopped after the 18 days, due to membrane clogging. Intermittent and continuous bleeding cultures were maintained the cell density around 1 x$10^7$ cell/ml. The vWF mAb titer was 45mg/l and 53mg/l for intermittent and continuous bleeding culture. As for volumetric productivity, the continuous bleeding culture was higher than intermittent bleeding culture by 41%. For glucose consumption rate (GCR), lactate production rate (LPR), glutamine consumption rate (GnCR), and ammonia production rate (APR), continuous bleeding culture showed 28%, 70%, 31% and 40% higher than intermittent bleeding culture. These results suggest that physiological activity of the continuous bleeding culture was higher than that of intermittent bleeding culture.

## 1. Introduction

Many industrial cultures of hybridoma are performed in either batch or fed-batch system. In those systems, the reactor size must be very large to produce lots of protein. Perfusion culture has the advantages of high volumetric productivity and efficient medium usage. But the process control and long-term operation of perfusion culture are difficult because of complicate operation system. Recently, many researches have been carried out in the field of process control, and it is possible to cultivate the hybridoma at high density with perfusion. Bleeding the cells is useful strategy to keep a high viability of the cell and to prevent from clogging of the membrane. The strategy of bleeding is complicate because it is influenced by growth rate, death rate, concentration of cell debris, perfusion rate, product titer and so on. In this study, we cultivated the hybridoma producing vWF antibody in a 10L bioreactor with two different bleeding modes, intermittent and continuous bleeding. None bleeding culture is also performed in order to compare with bleeding culture. Both the bleeding cultures are maintained the cell density around $10^7$cells/mL in serum-free media. To investigate the physiological state of each culture, we measured metabolite profiles and calculated the metabolite kinetics. Finally, we evaluate productivity at the unit cell and volume base in each culture.

## 2. Materials and Method

### 2.1 CELL LINE AND CULTURE MEDIA

The mouse-mouse hybridoma cell line vR8 was obtained by fusion using the technique described by Köhler and Milstein(Köhler, 1975). The cell line produces class IgG1 monoclonal antibody against human von Willebrand Factor(vWF). The culture media was a 1:1:1 mixture of DMEM, F12, and RPMI 1640 which contains 15mM HEPES, 2g/L $NaHCO_3$, 4.5g/L glucose, 5mM glutamine and 31.5 $\mu$g/mL of total protein.

### 2.2 BIOREACTOR OPERATION

Two 10L Biostat ED (B.BRAUN) reactors were used to conduct high cell density perfusion culture. The perfusion and bleeding rate were controlled by using the peristaltic pump. The cell density was maintained at around $10^7$cells/mL while the perfusion rate was 3-4 v/v/d. Intermittent bleeding was carried out by draining the culture broth through the sampling device.

### 2.3 ANALYTICAL METHODS

The cell concentration was counted on hemocytometer by trypan blue dye exclusion method. Glucose, glutamine, and lactate were measured using Biochemistry Analyzer (Model 2700, YSI). The vWF antibody concentration was determined by enzyme-linked immunosorbent assay (ELISA). The metabolite kinetics was calculated by using the equation below.

Specific metabolite consumption rate =

$$\frac{48(C_1-C_2)/(t_2-t_1)+2\{C_0-(C_1+C_2)/2\}\times R_P}{(x_1+x_2)/10^3}$$

Which, $C_O$ is initial chemical concentration of fresh media
$C_1$ is concentration of chemical at time $t_1$
$C_2$ is concentration of chemical at time $t_2$
$x_1$ is viable cell concentration at time $t_1$
$x_2$ is viable cell concentration at time $t_2$
$R_P$ is average perfusion rate from $t_1$ to $t_2$

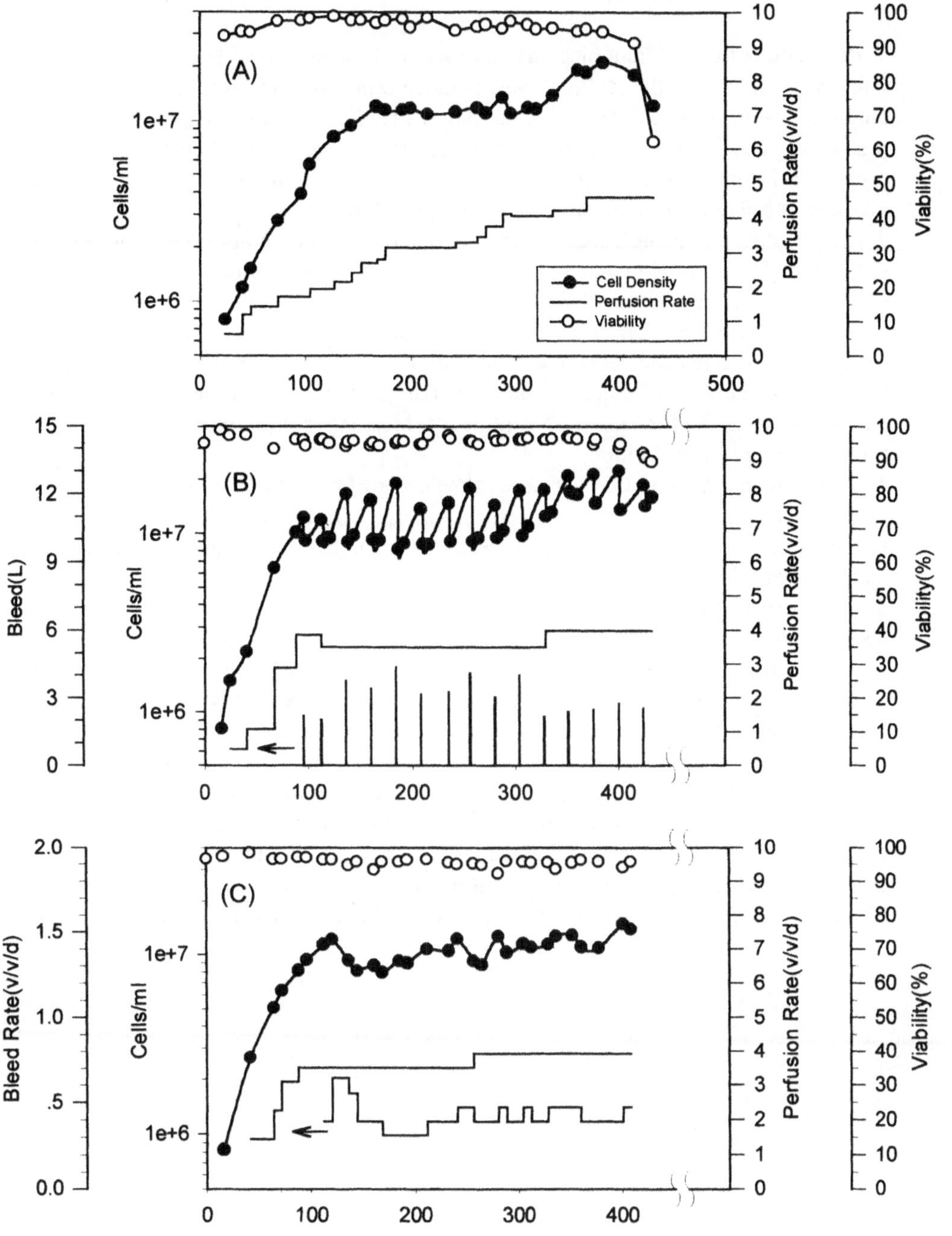


Fig.1 Culture Profiles of vR8 Hybridoma Cells in 10L Reactor.
(A)Without Bleed, (B)Intermittent Bleed, (C)Continuous Bleed

### 3. Results and Discussion

1. Perfusion culture without bleeding achieved a cell density of 2.1E7/ml during 18 days culturing. However, the culture was stopped after the 18days, due to membrane clogging. Both intermittent and continuous bleeding cultures maintained a similar profile of cell density and metabolite concentration. But intermittent bleeding procedure is not easy in operation, compared to the continuous bleeding because of possible introduction of contamination into the culture.
2. For all cultures, lactate concentrations were similar. But ammonia concentr-ation was slightly high (7 mM) in none and intermittent bleeding cultures, compared to the continuous bleeding culture (4-5 mM).
3. For GCR, LPR, GnCR and APR, continuous bleeding culture showed 21%, 12%, 31% and 17% higher than none bleeding culture, while it was 28%, 70%, 31%, and 40% higher than intermittent culture. The results suggest that physiological activity of the cultures is high in the order of continuous bleeding culture>none bleeding culture >intermittent bleeding culture.
4. YLac/Glc and YAmm/Gln decrease in the order of none bleeding culture, continuous bleeding culture and intermittent bleeding culture. On the other hand, OUR increases in the same order of the culture, suggesting that higher bleeding in culture enhances metabolic efficiency.
5. While vWF mAb titer increases continuously up to 110 mg/L during culturing, it decreases to 45 mg/L and 53 mg/L for intermittent and continuous bleeding cultures. As for volumetric productivity, the none bleeding culture was higher than the intermittent bleeding culture and continuous bleeding culture by 110% and 50%, respectively.
6. None bleeding culture produces high cell density in culture, but is prone to clog the membrane surface with the 20 days old culture. Therefore, it is very difficult to maintain the culture over 20days. Intermittent bleeding was not efficient in removing by-product, thereby its productivity was not as effective as the continuous culture. The continuous bleeding culture exhibited 40% higher in volumetric productivity than intermittent bleeding culture. For continuous bleeding culture, perfusion rate and bleeding rate must be optimal in order to maintain high cell density for up to 50 days.

Table 1. Comparison of metabolic parameters in different bleeding modes

| Parameters | Bleeding modes | | |
|---|---|---|---|
| | Without | Intermittent | Continuous |
| Glucose Consumption Rate(mM/$10^9$/h) | 0.102 | 0.096 | 0.123 |
| Lactate Production Rate(mM/$10^9$/h) | 0.158 | 0.104 | 0.177 |
| Glutamine Consumption Rate($\mu$M/$10^9$/h) | 35.28 | 35.36 | 46.3 |
| Ammonia Production Rate($\mu$M/$10^9$/h) | 56.35 | 47.14 | 66 |
| Oxygen Uptake Rate(mM/$10^9$/h) | 0.156 | 0.199 | 0.161 |
| Lactate Production Rate/ Glucose Consumption Rate | 1.55 | 1.08 | 1.44 |
| Ammonia Production Rate/ Glutamine Consumption Rate | 1.60 | 1.33 | 1.43 |
| Oxygen Uptake Rate/ Glucose Consumption Rate | 1.53 | 2.07 | 1.31 |
| Product Titer(mg/L) | 78.81 | 44.82 | 52.54 |
| Specific Productivity($\mu$g/$10^6$/d) | 22.87 | 13.12 | 18.23 |
| Volumetric Productivity(mg/L/d) | 294.8 | 138.3 | 195 |

*Each parameters are average value acquired after 100hrs of culture period.

## References

Seamans, T.C. and Hu, W.S. (1990) Kinetics of growth and antibody production by a hybridoma cell line in a perfusion culture, *J. Fer. and Bioeng.* **70**, 241-245.

Tokashiki, M. and Takamatsu, H. (1993) Perfusion culture apparatus for suspended mammalian cells, *cytotechnology* **13**, 149-159.

Hiller, G.W., Clark, D.S., and Blanch, H.W. (1993) Cell retention-chemostat studies of hybridoma growth and metabolism in continuous suspension culture on serum-free medium, *Biotechnol. Bioeng.* **42**, 185-195.

Zamboni, A., Giuntini, I., Gianesello, D., Maddalena, F., Rognoni, F., and Herbst, D., (1994) Production of mouse monoclonal antibodies suing a continuous cell culture fermenter and protein G affinity chromatography, *Cytotechnology* **16**, 79-87.

Banik, G.G., and Heath, C.A., (1995) Hybridoma growth and antibody production as a function of cell density and specific growth rate in perfusion culture, *Biotechnol. Bioeng.* **48**, 289-300.

# METABOLIC FLUX DISTRIBUTIONS IN HYBRIDOMA CELLS AT DIFFERENT METABOLIC STATES

Peng-Cheng Fu, Anna Europa, Anshu Gambhir and Wei-Shou Hu
*Department of Chemical Engineering & Materials Science,*
*University of Minnesota, Minneapolis, MN 55455 USA*

## ABSTRACT

The metabolic state, specifically the conversion of nutrients to lactate and other metabolites, of hybridoma cells can be manipulated in a fed-batch culture by controlling the level of glucose. When cultivated in continuous cultures, these different metabolic states result in multiple steady states marked by different cell, residual nutrient and metabolite concentrations. Most notably, the ratio of lactate produced to glucose consumed was markedly different. To better understand the underlying mechanisms of these metabolic states, metabolic flux analysis was performed. The intracellular fluxes are greatly different in the glycolytic pathway and amino acid catabolism among these steady states. The fluxes in the high lactate producing state were much greater than in the "efficient" state. The comparative analysis of intracellular fluxes lends credence to the idea of metabolic overflow in the excessive production of the metabolic byproducts: lactate and ammonia.

## INTRODUCTION

Animal cells in culture typically convert most of the glucose they consume into lactate. The accumulation of lactate, however, is commonly cited as one of the factors that inhibit cell growth and limit the maximum cell concentration that can be achieved in culture. The production of lactate and the amount of glucose converted to lactate can be reduced when cells are grown in a fed-batch culture in which the residual glucose concentration is maintained at low levels.(Zhou, Rehm et al. 1995; Zhou, Rehm et al. 1997) Such a fed-batch culture was used to grow and adapt hybridoma cells into a low-lactate producing state before changing into continuous culture. The cells reached and maintained a high viable cell concentration at steady state. In a similar manner, cells that were initially grown in batch culture and a glucose-rich environment reached a steady state with a cell concentration that is much lower. We have performed dozens of such experiments at dilution rates ranging from 0.028–0.033 $hr^{-1}$ in the past year. Starting from a batch culture, the resulting steady states were always in the high lactate producing metabolic state. While starting from a fed-batch culture by controlling the residual glucose at a low level of ~0.05 g/l, the resulting steady states fall into two categories: one with nearly no lactate production, the other with drastically reduced but still appreciable lactate production. The feed composition rates for all these cultures were the same. Many of the cultures were operated at the same dilution rate, suggesting steady state multiplicity. From a processing perspective, the desired steady state among those is the one with the least metabolite production. At such steady state nutrient concentration in the feed can be further increased to increase cell and product concentrations without causing the metabolite inhibitory effect typically seen in a cell culture. Controlling cell metabolism in a continuous culture to reduce or eliminate waste metabolite production may significantly improve the productivity of mammalian cell culture processes.

To better understand the underlying mechanism governing these different metabolic states associated with different steady states, we performed material balances on the key intermediate involved in the energy metabolism of glucose and amino acids. Such metabolic flux analysis is a powerful tool to elucidate cellular metabolism; however, its application is often obscured by the inability to account for material balance of medium components and biomass. Since our continuous cultures were carried out in a defined medium and the elemental composition of the biomass and products were known, a high degree of confidence can be attained with our experimental results. This paper presents the results of our metabolic flux analysis.

## METABOLIC REACTIONS AND CALCULATION

The metabolic reactions considered are shown in Table 1. A simplified presentation is shown in Figure 1. Three sets of steady state data associated with three different metabolic states with a molar stoichiometric ratio of lactate to glucose (L/G) of 1.40, 0.12 and 0.03 respectively were used (Table 2). Each set of data includes the experimentally derived specific consumption or production rates of glucose, lactate, seventeen amino acids, ammonia and biomass. The organic medium components which are not included in the analysis are insulin, transferrin, ethanolamine, vitamins, linoleic acid, thymidine and the three amino acids (cystine, tryptophan and proline) not measured in our HPLC analysis. The consumption rates of these compounds were relatively small. Their exclusion from the analysis does not affect the results. The biomass formula was derived from elemental analysis of C, N, H, O. The stoichiometric coefficients of amino acids (except glutamine) shown on the left hand side of the equation (Eq. 12 in Table 1) were obtained by first plotting their respective specific consumption rates from all cultures vs. their corresponding specific lactate production rates and then extrapolating to zero lactate production. In our analysis, only carbon and nitrogen balances are considered. In other words, energy balance (NAD, NADP, ATP, etc.) is not considered.

A general matrix representation of cellular metabolism is shown as follows:

$$Ax(t) = r(t) \tag{1}$$

where A represents the stoichiometric coefficients, x(t) the intracellular fluxes, and r(t) the vector for the accumulation rate of each metabolite in the network. With a pseudo-steady state assumption, the value for all intracellular compounds in r(t) becomes nil, while those that are present in the medium have the values of the corresponding specific consumption or production rates. The standard deviation of each measured specific rate was also included in the analysis. The algorithm also allows the solution to be constrained as follows:

## TABLE 1. Biochemical reactions in hybridoma cells

**A Glycolysis:**
1) GLC → 2 PYR
2) PYR → LAC

**B TCA cycle**
3) PYR → AcCoA + $CO_2$
4) AcCoA + OAA → αKG + $CO_2$
5) αKG + NAD → SUCCOA + $CO_2$
6) SucCoA → FUM
7) FUM → MAL
8) MAL → OAA

**C Glutaminolysis:**
9) GLN → GLU + $NH_3$
10) GLU → αKG + $NH_3$
11) OAA → PYR + $CO_2$

**D Biomass Synthesis:**
12) 0.1016 GLC + 0.031 GLN + 0.008 ARG + 0.0003 ASN + 0.001 GLU + 0.0038 GLY + 0.0028 HIS + 0.0071 ILE + 0.008 LEU + 0.0043 LYS + 0.001 MET + 0.0152 THR + 0.0051 VAL → $CH_{1.975}N_{0.1401}O_{0.489}$

**E Amino Acid Metabolism**
13) PYR + GLU → ALA + αKG
14) SER → PYR + $NH_3$
15) GLY → SER
16) CYS → PYR + $NH_3$
17) ASP + αKG → OAA + GLU
18) ASN → ASP + $NH_3$
19) HIS → GLU + $NH_3$
20) ARG + αKG → 2 GLU
21) PRO → GLU
22) ILE + αKG → SUCCoA + AcCoA + GLU
23) VAL + αKG → GLU + $CO_2$ + SUCCoA
24) MET → SUCCoA
25) THR → SUCCoA + $NH_3$
26) PHE → TYR
27) TYR + αKG → GLU + FUM + 2 AcCoA
28) LYS + 2 αKG → 2 GLU + 2 $CO_2$ + 2 AcCoA
29) LEU + αKG → GLU + 3 AcCoA

**AcCOA**: Acetyl coenzyme A; **αKG**: α-ketoglutarate; **ALA**: Alanine; **ARG**: Arginine; **ASN**: Asparagine; **ASP**: Aspartate; **BIOMAS**: Biomass; **CYS**: Cysteine; **$CO_2$**: Carbon dioxide; **FUM**: Fumarate; **GLC**: Glucose; **GLN**: Glutamine; **GLU**: Glutamate; **GLY**: Glycine; **HIS:** Histidine; **ILE**: Isoleucine; **LAC**: Lactate; **LEU**: Leucine; **LYS**: Lysine; **MAL**: Malate; **MET**: Methionine; **$NH_3$**: Ammonia; **$O_2$**: Oxygen; **OAA**: Oxaloacetate; **PHE**: Phenylalanine; **PRO**: Proline; **PYR**: Pyruvate; **SER**: Serine; **SUCCOA**: Succinate coenzyme A; **THR**: Theonine; **TYR**: Tyrosine; **VAL**: Valine

## TABLE 2. Values of key variables associated with the three different steady states

| | I | II | III |
|---|---|---|---|
| Cell concentration ($10^6$ cells/mL) | 0.88 | 4.30 | 5.24 |
| Residual glucose (mM) | 0.40 | 0.22 | 0.16 |
| Residual glutamine (mM) | 0.10 | 0.002 | 0.04 |
| Lactate (mM) | 7.19 | 1.43 | 0.13 |
| Ammonium (Mm) | 1.27 | 2.59 | 3.22 |
| Lactate production/glucose consumption (mole/mole) | 1.40 | 0.12 | 0.02 |
| Specific glucose consumption rate (mmole/cell-hr) | 0.18 | 0.04 | 0.03 |
| Specific glutamine consumption rate (mmole/cell-hr) | 0.10 | 0.03 | 0.02 |

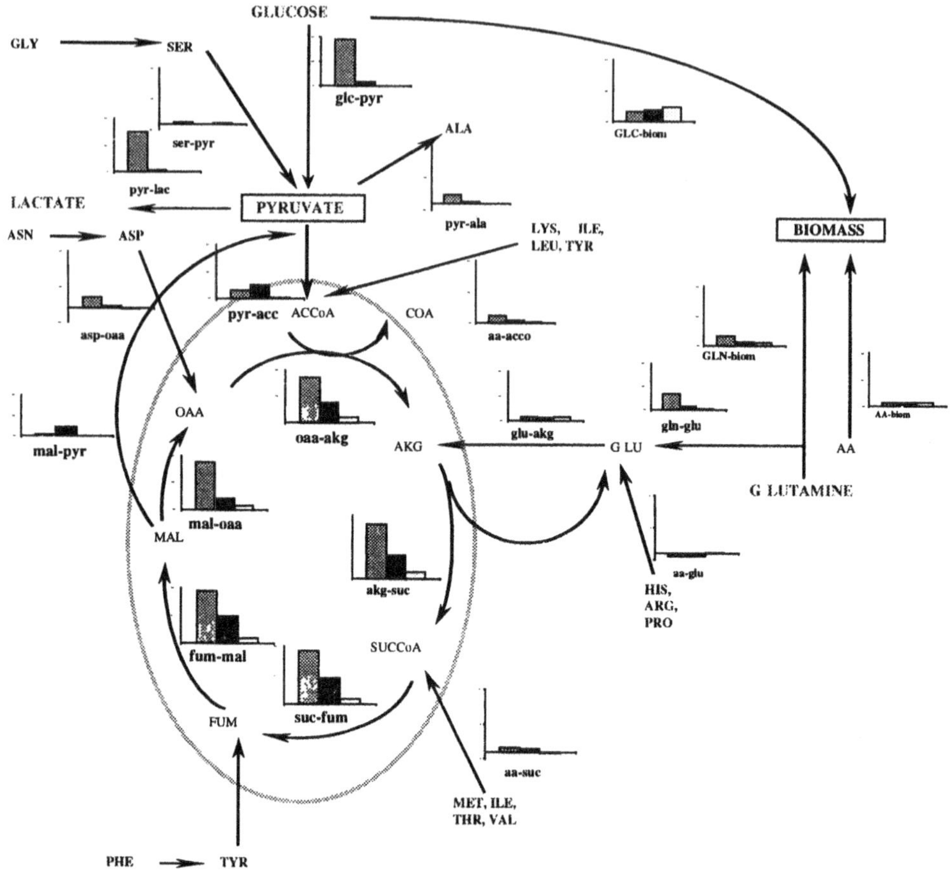


**FIGURE 1. Intracellular flux distribution for hybridoma cells at three different metabolic states. The three bars, from left to right, are for L/G of 1.4, 0.12 and 0.03 respectively.**

$$Cx \leq \mathbf{b} \quad (2)$$

where C is the constraint matrix that specifies the reactions to be constrained to be greater than or equal to **b**. **b** is usually taken as the zero vector. Once a nonsingular, well-conditioned stoichiometry matrix A is constructed, and the constraint (2) is determined, the flux estimates, x, can be calculated from the measurements r via least squares method:

$$x(t) = (A^T \Psi^{-1} A)^{-1} A^T \Psi^{-1} r(t) \quad (3)$$

where $\Psi$ is the measurement noise variance matrix associated with r.

## INTRACELLULAR FLUXES

The carbon flows for the three metabolic states analyzed are shown in Figure 1. The metabolic flux analysis show that under conditions of the same feed and the same dilution rates, multiple steady states exist. At the high lactate-producing state (L/G 1.4), the carbon flux is the highest in almost all the intracellular reactions. In addition to a higher diversion of pyruvate to lactate, the fluxes through the TCA cycle are also higher. In the intermediate state in which less lactate is produced, the flux through most reaction steps is lower than that in the high lactate producing state. Note that one of the exceptions is the flux from pyruvate to acetyl CoA: the flux in the intermediate state is greater than that in the glycolytic state. In the metabolic state associated with the lowest L/G, a very large fraction of the carbons from glucose is directed to the biomass formation, and the fluxes through the central metabolic pathways are significantly lower than the other two metabolic states. It is therefore a most "efficient" metabolic state for the cells.

The results of these metabolic flux analyses reveal that cells at the least "efficient" metabolic state have higher fluxes in both glycolysis and TCA cycles. This is paradoxical as it implies a higher energy production associated with this high lactate producing metabolic state. In our analysis, energy balance was not considered. It was felt that much biological phenomena (sustaining membrane potential, osmotic balance, and various transport, etc.) involving energy are unaccounted for in this metabolic flux analysis, thus a meaningful energy balance is elusive. However, if the results of our metabolic flux are correct, it implies that energetic aspects of cellular process are an important variable associated with different metabolic states.

## REFERENCES

Zhou, W. C., J. Rehm, et al. (1995). "High viable cell concentration fed-batch cultures of hybridoma cells through on-line nutrient feeding." Biotechnology & Bioengineering **46**(6): 579-587.

Zhou, W.-C., J. Rehm, et al. (1997). "Alteration of mammalian cell metabolism by dynamic nutrient feeding." Cytotechnology **24**: 99-108.

# ON-LINE DETECTION OF METABOLIC DEMAND AND CHANGES TO IT: HEAT FLUX IN CONTINUOUS CULTURES OF ANIMAL CELLS

Y.H. GUAN and R.B. KEMP
*Institute of Biological Sciences, The University of Wales*
*Penglais, Aberystwyth, SY23 3DA, Wales, UK*

**Abstract** Heat flow rate was used on-line and in real time to monitor continuous cultures of CHO320 cells for achievement of steady state conditions. The metabolic demand for the key substrates, glucose and glutamine, was shown by the link between metabolic activity (heat flow rate in this work) and the stoichiometry of the growth reaction. By relating these findings to those for batch culture, it was deduced that cell metabolism for most of the time in the latter is close to steady state. It was also shown in the continuous cultures that metabolic activity was a reflection of mitochondrial membrane potential.

## 1. Introduction

For many mechanistic studies of cell metabolism in relation to growth and for precise investigations relying on a strict thermodynamic explanation, it is necessary to achieve a specific steady state. For cells *in vitro*, the classical way to set such a state is to arrange a continuous culture at a particular dilution rate [1]. Information that this state has indeed been accomplished is traditionally furnished by off-line assays of cell numbers and the concentrations of major metabolites. These are important individual facets of the overall growth reaction but, of course, growth is an explicitly dynamic process and is described most properly by its rate. This requires two interrelated conditions: (i) the measurement should be on-line and (ii) an instantaneous rate should be found rather than an average relative rate. A clue to serving these demands can be ascertained when it is considered that the growth reaction is in fact a description of cellular metabolic activity. In terms of non-equilibrium thermodynamics (NET), this is characterised as the Gibbs energy of the substrates being dissipated largely as heat with a small fraction of it dissipated as entropy in the formation of biomass [2]. So, by definition, the rate of heat flow must be a function of metabolic activity and is measured as the instantaneous rate by heat conduction calorimetry [3]. For cell cultures in suspension, it has been found possible to monitor heat flow rate on-line and in real time [4]. The first aim of this study is to demonstrate that such a measurement can show the steady state of cells in culture, using CHO320 cells producing interferon-$\gamma$ as the example.

From our earlier results with batch cultures, it proved possible to construct a simplified growth reaction which was validated by the enthalpy balance method [5].

$$\nu_{S_1} C_6H_{12}O_6(\text{Glc}) + \nu_{S_2} C_5H_{10}N_2O_3(\text{Gln}) + \nu_O O_2 \rightarrow CH_\alpha O_\beta N_\gamma(\text{cell}) + \nu_L C_3H_6O_3(\text{Lac}) + \nu_C CO_2 + \nu_N NH_3 + \nu_H H_2O + \nu_x\left(\int_{t_1}^{t_2} J_{H,r}\,dt\right) \quad (1)$$

Calculations from experimental data, when applied to such an equation, can show the actual cellular demand for substrates as opposed to the amount of them simply provided in the medium. For the medium originally used in our studies [6], the ratio for the demand for glucose and glutamine was 3:1, in contrast to the 5:1 ratio supplied in it. The required proportion was incorporated into the recipe for an improved medium, which also had increased concentrations of amino acids and some other ingredients [7]. This was used in the present study. It was recognised however that the demand for substrates by cells in steady state may well be markedly different from that of cells in batch culture where there are constantly changing conditions. The second aim is, therefore, to calculate the stoichiometric coefficients for cells in continuous culture and compare them with those previously obtained for batch cultures.

With increased metabolic activity owing to the use of optimised media comes the possibility that mitochondrial capacity may limit energy provision for cellular growth and maintenance. Earlier investigations on these cells growing in batch culture in the original medium had shown 40% spare capacity [8] and 20% in the improved medium [7]. It seemed likely however that the greater metabolic activity induced by faster dilution rates would be a reflection of increased mitochondrial membrane potential. The third aim was to show the relationship between the glucose consumption at different dilution rates and mitochondrial activity by flow cytometry.

## 2. Experimental

In order to achieve the relatively high cell density required for continuous cultures, CHO320 cells [6] were grown initially as a batch culture in a controlled bioreactor [4] (Applicon Ltd., Tewksbury, Glous., UK) using an improved, RPMI-based medium with 24 mM bicarbonate buffer at pH 7.25 ± 0.02 and a dissolved oxygen concentration of 55%[7]. Once the chosen cell density was achieved, the continuous culture was set up by constantly drawing fresh medium at a given (dilution) rate into the culture using one of the pumps in the Applicon controller, while maintaining an unchanging bulk volume. Heat flow rate was measured on-line and ex situ [9] using a TAM microcalorimeter (Thermometric, AB, Järfälla, Sweden) with a flow module optimised for use with animal cells in suspension [10]. Off-line measurements were made of cell density and the concentrations of major metabolites as described previously [4]. Oxygen uptake rate was estimated using a Paar Oroboros high resolution respirometer [4,11] and mitochondrial membrane potential was shown by staining cells with rhodamine-123 and measuring its intensity by flow cytometry [12] using an Argus 100 instrument (Skatron Ltd., Newmarket, UK).

## 3. Results and Discussion

When the batch culture of CHO 320 cells had reached a high density (ca. $1\times10^6$ $cm^3$), the first of three sequentially lower dilution rates was set up at 0.020 $h^{-1}$ (see Figure 1). The results showed that heat flow rate and oxygen uptake rate both increased with the additional supply of substrates afforded by higher dilution rates. From an understanding of NET, the thermal data mean the metabolic activity was at least partly a function of substrate availability, though the exact form of the relationship and its connection to cellular demand will require further careful experimentation. Measurements of the concentrations of the two major substrates, glucose and glutamine, indicated that the latter was the rate-limiting metabolite (Figure 2). This is not surprising since it is the substance which effectively has the greatest control over the growth of cells in batch culture [4]. The amount of ammonia in the culture also remained constant over the different dilution rates, suggesting that the proportions of ammonia utilisation between the transamination, glutaminolysis and nucleic acid synthesis remained the same at the three steady states. There was, however, some indications that the production of lactate may not be the

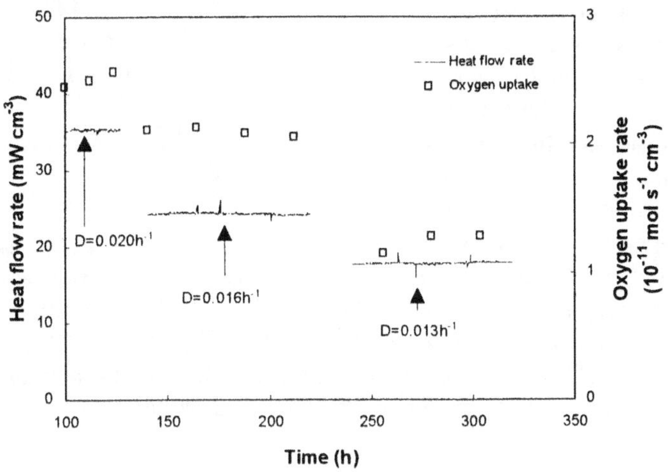


*Figure 1.* Heat flux is a reflection of steady states achieved in continuous culture.

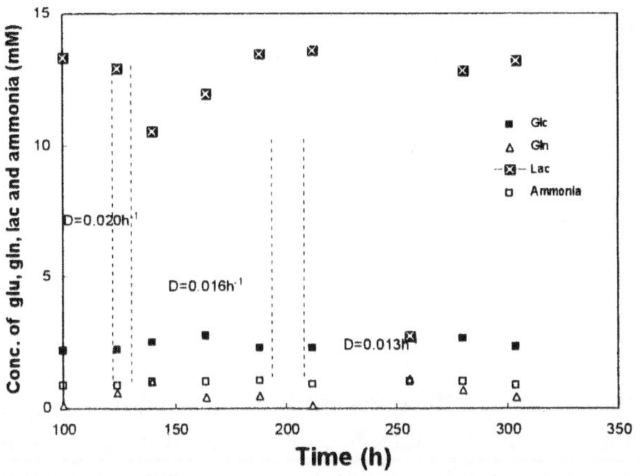


*Figure 2.* Variations in the concentrations of glucose, glutamine, lactate and ammonia when steady states were achieved in continuous culture. The results show that glutamine is the limiting substrate for cell growth.

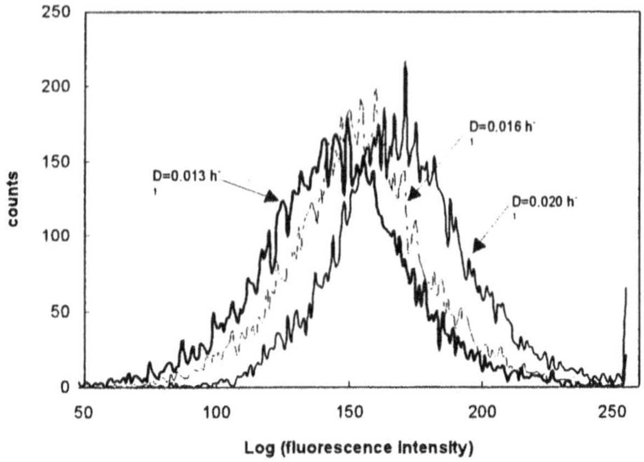


*Figure 3.* The cells growing at different dilution rates were stained by rhodamine 123 to reveal the change in mitochondrial membrane polarization with specific cell growth rate.

same throughout the range of continuous cultures and this raised the need for further experimentation to determine the fate of glucose in catabolism by these cells.

As stated earlier [2,4,5,7,9], the stoichiometric coefficients obtained from growth reactions can provide important information about the demands of cells for substrates (see the simplified equation, Eq. (1)). This methodology was applied to the present data for continuous cultures and the batch cultures preliminary to them (see Table 1). It is clear that the values for the

TABLE 1. Stoichiometric coefficients and heat flux obtained from continuous and batch culture

| Culture condition | $\nu_{S_1}$ (Glc) | $\nu_{S_2}$ (Gln) | $\nu_O$ ($O_2$) | $\nu_L$ (Lac) | $\nu_C$ ($CO_2$) | $\nu_N$ ($NH_3$) | $\nu_H$ ($H_2O$) | Heat flux (pW per cell) |
|---|---|---|---|---|---|---|---|---|
| Continuous culture at D=0.020 $h^{-1}$ | 0.842 | 0.307 | 1.881 | 1.233 | 1.32 | 0.69 | 0.50 | 35.8 |
| Continuous culture at D=0.016 $h^{-1}$ | 0.682 | 0.240 | 1.762 | 1.190 | 1.011 | 0.513 | 0.498 | 28.5 |
| Continuous culture at D=0.013 $h^{-1}$ | 0.54 | 0.20 | 1.48 | 0.91 | 0.87 | 0.47 | 0.39 | 21.9 |
| Batch culture at ave. $\mu$ = 0.018 $h^{-1}$ | 0.55 | 0.17 | 0.51 | 0.87 | 0.60 | 0.43 | 0.32 | 21.8 |

heat flow rate reflect the different demand for key substrates. Since all the sets of stoichiometric coefficients are similar and those for the continuous cultures are at steady state conditions, this means that batch cultures of CHO320 cells, at least under the typical conditions in this laboratory, were close to a steady state at any one particular time.

Since glucose flux cannot be determined easily from the concentration data, a set of flow cytometric experiments were arranged to determine the intensity of mitochondrial membrane potential at different dilution rates. The results are depicted in Figure 3. Although the relationship between the curve for fluorescence intensity by rhodamine-123 and mitochondrial membrane potential is complex, a simple approximation may be obtained from the position of the peak values. It can be seen from Figure 3 that the higher rate of glucose consumption with increasing dilution rates was, in part at least, due to the greater mitochondrial activity. It is not possible to state from these results if total mitochondrial capacity has been reached at the highest dilution rate applied in these experiments but care will be needed for future studies of this type.

## 4. Conclusions

The results show that the heat flow rate is an early and sensitive indicator of the changing metabolic state of cells in continuous as well as in batch culture. Since it has been shown in previous results that production of the target protein deteriorates with decreasing metabolism [4], in an industrial sense action must be taken to maintain productivity. Since data from continuous cultures prove the relationship between the stoichiometric coefficients and heat flow rate, the way to improve cell growth and the quality production of target protein is to feed the cells continuously at a rate determined by the specific heat flow rate (flux) and dependent on the fluxes of the major substrates in their correct molar ratio.

**Acknowledgments:** Y.H.G. was the recipient of a conference grant from the Royal Society. The work was funded by UK Biotechnology and Biological Sciences Research Committee grants 2/3680 and 2/TO3789.

## 5. References

1. Pirt, S.J. (1975) *Principles of Microbe and Cell Cultivation*, Blackwell Scientific Publ., Oxford, p.4.
2. Kemp, R.B. and Guan, Y. (1997) Heat flux and calorimetric-respirometric ratio as measures of catabolic flux in mammalian cells. *Thermochim. Acta* **300**, 199–211.
3. Kemp, R.B. (1998) Nonscanning calorimetry, in M.E. Brown (ed.), *Handbook of Thermal Analysis and Calorimetry*, Vol. 1, Elsevier, Amsterdam, pp. 577–675.
4. Guan, Y., Evans, P.M., Kemp, R.B., Heat flux: an on-line monitor and potential control variable of metabolic flux in animal cell culture that combines microcalorimetry with dielectric spectroscopy, Biotechnol. Bioeng. **58**, 464-477.
5. Guan, Y.H. and Kemp, R.B. (1998) Detection of dynamic substrate requirements of animal Cells by stoichiometric growth equations and enthalpy balance, submitted to *J. Biotechnol.*
6. Hayter, P.M., Curling, E.M.A., Baines, A.J., Jenkins, N., Salmon, I., Strange, P.G. and Bull, A.T. (1991) Chinese hamster ovary cell growth and interferon production kinetics in stirred batch culture. Appl. Microbiol. Biotechnol. **34**, 559-564.
7. Y.H. Guan and R.B. Kemp (1998) On-line heat flux measurements improve the culture medium for the growth and productivity of genetically engineered CHO cells. *Cytotechnology* (in press).
8. Kemp, R.B. and Guan, Y. (1998) Probing the metabolism of genetically-engineered mammalian cells by heat flux. *Thermochimica Acta* **309**, 63-78
9. Guan, Y., Lloyd, P.C., Evans, P.M. and Kemp, R.B. (1997) A modified continuous flow microcalorimeter for measuring heat dissipation by mammalian cells in batch culture, *J. Thermal Anal.*, **49**, 785-794.
10. Guan, Y.H. and Kemp, R.B. (1998) A calorimetric flow vessel optimised for measuring the metabolic activity of animal cells, *Thermochim. Acta* (in press).
11. Gnaiger, E., Steinlechner-Maran, R., Méndez, G., Eberl, T. and Margreiter, R. (1995) Control of mitochondrial and cellular respiration by oxygen. J. Bioenerg. Biomembr. **27**, 583-596.
12. Kidane, A., Guan, Y., Evans, P.M., Kaderbhai, M.A. & R.B. Kemp. (1997). Comparison of Heat Yields in Wild and Genetically Engineered Chinese Hamster Ovary Cells. *J. Thermal Anal.* 49, 771–784.

# GROWTH STIMULATION OF CHO CELLS BY PHOSPHOLIPIDS IN SERUM-FREE CULTURE

KENTARO SAKAI,[1] TOSHIO MATSUNAGA,[1] HIDEKI YAMAJI,[1]* AND HIDEKI FUKUDA[2]
[1]*Department of Chemical Science and Engineering, Faculty of Engineering, and* [2]*Division of Molecular Science, Graduate School of Science and Technology, Kobe University, 1-1 Rokkodai, Nada, Kobe, Hyogo 657-8501, Japan.* *Corresponding author.

**Abstract.** The effects of phosphatidic acid (PA) and lysophosphatidic acid (LPA) on the growth of Chinese hamster ovary (CHO) cells in serum-free media were evaluated by MTT assay. Supplementing PA or LPA to a protein-free basal synthetic medium or a serum-free medium containing insulin and transferrin markedly stimulated the growth of CHO cells, suggesting that PA and LPA are promising growth-promoting supplements for use in constituting a low-protein serum-free medium.

## 1. Introduction

For effective mass production of clinically and diagnostically important proteins, serum-free culture of animal cells is increasingly desired since it can facilitate the purification of cellular products and thereby reduce production costs. For serum-free culture, a serum-free medium is generally prepared by supplementing a basal synthetic medium with various hormones and growth factors, large amounts of serum proteins, and other substances. However, addition of such proteins, especially those from animal sources, should as far as possible be reduced for effective purification of cellular products and to minimize potential contamination by viruses, mycoplasmas, and/or other infectious agents.

In this connection, exogenous phospholipids like phosphatidic acid (PA) and lysophosphatidic acid (LPA) have been reported to act as mediators in signal transduction, leading to the stimulation of DNA synthesis in fibroblast cells (van Corven *et al.*, 1989; Moolenaar, 1994). In this study, we examine the efficacy of PA and LPA as possible growth-promoting supplements for use in developing a low-protein serum-free medium for Chinese hamster ovary (CHO) cells.

## 2. Materials and Methods

### 2.1. CELL LINE AND ROUTINE MAINTENANCE

The anchorage-dependent CHO DXB11 cell line, a dihydrofolate reductase-deficient mutant (Urlaub and Chasin, 1980), was used. The cells were maintained in α-MEM with nucleosides (Life Technologies, Grand Island, NY, USA) supplemented with 5% fetal bovine serum (FBS; Life Technologies) in T-flasks kept in a $CO_2$ incubator (37℃, 5% $CO_2$).

### 2.2. SERUM-FREE MEDIA

To examine effects of the phospholipids on the growth of the CHO cells, two kinds of basal synthetic medium, α-MEM and UC212 (Nissui Pharmaceutical, Tokyo, Japan), were used as protein-free media. UC212 is a modification of α-MEM formulated for growing CHO cells under serum-free conditions and for the production of recombinant protein (Mizuguchi *et al.*, 1993). The supplementation of these media with ITES (10 mg/*l* insulin, 5.5 mg/*l* transferrin, 2 mg/*l* ethanolamine, and 6.7 μg/*l* sodium selenite; Murakami *et al.*, 1982) was also examined.

### 2.3. PHOSPHOLIPIDS

PA from egg yolk lecithin, dimyristoyl-PA (C14:0), dipalmitoyl-PA (C16:0), dioleoyl-PA (C18:1, [*cis*]-9), distearoyl-PA (C18:0), and 1-oleoyl-LPA (C18:1, [*cis*]-9) were purchased from Sigma, St. Louis, MO, USA. Each phospholipid was suspended at 0.4–1 g/*l* in HEPES buffered saline (15 mM HEPES and 8 g/*l* NaCl, pH 7.4) by sonication with a probe sonicator and was added to the serum-free medium to give the final concentrations indicated.

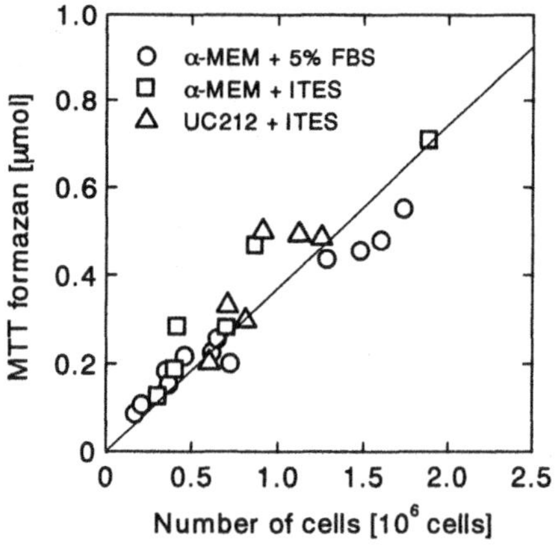


*Figure 1.* Relationship between number of viable CHO DXB11 cells cultured in serum-containing or serum-free medium and production of MTT formazan in MTT assay.

## 2.4. ASSAY FOR EVALUATING CELL GROWTH

Cells were allowed to grow in α-MEM supplemented with 5% FBS before the medium was replaced with protein-free α-MEM. After serum starvation of the protein-free α-MEM for 2 d, cells were inoculated at the same cell density in 12-well plates containing the serum-free medium to be tested. After 3-d incubation in the serum-free medium, the number of viable cells in each well was evaluated by MTT (3-(4,5-dimethylthiazol-2-yl)-2,5-diphenyl tetrazolium bromide) assay (Mosmann, 1983; Yamaji and Fukuda, 1992).

# 3. Results and Discussion

Figure 1 shows the relationship between number of viable CHO DXB11 cells in a monolayer culture using serum-containing or serum-free medium and production of MTT formazan in the MTT assay. The MTT formazan produced increased linearly with the number of cells, suggesting that the specific activity of MTT reduction by the cells was almost constant irrespective of the medium. The number of viable cells after 3-d incubation in the serum-free medium being tested was then quantified spectrophotometrically by MTT assay.

Figure 2 (a) shows the effect of PA from egg yolk lecithin on the growth of CHO DXB11 cells in protein-free α-MEM. The optical density on the ordinate reflects the number of viable cells in one well of a 12-well plate after 3-d incubation in the tested medium as described above. As shown in the figure, supplementing PA from egg yolk lecithin to protein-free α-MEM markedly promoted the growth of CHO DXB11 cells. A PA dose-dependent increase in number of the cells was observed in the range 1–40 mg/*l*. Figure 2 (b) shows the effect of PA on the

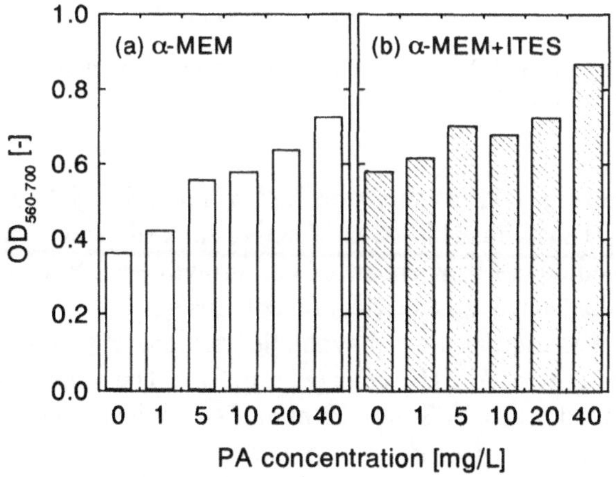


*Figure 2.* Effect of PA from egg yolk lecithin on growth of CHO DXB11 cells in (a) protein-free α-MEM and (b) α-MEM supplemented with ITES. $OD_{560-700}$, optical density at 560–700 nm in MTT assay. Each bar represents the mean of values from duplicate wells in 12-well plates.

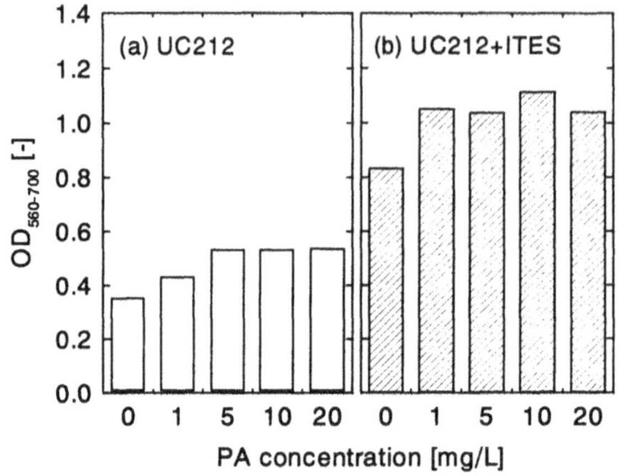


*Figure 3.* Effect of PA from egg yolk lecithin on growth of CHO cells in (a) protein-free UC212 and (b) UC212 supplemented with ITES.

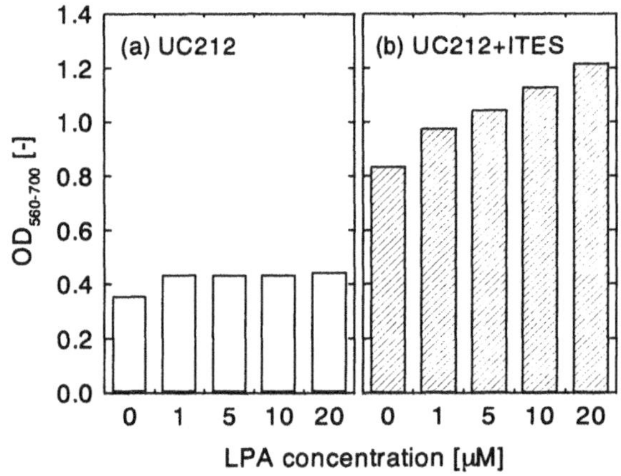


*Figure 4.* Effect of 1-oleoyl-LPA on growth of CHO cells in (a) protein-free UC212 and (b) UC212 supplemented with ITES.

growth of CHO cells in α-MEM supplemented with ITES. While supplementation of protein-free α-MEM with ITES was effective for cell growth, addition of PA to the ITES-supplemented α-MEM enhanced cell growth even further. This suggests that ITES and PA had a synergistic effect on cell growth.

When UC212 was used as a basal synthetic medium, PA from egg yolk lecithin also stimulated the growth of CHO cells (Figure 3). The number of viable cells in UC212 supplemented with ITES and 10 mg/*l* PA reached over 70% of that in α-MEM supplemented with 5% FBS. Supplementation of serum-free UC212 with 1-oleoyl-LPA also accelerated the growth of CHO cells (Figure 4). The results shown in Figures 2–4 thus suggest that both PA and LPA are promising as a growth-promoting supplements for use in the serum-free culture of CHO cells.

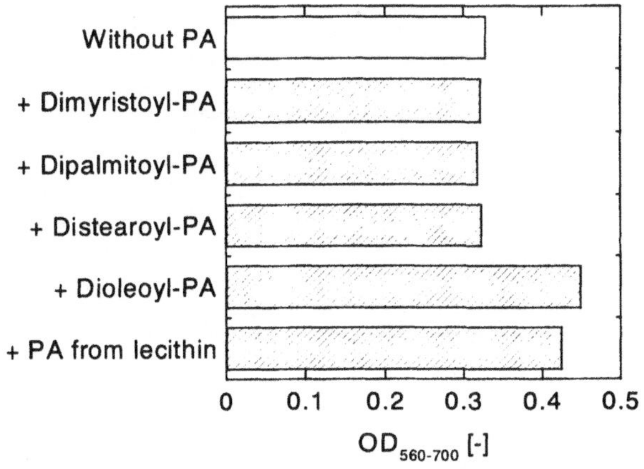


*Figure 5.* Effect of PA analogues with different acyl chains in protein-free α-MEM. Concentration of dimyristoyl-PA (C14:0), dipalmitoyl-PA (C16:0), distearoyl-PA (C18:0), and dioleoyl-PA (C18:1), 15 μM. Concentration of PA from egg yolk lecithin, 10 mg/*l*.

To determine how the nature and length of the PA acyl chain affects cell growth, we tested various PA analogues with different acyl chains. Of the analogues examined, only dioleoyl-PA (C18:1, [*cis*]-9) was potent in stimulating the growth of CHO cells (Figure 5). In this study, the phospholipids were suspended directly in the serum-free medium. Since PAs with unsaturated acyl chains, like dioleoyl-PA, are more water-soluble than those with saturated acyl chains, the above result may indicate that the availability to the cells of the former type is superior to that of the latter. Hence, a suitable PA analogue should be selected depending on the method employed for adding the PA to the medium.

## 4. References


Mizuguchi, M., Matsumoto, K., and Onodera, K. (1993) Production of the human growth hormone in serum-free UC203 medium, in S. Kaminogawa *et al.* (eds.) *Animal Cell Technology: Basic & Applied Aspects* **5**, Kluwer Academic Publishers, Dordrecht, pp. 57-61.

Moolenaar, W.H. (1994) Lysophosphatidic acid, a multifunctional phospholipid messenger, *J. Biol. Chem.* **270**, 12949-12952.

Mosmann, T. (1983) Rapid colorimetric assay for cellular growth and survival: application to proliferation and cytotoxicity assays, *J. Immunol. Methods* **65**, 55-63.

Murakami, H., Masui, H., Sato, G.H., Sueoka, N., Chow, T.P., and Kano-Sueoka, T. (1982) Growth of hybridoma cells in serum-free medium: ethanolamine is an essential component, *Proc. Natl. Acad. Sci. USA* **79**, 1158-1162.

Urlaub, G. and Chasin L.A. (1980) Isolation of Chinese hamster cell mutants deficient in dihydrofolate reductase activity, *Proc. Natl. Acad. Sci. USA* **77**, 4216-4220.

van Corven, E.J., Groenink, A., Jalink, K., Eichholtz, T., Moolenaar, W.H. (1989) Lysophosphatidate-induced cell proliferation: identification and dissection of signaling pathways mediated by G proteins, *Cell* **59**, 45-54.

Yamaji, H. and Fukuda, H. (1992) Growth and death behaviour of anchorage-independent animal cells immobilized within porous support matrices, *Appl. Microbiol. Biotechnol.* **37**, 244-251.

# A KINETIC MODEL OF MACROMOLECULAR METABOLISM IN BATCH AND FED-BATCH CULTURES OF HYBRIDOMA CELLS PRODUCING MONOCLONAL ANTIBODY

J.D. JANG AND J.P BARFORD*

Department of Chemical Engineering,
University of Sydney 2006 NSW Australia

**Keywords: hybridoma; fed-batch; monoclonal antibody; kinetics; modelling**

**Abstract**

Growth profiles of the batch and fed-batch culture of hybridoma cells producing monoclonal antibody were simulated using a structured model previously developed. The model describes the production of cellular macromolecules and monoclonal antibody, the metabolism of glucose and glutamine with the production of lactate and ammonia, and the profiles of cell growth in batch and fed-batch culture. Equations describing the cells arrested in G1 phase (Linardos et al., 1992; Suzuki et al., 1989) was included in this model to describe the increase of the specific antibody productivity in the near-zero specific growth rate, which was observed in the recent experiments in fed-batch cultures of a previous study. This model well predicted the increase of specific antibody production rate and the decline of the specific production rate of cellular macromolecules such as DNA, RNA, protein, and polysaccharide with lower specific growth rate in the fed-batch culture.

## 1. Introduction

Antibody production by hybridoma cells requires all the cellular machinery for protein synthesis including transcription, translation, posttranslational modification, and secretion because antibodies are proteins or glycoproteins. On the other hand, antibody is not an essential product for hybridoma cell growth, thus its production is expected to be less favored by fast growing cells when compared to the production of cellular proteins. Therefore antibody productivity is expected to increase by limiting cell growth in batch, fed-batch, or continuous culture. It has been reported that the specific antibody production rate is higher in the G1 phase of cell cycle or in the cells arrested in G1 phase (Garatun-Tjeldstø et al., 1976; Linardos et al., 1992). It has been suggested that the fraction of G1-arrested cells increase as decreasing the dilution rate in the cell cycle model of Suzuki et al. (1989).
Mathematical models are useful in predicting the behavior of animal cell culture and in optimizing culture condition. A structured and compartmented model including most detailed biochemical pathways of important metabolites in the hybridoma cell

metabolism has been developed by our group (Sanderson et al., 1995; Sanderson, 1996). In this study, the model has been upgraded to include growth kinetics in terms of macromolecules and antibody production from G1-arrested cells. The model simulation was compared with the experimental results from a batch culture and a set of fed-batch culture of a murine hybridoma cell line, AFP-27.

## 2. Model development

The model consists of 5 basic compartments: feed, harvest, medium, cytoplasm, and mitochondria. All cells are represented as cytoplasm which are divided into 3 cell population: viable cycling, G0 (viable cells arrested in G1 phase), and dead cells. Specific antibody production rate was assumed to be higher in G0 cells than in normal cycling cells. Glucose metabolism, amino acid metabolism, TCA cycle, and antibody production were included within cytoplasm or mitochondria (Figure 1).

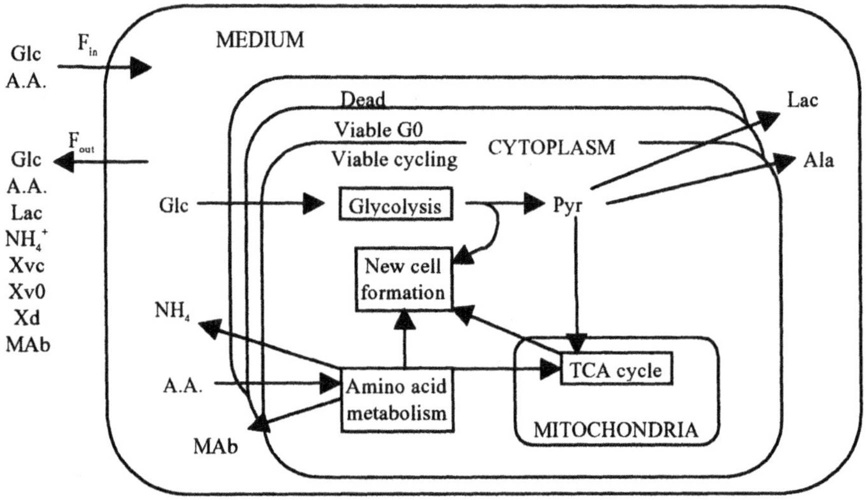


Figure 1. Simplified diagram of the compartmented structure in the model.

The equations for the cell growth and cellular metabolism were based on Michaelis-Menten equation with multi-limitation and multi-inhibition terms. For example, specific cell growth rate ($\mu$) is determined in cytoplasm as a function of limiting substrates ($[S_i]$) and inhibitors ($[I_i]$).

$$\mu = \mu_{max} \prod_{i=1}^{n} \left( \frac{[S_i]_{cyt}}{K_S + [S_i]_{cyt}} \right) \prod_{i=1}^{n} \left( \frac{K_I}{K_I + [I_i]_{cyt}} \right)^2 - \mu_d$$

where $K_S$ and $K_I$ represent limitation and inhibition constant, respectively.

Production of cellular macromolecules was assumed to occur in cytoplasm, in which the production rate is limited by availability of resources for building blocks. DNA and RNA were assumed to be required as a template for the synthesis of RNA and protein, respectively. All the cellular macromolecules were assumed to be degraded

with first-order kinetics. The degradation rate of cellular RNA was assumed to be different between stable and unstable species.
Maximum specific antibody production rate ($V^{ab}_{m}$) was described as a function of G0 fraction ($f_{G0}$), which was assumed to be determined by specific death rate in cytoplasm.

$$V_m^{ab} = V_{m1}\left(1 - f_{G0}\right) + V_{m2} f_{G0}$$

## 3. Results and discussions

Batch and fed-batch cultures of hybridoma cells producing monoclonal antibodies were simulated using the structured model in dynamic mode.
Simulated data of batch culture was relatively well fitted to the experimental data with 25 mM initial glucose and 4 mM initial glutamine (Figure 2). The synthesis of cellular DNA, RNA, and protein are clearly shown to be interrelated each other in the simulation as expected from the flow of genetic information.

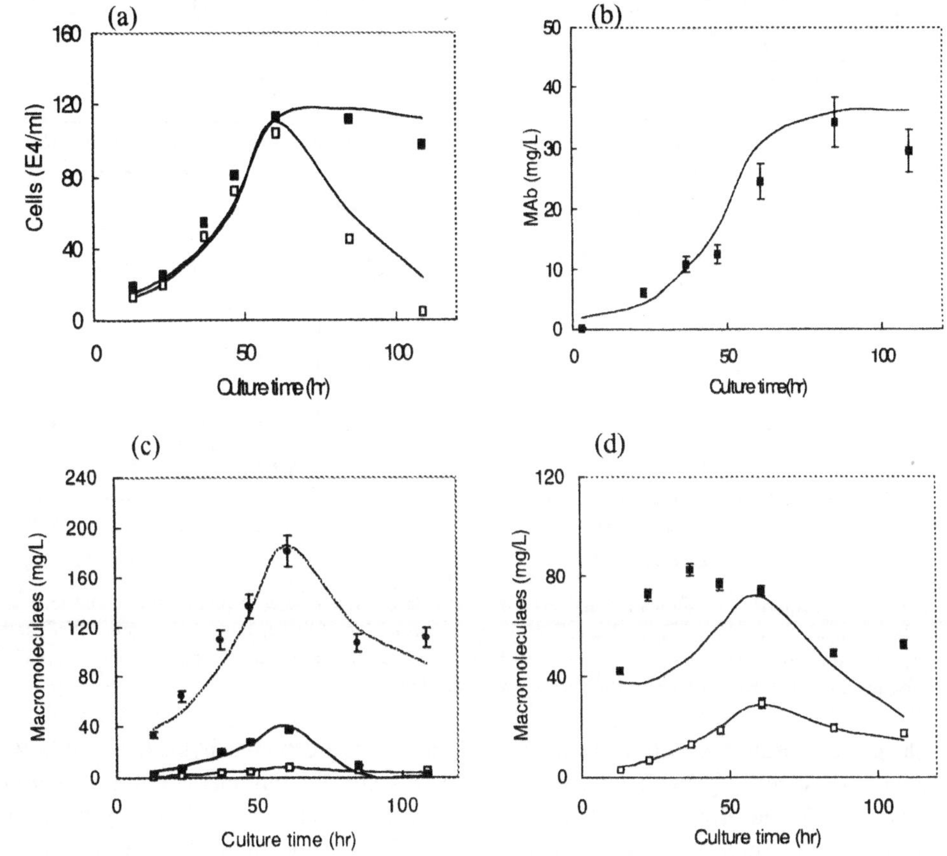


Figure. 2. Comparison of simulation and experimental data of (a) viable cells (open square) and total cells (filled square), (b) antibodies, (c) cellular DNA (open square), RNA (filled square), and protein (filled circle),

(d) cellular lipid (open square) and total carbohydrate (filled square) in the batch culture of AFP-27 producing monoclonal antibodies.

The result of simulation with specific growth rate 0.01 $hr^{-1}$ are presented in Figure 3, though several set of specific growth rate were tried in the simulation of the model in fed-batch mode (data not shown for other specific growth rates). By including the feature of cell arrest in G1 phase, it was possible to predict increasing total cell concentration and nearly constant viable cell concentration as well as continuing production of monoclonal antibodies during pseudo-steady state.

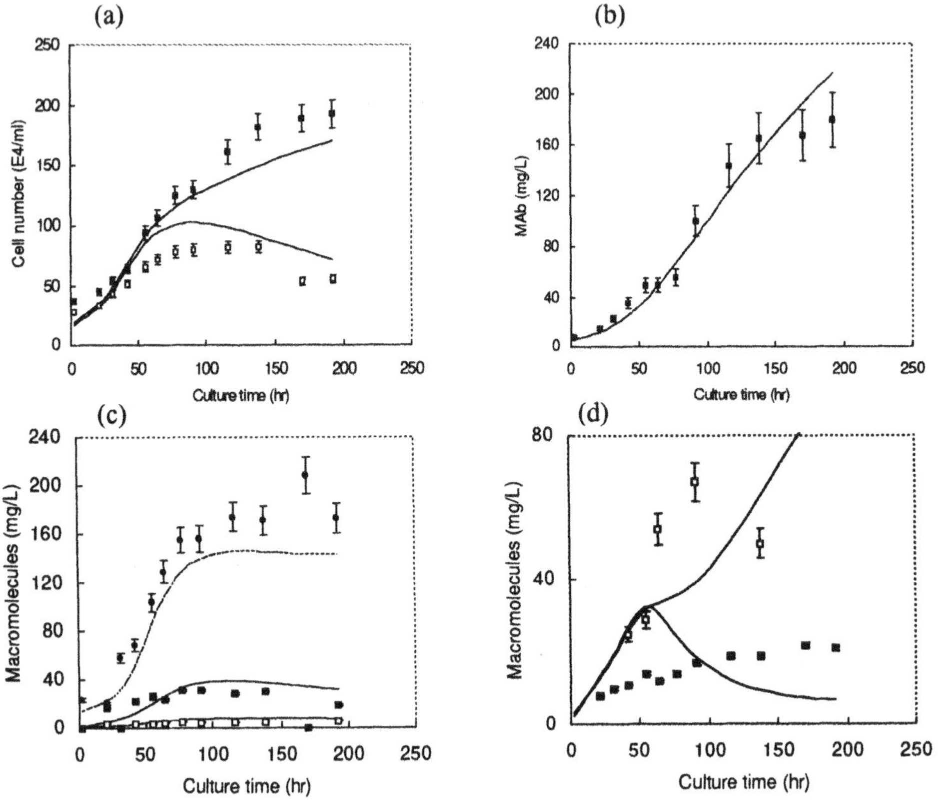


Figure. 3. Comparison of simulation and experimental data of (a) viable cells (open square) and total cells (filled square), (b) antibodies, (c) cellular DNA (open square), RNA (filled square), and protein (filled circle), (d) cellular lipid (open square) and total carbohydrate (filled square) in the fed-batch culture of AFP-27 producing monoclonal antibodies at specific growth rate of 0.01 $hr^{-1}$.

During the pseudo-steady state in the fed-batch culture of the hybridoma cells with near-zero specific growth rate, cell growth was strictly limited in terms of cellular DNA, RNA and proteins, while antibody production was continued during that period as shown in experiment and simulation results. This might imply that more resources for protein synthesis were diverted from cellular proteins to antibodies. Increased fraction of G0 cells in the fed-batch culture of near-zero specific growth rate might be partly responsible for low viability and high specific antibody productivity in the culture of near-zero specific growth rate as proposed by Suzuki et al. (1989) and Linardos et al. (1992).

## 4. Conclusions

A structured, compartmented model of hybridoma cell culture including macromolecular metabolism has been developed by our group. Hybridoma cells in culture were supposed to have G0 phase cells (arrested in G1phase) which might have higher antibody production rate.

Simulated results of hybridoma cell growth, MAb production, and macromolecular profiles fitted well with experimental data in batch mode, while those in fed-batch mode of low specific growth rate (0.01 $hr^{-1}$) showed more deviation from experimental data.

Cellular metabolism in fed-batch culture of low specific growth rate (0.01 $hr^{-1}$) might be concluded to be much different from that of batch case. Therefore more sophisticated modelling approaches are needed to simulate fed-batch with low specific growth rate.

## References

Garatun-Tjeldstø O., Pryme, I.F., Weltman, J.K., and Dowben, R.M. (1976) Synthesis and secretion of light-chain immunoglobulin in two successive cycles of synchronized plasmacytoma cells. *J. Cell. Biol.* **68**, 232-239.

Linardos, T.I., Kalogerakis, N., and Behie, L.A. (1992) Kinetic analysis of hybridoma growth and monoclonal antibody production in semi-continuous culture. *Biotech. Bioeng.* **40**, 359-368.

Sanderson, C.S. (1996) The development and application of a structured model for animal cell metabolism. Ph.D. Thesis, University of Sydney.

Sanderson, C.S., Barton, G.W., and Barford, J.P. (1995) Optimization of animal cell culture media using dynamic simulation. *Computers Chem. Eng.* **19**, s681-s686

Suzuki, E. and Ollis, D.F., 1989. Cell cycle model for antibody production kinetics. *Biotech. Bioeng.* **34**, 1398-1402.

# ENHANCEMENT OF ANTIBODY PRODUCTIVITY IN THE NEAR-ZERO SPECIFIC GROWTH RATE DURING GLUCOSE AND GLUTAMINE-LIMITED FED-BATCH CULTURE OF HYBRIDOMA CELLS.

J.D.JANG AND J.P BARFORD*

Department of Chemical Engineering,
University of Sydney 2006 NSW Australia

**Keywords: hybridoma; fed-batch; monoclonal antibody; near-zero growth; substrate limitation**

## Abstract

A set of fed-batch cultures feeding concentrated nutrients was carried out to find out the optimal feeding policy to give higher antibody productivity in the fed-batch culture. From the experiments, it was found that there were two different trends in the relationship between the antibody productivity and the specific growth rate. The specific antibody productivity (g/gCell/hr) was found to significantly increase (327 % over the average of other growth rates) when the apparent specific growth rate was lowered to 0.01 $hr^{-1}$. This might be caused by the increased fraction of G1-arrested cells in this low specific growth rate in the fed-batch culture as proposed by Linardos et al.(1992) and Suzuki et al.(1989) in their cell cycle model. On the other hand, there was little difference in the specific antibody productivity beyond this specific growth rate. The final antibody titer was higher in the higher specific growth rates beyond 0.01 $hr^{-1}$ (150 % higher at 0.04 $hr^{-1}$ than at 0.0175 $hr^{-1}$), possibly due to the higher viable cell density. Therefore, to get higher antibody productivity, it might be better to start with or grow to the higher viable cell density and then the specific growth rate should be maintained in the near-zero (0.01 $hr^{-1}$ in this study) to get highest concentration of G1-arrested cells as possible. In this paper, fed-batch data at near-zero specific growth rate are compared with data of batch culture and fed-batch culture at high growth rate.

## 1. Introduction

Antibodies have two important features to be considered for large-scale production. Since they are proteins, their production requires all the cellular machinery for protein synthesis including transcription, translation, posttranslational modification, and secretion. Since they are not essential products for cell growth, their production is expected to compete with the production of cellular proteins. Therefore, to achieve mass production of antibodies, the relationship between cell growth in terms of macromolecules and antibody production should be elucidated.

A specific antibody production rate has been reported to be not growth associated and to be higher in lower specific growth rate in many cases (Ray et al., 1989; Miller et al., 1988; Bibila et al., 1991; Hayter et al., 1992).. In this study, therefore, a fed-batch cultures of murine hybridoma cell at near-zero specific growth rate was performed and compared with those with different specific growth rates. The metabolic characteristics of near-zero specific growth rate including antibody productivity were also compared with those of other rates and possible implications of the results were discussed.

## 2. Materials and Methods

The cell line used in this study is a murine hybridoma AFT-27, producing an IgG1 antibody that reacts with human α-fetoprotein. The medium used in this study was based on DR medium (1:1 mixture of DMEM and RPMI) or eDR medium (DR with additional amino acids) with various concentration of glucose and glutamine. For the inoculum of fed-batch culture, subculture volume was doubled daily from 20 ml up to 1 liter with eDR medium with 5 mM glucose and 2 mM glutamine (eDR5/2) to inoculate 1.5 liter of fresh eDR5/2 medium. 60 mM glucose and 20 mM glutamine in eDR medium was used for feed in fed-batch cultures. Fetal calf serum (2 %, v/v) was added to the main culture medium and feed medium.

The fed-batch culture was performed in a stirred tank reactor in which pH and temperature were automatically controlled to 7.1 and 37°C, respectively. DO was controlled automatically to be maintained between 10 and 50 % by direct sparging of compressed air. Feed rate (F, liter/hr) was calculated from target specific growth rate ($\mu$), cell density (X, cell/liter), running volume (V, liter), cell yield from glucose consumed ($Y_{xs}$, cell/mmole glucose), and glucose concentration in feed medium ($F_{glc}$, mM) according to following equation:

$$F = \frac{\mu X V}{Y_{xs} F_{glc}} \tag{1}$$

Feed rate was recalculated after every sampling with updated cell numbers and running volume. Sample of about 100 ml was withdrawn from the culture every 12 hours for assay.

Glucose, lactate, and ammonia were assayed using an enzymatic technique. Total amount of antibody was measured using enzyme-linked immunosorbent assay (ELISA). Pre-column derivatization technique using FDNDEA (N,N, diethyl-2,4-dinitro-5-fluoro-aniline, Fluka, Buchs, Switzerland) was used to assay the concentration of amino acids in the cell culture supernatant (Fermo *et al.*, 1988).

### 3. Results and Discussion

Culture profiles in fed-batch culture at near-zero specific growth rate are shown in Figure 1(a). In the figure, viable cell concentration was maintained at nearly constant value during a pseudo-steady state, while total cell concentration increased during that period. This result agrees with previous reports by others (Miller et al., 1988; Ray et al., 1989) that viability decreases as specific growth rate decreases. No increase in the viable cell concentration means that the rate of cell death is equal to the rate of viable cell growth to give net near-zero growth. In normal batch culture, when the cell start to die by certain reasons such as nutrient exhaustion or inhibitory product accumulation, the concentration of viable cells decreases as that of dead cells increases to give constant total cell concentration. Therefore, in the fed-batch culture at near-zero growth rate, the specific death rate might not be directly related to the specific growth rate of viable cell, but other mechanism could be involved. One of the possible explanations could be an increased fraction of non-cycling viable cells (G0 cells in hybridoma cells) in near-zero growth, as proposed by Linardos et al. (1992) and Suzuki et al. (1989). In such case, some portion of cells that are not stained with trypan blue could be the fraction of G0 cells. G0 cells may not contribute to multiplying the viable cells, while they may die with certain specific death rate, giving increase only in the concentration of dead and total cells, but not in that of viable cells. Therefore, the increase of total cell concentration without increase in viable cell concentration might imply that there could be some portion of non-cycling viable (G0) cells.

As shown in Figure 1(b), lactate concentration was maintained at nearly constant level while ammonia level increased progressively during the pseudo-steady state of the fed-batch culture with near-zero growth. On the other hand, the concentrations of glucose and glutamine were maintained at minimum level during the period, indicating both nutrients were limiting to the cell growth. Since feed solution was supplied continuously, minimum levels of glucose and glutamine during the pseudo-steady state might imply that most of glucose and glutamine were consumed as they were supplied, which could be the criteria of nutrient limitation and control of specific growth rate as targeted.

Final Mab titer and Mab per cell mass were 2.5 and 4.5 fold higher, respectively in near-zero specific growth rate(Figure 2). As the cells with near-zero specific growth rate utilised glucose and glutamine more efficiently than those with higher specific growth rate, the level of alanine and proline as well as lactate showed net consumption as a whole rather than accumulation. This might be due to reduction of pyruvate pool and more effective utilisation of glutamine and glutamate in the cell.

Since fed-batch culture with near-zero specific growth rate limits cell growth so strictly that more resources of protein synthesis might be available for the synthesis of monoclonal antibodies. In addition, many charactristics of fast growing hybridoma cells such as high consumption rate of glucose and gluatmine with concomitant massive production of lactate and ammonia might be greatly reduced or turned off.

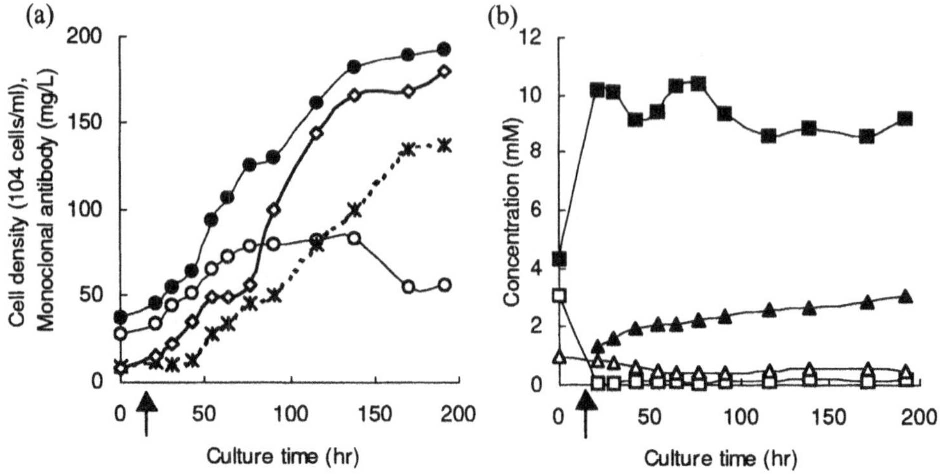


Figure 1. Culture performance in fed-batch culture at near-zero specific growth rate. (a) Concentrations of viable cell (open circle), dead cell (asterisk), total cell (closed circle), and monoclonal antibody (open diamond). (b) Concentrations of glucose (open square), glutamine (open triangle), lactate (closed square), and ammonia (closed triangle). Arrows indicate start time of feeding.

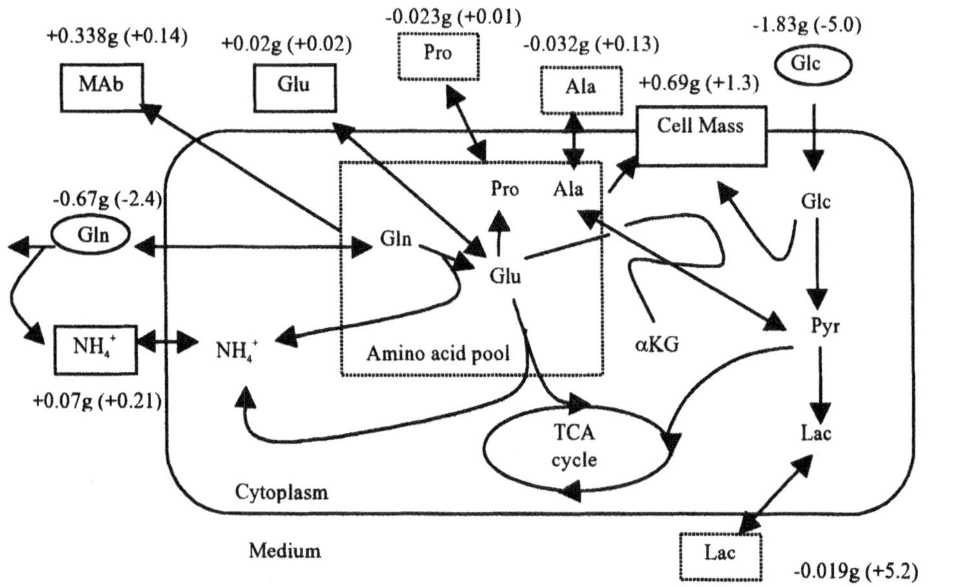


Figure 2. Metabolic pathways showing production or consumption of the medium components and products during the fed-batch culture of AFP-27 with specific growth rate of near-zero and 0.04 $hr^{-1}$ (values in parenthesis). Positive and negative value represents production and consumption, respectively. Components in the oval represent net consumption in both cultures and those in dotted box for different trend between two cultures.

## 4. Conclusions

Fed-batch culture of near-zero specific growth rate was performed to increase antibody productivity. The specific production rate and maximum titer of monoclonal antibodies in the fed-batch culture was found to be more than 5 fold of batch case, while the maximum viable cell density was similar to batch case. From the average data of the quasi-steady state, some conclusions could be drawn.

(1) Antibody productivity could be significantly enhanced by maintaining near-zero specific growth rate in fed-batch culture with exponential feeding of concentrated feed medium.
(2) Specific consumption rates of glutamine and glucose could be significantly decreased by lowering the specific growth rate in the fed-batch culture.
(3) Specific production rate of ammonia and lactate also could be significantly decreased by slow feeding of glutamine and glucose, respectively.
(4) Patterns of some amino acid metabolism might be changed at the near-zero specific growth rate in the fed-batch culture.

In conclusion, this study suggests that high antibody productivity could be obtained by controlling the specific growth rate in fed-batch culture of hybridoma cells. It may also be suggested from this study that fed-batch culture could be a useful strategy for the production of monoclonal antibodies in large scale.

## References

Bibila, T. and Flickinger, M.C. (1991) A structured model for monoclonal antibody synthesis in exponetially growing and stationary phase hybridoma cells. *Biotech. Bioeng.* **37,** 210-226

Hayter, P.M., Kirkby, N.F., and Spier, R.E. (1992) Relationship between hybridoma growth and monoclonal antibody production. *Enzyme Microb. Technol.* **14,** 454-461.

Linardos, T.I., Kalogerakis, N., Behie, L.A. (1992) Kinetic analysis of hybridoma growth and monoclonal antibody production in semi-continuous culture. *Biotech. Bioeng.* **40,** 359-368.

Miller W.M., Blanch, H.W., and Wilke, C.R. (1988) A kinetic analysis of hybridoma growth and metabolism in batch and continuoous suspension culture: Effect of nutrient concentration, dilution rate, and pH. *Biotech. Bioeng.* **32,** 947-965

Ray, N.G., Karkare, S.B., and Runstadler, Jr. P.W. (1989) Cultivation of hybridoma cells in continuous cultures: Kinetics of growth and product formation. *Biotech. Bioeng.* **33,** 724-730.

Suzuki, E. and Ollis, D.F. (1989) Cell cycle model for antibody production kinetics. *Biotech. Bioeng.* **34,** 1398-1402.

# LONG-TERM CULTURE OF CHO CELLS ON POROUS MICROCARRIERS IN A STIRRED TANK EQUIPPED WITH A MODIFIED CELL RETENTION SYSTEM

XIANWEN HU, CHENGZU XIAO, ZICAI HUANG&ZHIXIA GUO
*Department of Cell Engineering, Institute of Biotechnology, 20 Dongdajie, Fengtai, Beijing, 100071 , P.R. China*

**Abstract** A recombinant DNA CHO cell line secreting urokinase-type plasminogen activitor (uPA) was cultivated with Cytopore cellulose porous microcarriers in a 20L Biostat UC20 stirred tank reactor. The reactor was equipped with bubble-free silicone tubing oxygenator, and a modified perfusion system which can be washed-out backward was substitute for spin-filter to prevent filter clogging. During 100days culture with 0.1% fetal bovine serum (FBS) medium , the maximal cell density was over $10^7$/ml , the maximal uPA activity in supernatant reached 6250 IU/ml , and 1604L supernatant contained about 51g uPA (about 70% was single chain form) was harvested . Recombinant CHO cells can move from seed porous microcarriers occupied by rCHO cells to vacant microcarriers spontaneously without trypsinization, and continue to grow until all the microcarriers contained rCHO cells. It shows that Cytopore porous microcarriers is very useful and convenient for enlarging cell culture scale step by step.

**Keywords:** cell culture porous microcarriers prourokinase

## 1. Introduction

Cytopore is one of excellent porous microcarriers made of cellulose for animal cell culture , which almost possesses all of characteristics of an ideal cell culture support described by Griffiths(1). It can be used for both anchorage dependent and suspension cells, is capable of long-term continuous operation with low-serum/serum-free media, has good mechanical stability and considerable volumetric scale-up potential (2). In the previous study, we reported that a rCHO cell line which express prourokinase was cultivated on Cytopore porous microcarriers in a 20L Biostat UC stirred tank reactor with spin-filter to retain cells, the cell density could reach $1.33\times10^7$/ml, the maximal uPA activity was over 7000IU/ml, but the spin-filter was completely fouled by cells and microcarriers after 21days culture, and had to stop running the reactor (3). So we developed a new cell retention system to overcome this problem, and continuously cultured rCHO cells over 100days without serious clogging.

## 2. Materials and Methods

### 2.1. CELL LINE AND MEDIA AND MICROCARRIERS

GL-11G is a new genetically-engineered CHO cell line secreting prourokinase. The growth medium was DMEM(low glucose):F12 (1:1) , supplemented with 0.1% FBS, 4.5g/L glucose, 0.5g/L glutamine, 1.8g/L Hepes , 1g/L peptone, 20KIU/ml aprotinin , and plus a low-serum medium supplement-BIGBEF-3(4) .
The porous microcarriers, Cytopore, were hydrated in 0.1M phosphate-buffered saline(PBS), prewashed and autoclaved for 30min at 121℃ according to manufacturer's recommendation .

## 2.2. BIOREACTOR AND CELL CULTURE

A 20L (working volume) Biostat UC stirred tank reactor (B.Braun , German) , was equipped with bubble-free silicone tubing oxygenator. The modified cell retention system composed of pumps, stainless steel filter and a time-controller. In order to prevent filter fouling, the filter with 200 mesh openings stainless steel screen could be washed-out backward once a time controlled by time-controller .(Figure 1).

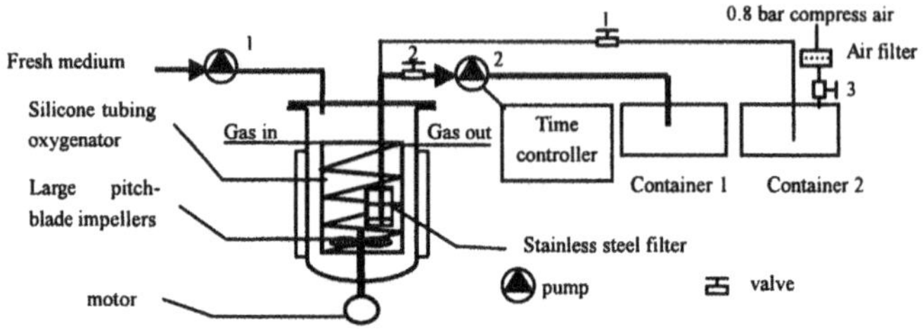


*Fig. 1* **Schematic diagram of a 20L Biostat UC reactor equipped with a modified cell retention system**

The modified cell retention system could work in two mode:
**Mode 1**: Semi-continuous culture, harvest pump and time-controller is unnecessary, when harvesting supernatant, close valve 1 and 3, open valve 2, and keep pressure of reactor headspace at 0.2bar~0.3bar, supernatant flows into container 1 automatically. If replenishing medium rate is over 50% working volume per day, the medium can be replaced twice or three times a day. The fresh medium equivalent to removed supernatant is added to keep constant working volume in the reactor. When washing-out filter backward, open valve 1 and 3, close valve 2, owing to overpressure, the liquid in the container 2 flows into the filter to prevent filter clogging.
**Mode 2**: Perfusion culture, close valve 1and 3, open valve2, in a cycle of harvesting and washing-out controlled by time controller, the harvest pump runs forwardly for 10 min to harvest supernatant, and runs in reverse for 1min to wash-out the filter.

The agitate rate was 70rpm.Temperature was maintained at 37±0.2°C. The pH was kept at 7.1± 0.1 by automatic addition of 1M NaOH solution. Dissolved oxygen (DO) was maintained at 40± 5% of air saturation. The concentration of microcarriers was 2g/L.The seed microcarriers with GL-11G cells obtained from 1000ml spinner flasks culture were transferred directly to the Biostat UC reactor via tube without trypsinization. The frequency and extent of medium replenishment depended on the residual glucose concentration. The apparent specific growth rate ($\mu_{app}$) is determined by linear regression of the exponential increase of cell concentration, C, by plotting the natural log of C vs. time, t (7).

## 2.3. ASSAYS

Cells were enumerated by counting nuclei stained with 0.1% crystal violet(w/v) in 0.1M critic acid solution. The cell adherent ratio of microcarriers was determined by dyeing with MTT. Viable cell counts were performed with exclusion trypan blue method after trypsinization. Glucose concentration was measured using a clinical glucose analyzer. For distinguishing the proportion of single-chain and two-chain uPA in supernatant, the samples were analyzed both in non-reducing and reducing condition with western blot analysis (5).

## 2. Result

## 2.1. CELL GROWTH AND REPLENISHMENT OF MEDIUM

GL-11G cells had been successfully cultivated on Cytopore cellulose porous microcarriers in a 20L Biostat UC stirred tank reactor for 100 days. The culture process could be divided into five stages (Figure 2). The seed cells were cultivated on Cytopore porous microcarriers in a 1000ml Spinner Flasks, and directly transferred into the 20L reactor without trypsinization, the scale-up rate was 1:20, the initial cell density was $2.0\times10^5$/ml. After an expanded lag time of 3 days, the cells began to proliferate and move spontaneously to vacant microcarriers.

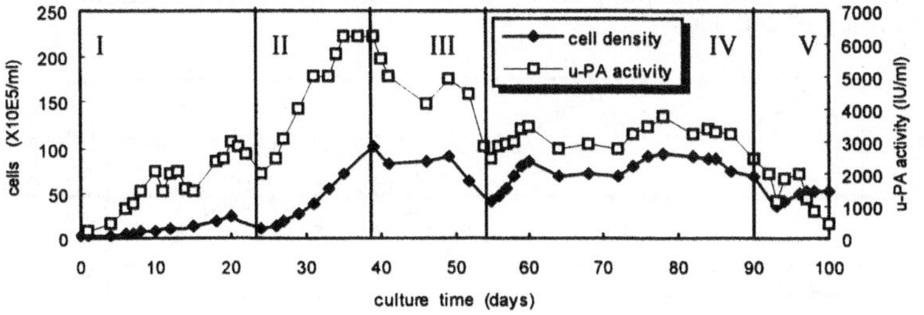


*Fig.2* **Growth curve and u-PA activity of GL-11G cells cultivated on Cytopore porous microcarriers in a 20L Biostat UC bioreactor with a modified cell retention system**

stage I : semi-continuous culture with 0.1% FBS growth medium at initial cell density of $2.0\times10^5$/ml (0day-22day);

stage II : semi-continuous culture with 0.1% FBS growth medium at initial cell density of $9.8\times10^5$/ml after electricity broke down (24day-39day);

stageIII: perfusion culture at stationary phase (40day-54day);

stageIV: semi-continuous culture after the clogging filter was completely washed-out(55day-90day);

stage V : cell decline phase, the pH in the reactor was not properly controlled(91day-100day);

During stage I, the $\mu_{app}$ was 0.112d$^{-1}$, and cell density increased up to $2.38\times10^6$/ml after 20 days culture. In our previous study (3), the $\mu_{app}$ was 0.182d$^{-1}$ at the initial cell density of $3.5\times10^5$/ml when 1% FBS culture medium was used. The expanded lag phase and relatively lower specific growth rate were due to (1) the inoculated cell density was low; (2) 0.1% FBS growth medium was not very suitable at the beginning of cell subculture, especially at low inoculated cell density. The replenishing medium rate was about 7L~12L/day to maintain glucose concentration at 3.0~3.3g/L, which depended on the residual glucose concentration in the reactor. .

The culture process of stage II started from the viable cell density of $9.8\times10^5$/ml, and the cell density reached $1.01\times10^7$/ml after another 15 days culture. The $\mu_{app}$ at stage II was 0.153d$^{-1}$, higher than that of stageI. It indicated that higher inoculated cell density was beneficial to shorten lag phase and speed up cell proliferation. The replenishing medium rate increased gradually from 7L/day to 23L/day, and the residual glucose concentration oscillated around 2.0g/L.

During stage III, cell growth was at stationary phase, and perfusion culture mode was adopted. The cell density was around $9.0\times10^6$/ml, 20~23 liters medium was replaced per day to keep glucose concentration at a level around 2.0g/L. After 15 days perfusion culture, the filter was clogged seriously, and the replacing medium rate was only 12L/d, the residual glucose concentration decreased down to 1.2g/L, the culture medium couldn't supply enough nutrient to support cell growth, and a large number of cells died due to undernourishment.

In order to wash-out the filter completely, the working mode 1 of cell retention system was adopted. During the early period of stage IV, the $\mu_{app}$ was 0.172d$^{-1}$ at the initial cell density of $4.0\times10^6$/ml, showing that $\mu_{app}$ increased with inoculation density (Table 1). Comparing to the $\mu_{app}$ of 0.182d$^{-1}$ in 1% FBS culture medium, low-serum growth medium seemed to have a little effect on cell growth at relatively high inoculation density. The replacing medium rate maintained 20L/d ~ 22L/d to keep residual glucose concentration at the range of 2.3~2.7g/L. The sample of the 87$^{th}$ day was enumerated with exclusion trypan blue method after trypsinization, the cell viability was about 70%. It seemed to indicate that the cells in the center of microcarriers may encounter mass transfer limitation, especially under the circumstance which the cells in the porous microcarriers were very dense after long term culture.

## 2.2. BIOREACTOR PRODUCTIVITY AND PRODUCT QUALITY

The uPA activity curve was shown in Figure 2. The uPA activity reached its vertex of 6250 IU/ml in stage II, and reduced to 4500 ~ 5500IU/ml during perfusion culture process. During stageIV, cell density and product expressing was in relatively stead state, the uPA activity fluctuated about 2800IU/ml~3600IU/ml. In decline phase, the uPA activity decreased from 2500IU/ml to 500IU/ml as a result of cell death.

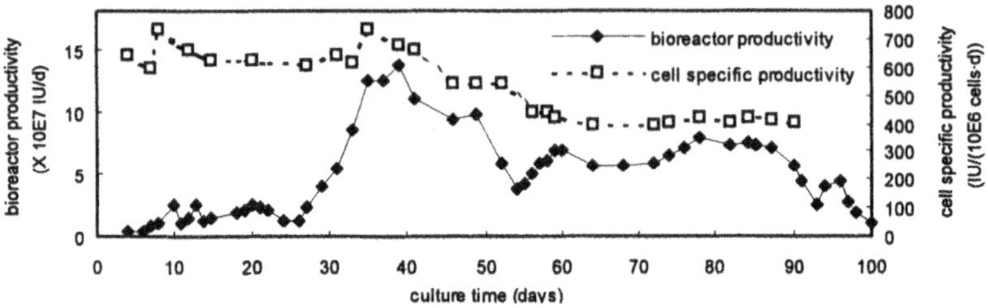


*Fig. 3* **Bioreactor productivity and cell specific productivity**

The bioreactor productivity and cell specific productivity was given in Figure 3 and Table 1. The maximum bioreactor productivity could reach $1.375\times10^8$IU/day, about 1.32g uPA/day calculated according to uPA specific activity=10400 IU/mg, and the uPA yield of 100 days culture added up to 51g more or less. Cell specific productivity in stage I and II was around 600~730IU/$10^6$cells •d, a little higher than that in stage III(640~540 IU/$10^6$cells • d). It appears that comparing to perfusion culture, batch replenishing medium culture has some advantages in utilizing culture medium and cell expressing product. This may be attributed to that the environment after replenishing medium was more suitable for cell growth like fresh medium, while some fresh medium would flow out of reactor without utilizing and the environment always keep at not very perfect state in perfusion culture. In stage IV, cell specific productivity maintained in the range of 400~440 IU/$10^6$cells • d, the relatively low value was due to cell degeneration after long term culture and not taking dead cells into account when calculating cell specific productivity.

**Table 1 rCHO cells cultivated on porous microcarriers for 100days**

| | stageI | stageII | stageIII | stageIV | stageV |
|---|---|---|---|---|---|
| Cell density ($\times10^5$/ml) | 2.0~23.8 | 9.8~101.2 | 100~90 | 40~93 | ~45 |
| $\mu_{app}$ ($d^{-1}$) | 0.112 | 0.153 | - | 0.172 | - |
| Replenishing medium rate (L/d) | 7~12 | 7~22 | 20~23 | 20~22 | 22 |
| UPA activity (IU/ml) | 1000~2980 | 2000~6500 | 4000~5500 | 2500~3700 | 2000~500 |
| ScuPA ratio | >90% | ~85% | 80%~30% | 70%~80% | ~70% |
| Cell specific productivity (IU/$10^6$cell • d) | 620~730 | 640~730 | 540~660 | 400~440 | - |
| Bioreactor productivity (g uPA/d) | 0.05~0.23 | 0.118~1.32 | ~1.0 | 0.54~0.75 | 0.1~0.42 |
| Total productivity (g uPA) | 2.73 | 10.82 | 12.42 | 22.12 | 2.8 |

Single chain urokinase-type plasminogen activator(scuPA, i.e. prourokinase)is a potential thrombolytic drug with fewer side effect than two chain urokinase-type plasminogen activator(tcuPA, i.e. urokinase) (8), so the quality of uPA product depends on the ratio of scuPA. ScuPA is easily activated by plasmin, trypsin and plasma kallekrein, resulting in the conversion of single chain prourokinase into two chain urokinase.(9). The main problem in prourokinase production by cell culture is how to keep unsteady prourokinase in single chain form. In stage I and II, 85%~90% of uPA was maintained in single chain form in the harvested supernatant. During perfusion culture, scuPA ratio decreased from about 80% to 60% gradually along with culture process, and even dropped to 30%~40% when large number of cells died due to filter clogging. In stage IV and V, the ratio of scuPA got back to 70%~80% after batch replenishing culture adopted,. This suggested that cell death in large quantities would release some protease to convert scuPA into two chain form, and semi-continuous culture could harvest supernatant with relatively high ratio of scuPA due to the short residence time of the product after secretion.

## 3. Discussion

A modified cell retention system was developed to prevent clogging in long-term and large-scale continuous cell culture. After 100 days culture, the filter worked excellent and 1604L supernatant containing 51g uPA was harvested.

Batch replenishing medium culture has advantages of batch culture and perfusion culture when replacing medium rate was about 1~2 working volume per day, like batch culture, it could utilize medium more efficient and obtain supernatant with relatively high concentration product. On the other hand, it reduced accumulation of toxic metabolites and operation expenses to raise utilizing efficiency of equipment. Moreover, when the product was unsteady in 37℃ environment such as prourokinase, batch replenishing medium culture will be a ideal choose for product quality assurance.

Cytopore porous microcarrier was very useful for long-term cell culture with low-serum medium, and subculture method was simple and safe, there is no fear for contamination during inoculation process because cells can move from seed microcarriers to vacant ones automatically without trypsinization(6), this was convenient for preparing seed cells and enlarging culture scale. It can reduce serum concentration in culture medium that can't result in cells detaching from microcarries. Low-serum culture medium seemed to have no significant effect on cell growth and product secreting. The $\mu_{app}$ in 0.1% FBS culture medium was a little lower than that in 1% FBS culture medium, while the cell specific productivity in 0.1% FBS medium was 600~700 IU/($10^6$cells • d), a little higher than 450~500 IU/($10^6$cells • d) in 1% FBS growth medium(3).

## 4. Reference


1. Griffiths J.B. (1990) Immobilization of animal cells in porous carriers culture, *TIBTECH* **8**: 204.
2. Shirokaze J., Nogawa M. and Ogura R. (1994) High density culture using macroporous microcarrier, in Spier R.E. and Griffiths J.B.(eds.), *Animal cell technology: products for today, prospects for tomorrow,* Butterworth-Heinemann, Oxford: 261.
3. Hu Xianwen, Xiao Chengzu and Huang Zicai (1998) Large scale culture of rCHO cells using porous microcarriers, *Chin. J. Biotechnol.* **14** (3): 348.
4. Xiao Chengzu, Huang Zicai and Zhuang Zhengguang (1995) Production of recombinant prourokinase by CHO cells in a perfusion biosilon microcarrier culture system, in Beuvery EC et al.(eds.), *Animal cell technology: Developments towards the 21st century*, Kluwer Academic Publishers, Netherlands: 97.
5. Han Suwen, Yu Weiyuan and Cheng Dusheng (1997) The construction and expression of human prourokinase cDNA coding deleted mutant at 150~156 amino acids, *Chin. J. Biotechnol.* **13**(2): 127.
6. Kamiya K., Yanagida K. and Shirokaze J. (1995) Subculture method for large scale cell culture using macroporous microcarrier, in Beuvery EC et al.(eds.), *Animal cell technology: Developments towards the 21st century*, Kluwer Academic Publishers, Netherlands: 759.
7. Michaels J.D., Petersen J.F. and McIntire L.V. (1991) Protection mechanisms of freely suspended animal cells from fluid-mechanical injury, *Biotechnol. Bioeng.* **38**:169.
8. PRIMI Trial Study Group (1989) Randomized double-blind trial of recombinant prourokinase against streptokinase in acute myocardial infarction, *Lancet.* **1**:863.
9. Ichinose A., Fujikawa K. and Suyama T. (1986) The activation of prourokinase by plasma kallikrein and its inactivation by thrombin, *J. Biol. Chem.*,**261**: 3486.

# Continuous production and recovery of recombinant $Ca^{++}$ binding receptor from HEK 293 cells using perfusion through a packed bed bioreactor

J.B. KAUFMAN[1], G. WANG[2], W. ZHANG[2], M.A. VALLE[1], J. SHILOACH[1*]
[1] *Biotechnology Unit, LCDB, NIDDK, NIH, Bdg. 6, Rm. B1-33, Bethesda, MD 20892 USA;* [2] *New Brunswick Scientific Co., Inc., Edison, NJ 08818-4005 USA;*
* *Corresponding author (phone: 301-496-9720, fax: 301-496-5239)*

**Keywords:** perfusion, packed bed bioreactor, HEK 293

The extracellular domain of human parathyroid $Ca^{++}$ receptor was needed for structural and clinical studies. The desired protein was produced by immobilizing the transformed HEK 293 cells in a packed bed configuration using a 2.2 liter bioreactor equipped with a vertical mixing impeller assembly and an internal basket. The production was done in two stages: a propagation stage followed by a production stage. In the first stage, which lasted approximately 140 hours, the bed was perfused with serum-containing medium, allowing the cells to grow at a constant growth rate. When the total cell population reached approximately $3\times10^{10}$, the second stage was begun. Serum-free medium replaced the serum containing medium, and the perfusion process was continued for an additional 400 hours. During this time, the medium was pumped through the packed bed at a rate of 4-6 liters per day, keeping the residual glucose concentration at 1 $gl^{-1}$.

## 1. Introduction

The packed bed bioreactor is an efficient and convenient system for the production of extracellular biomolecules from trapped or immobilized microorganisms. In the last few years, this arrangement has gained popularity for the production of proteins from anchorage-dependent as well as from suspension mammalian cells. The development of various packed bed supports, such as porous ceramic beads, glass fibers, porous glass beads , cylinder-shaped sintered glass carriers and polyester disks (Racher *et al.*, 1995) , together with a bioreactor configuration suitable for the fragile cells, is the main reasons for the increasing interests in this production system. Cell concentrations between $10^7$ and $5\times10^8$ per milliliter of bed have been obtained during the production of proteins such as insulin, $\gamma$-interferon, IL 6 and monoclonal antibodies from AtT-20, CHO, BHK and hybridoma cells.

Packed bed bioreactors are easy to use, and they allow high cell densities due to the large surface area and the easy perfusion operation. On the other hand, compared with fluidized bed or stirred tank reactors, the packed bed is inhomogeneous and sampling of cells is not possible. In the present work, we report the use of a packed bed of polyester disks for the production of human parathyroid $Ca^{++}$ receptor from human embryonic kidney (HEK 293) cells.

## 2. Materials and Methods

### 2.1. CELL LINES AND MEDIA

Human embryonic kidney cells (HEK 293), were stably transformed to produce the extracellular domain of the human parathyroid $Ca^{++}$ receptor. Growth medium consisted of DMEM supplemented with 10% fetal bovine serum, 200 U/ml Hygromicin B and 0.1% Pluronic F-68 (Gibco, USA). Production medium consisted of DMEM without phenol red (Biofluids) containing 293 Serum-Free Supplement (Kemp Biotechnologies, USA)(Shiloach *et al.*, 1996), 2 mM glutamine, 200 U/ml Hygromicin B and 0.1% Pluronic F-68.

### 2.2. REACTOR SYSTEM AND CULTIVATION CONDITIONS

#### 2.2.1. *Perfusion Process*

Cells grown on tissue culture flasks were transferred to a 2.2 liter bioreactor (1.6 liter working volume) equipped with an internal retention device (basket), containing 60 grams of Fibra-Cel$^{TM}$ disks made of polyester non-woven fabric laminated to a polypropylene screen and treated and pre-coated with poly-D-lysine, and a vertical mixing system. The dissolved oxygen concentration was kept above 30% saturation through sparging with oxygen. The pH was kept at 7.0 using a four gas mixing system (air, nitrogen, oxygen and carbon dioxide) and $NaHCO_3$. The agitation was kept at 60 RPM and the temperature at 37°C . Initially, the reactor was operated in batch mode until the glucose concentration reached 1.0 $gl^{-1}$, at which point growth medium was perfused to keep the glucose level at 1.0 $gl^{-1}$. Spent medium was collected into an intermediate reservoir from where it was pumped into an ion-exchange column. The overall production scheme is shown in Figure 1.

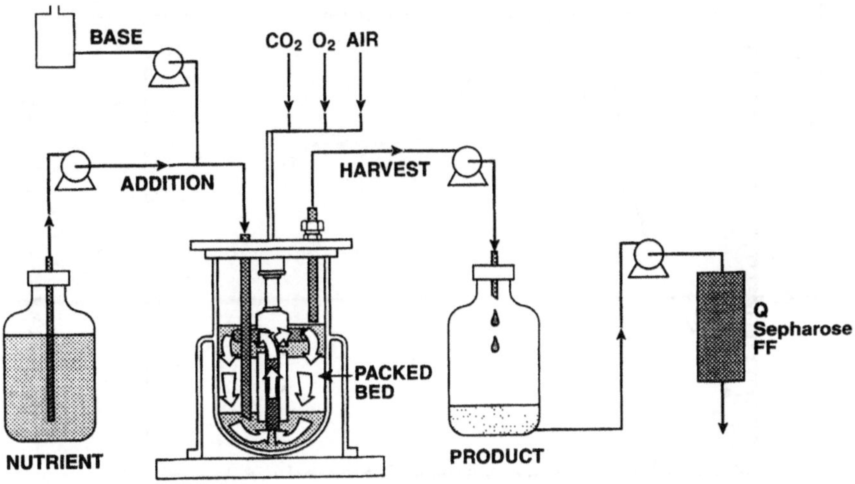


*Figure 1.* Overall production scheme for a perfused packed bed process.

## 3. Results

3.0.2. *Perfusion Process*

The results from a fermentation using continuous perfusion are presented in Figures 2 and 3. The 2.2 liter reactor was inoculated with $7\times10^8$ cells. During the first stage, which lasted approximately 140 hours, the bed was perfused with serum-containing medium and the cells grew at a constant growth rate of 0.0126 $hr^{-1}$ (Figure 2), as estimated from the glucose consumption rate and a constant specific glucose consumption rate of 25 $\mu g/10^6$ cells/hr. When the total cell number reached $3\times10^{10}$, the second stage was started and serum-fee medium replaced serum-containing medium. During this stage, which lasted approximately 400 hours, the cell number declined linearly at a rate of $2.8\times10^7$ cells/hr to $2\times10^{10}$ cells (Figure 2). Figure 3 shows the residual glucose concentration and the amount of medium used.

## 4. Discussion and Conclusions

Extracellular recombinant proteins can be produced from anchorage-dependent mammalian cells in several configurations. Among these are microcarriers in stirred tank reactors, macroporous microcarriers fluidized bed reactors, various adsorbent materials in packed bed reactors, hollow fiber reactors and flat surface arrangements. As was pointed

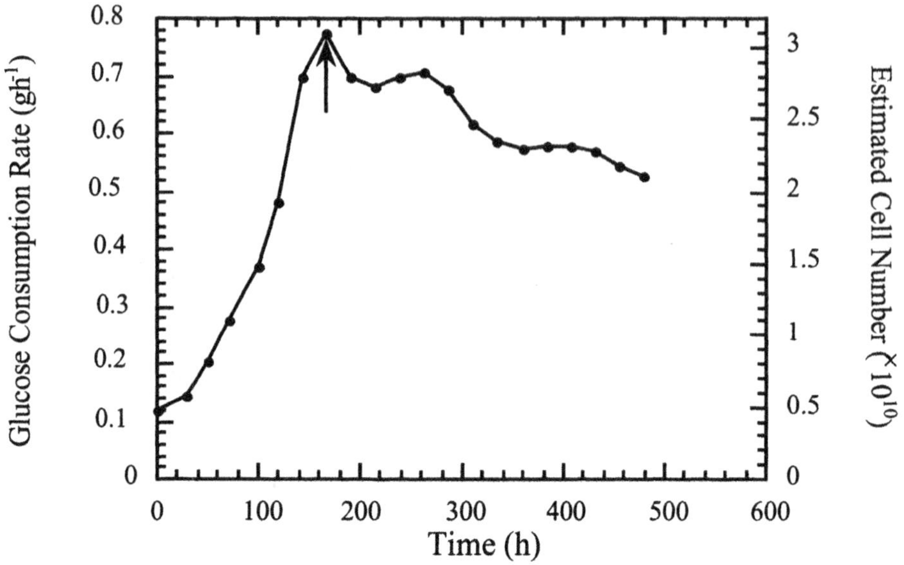


*Figure 2.* Glucose consumption rate and estimated cell number for a perfused packed bed process; arrow indicates change from the propagation stage to the production stage.

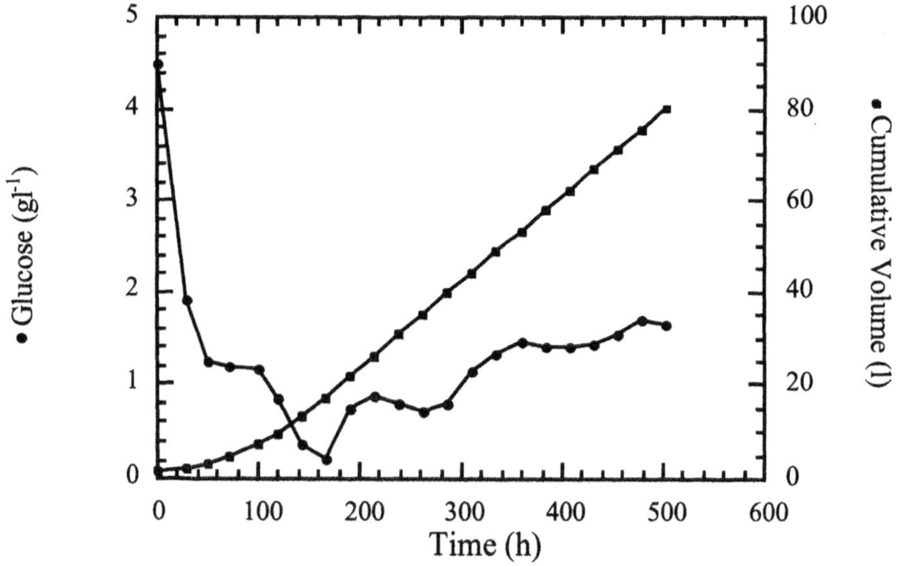


*Figure 3.* Glucose concentration and cumulative volume for a perfused packed bed process.

out in previous work (Shiloach *et al.*, 1996), the packed bed bioreactor is a very convenient arrangement for producing extracellular proteins from both anchorage-dependent and from suspension cells. It provides a large surface area in a relatively small volume, and it allows for easy medium replacement and continuous perfusion. The aeration and pH are controlled by direct aeration of the vertical mixing bioreactor, an arrangement that is less complicated than continuous flow across flat surfaces or the hollow fiber layout. Because the cells are immobilized on the paced bed, it is much easier to replace the media than in a stirred tank reactor containing microcarriers

Extracellular proteins are preferably produced from mammalian cells in serum-free medium to simplify the recovery and purification, and to insure the consistent quality of the product. In a two-stage operation, after achieving the proper biomass concentration using serum-containing medium, the medium is replaced with serum-free medium. This medium cannot support cell growth and, as a result, the cell number declines. In the described experiment, approximately, 30% of the cells were lost in 400 hours, indicating that serum-containing medium is needed periodically to restore the viability of the adherent cells.

The continuous mode operation is preferred to the repeated batch mode because it keeps the cells in a homeostatic environment like that experienced in vivo. In addition, processing the perfused medium by pumping it directly in an ion-exchange column simplifies the recovery process and improves the product stability. The observed decline in cell number can be eliminated by devising a strategy that incorporates periodic changes between serum-free medium and serum-containing medium, or by developing of a better serum-free medium.

## References


Racher, A.J., Fooks, A.R. and Griffiths, J.B. (1995) Culture of 293 cells in different culture systems: Cell growth and recombinant adenovirus production. *Biotechnology Techniques* **9**: 169-174.

Shiloach, J., Kaufman, J., Trinh, L. and Kemp, K. (1996) Continuous production of the extracellular domain of recombinant $Ca^{++}$ receptor from HEK 293 cells using novel serum free medium, in Carrondo, J.T., Griffiths, B. and Moreira, J.L.P. (eds.) *Animal Cell Technology* (pp. 535-540). Kluwer Academic Publishers, Dordrecht.

# CHARACTERIZATION OF RECOMBINANT BOVINE LACTOPEROXIDASE PRODUCED BY CHO CELLS.

*Recombinant bovine lactoperoxidase*

S. Watanabe[1, 2], K. Shimazaki[1], A. Bollen[2], and N. Moguilevsky[2]
[1]*Dairy Science Laboratory, Animal Science Department, Faculty of Agriculture, Hokkaido University, Japan.* [2] *Applied Genetics, Universit Libre de Bruxelles, 24 rue de l'Industrie, B-1400 Nivelles, Belgium.*

## Abstract

A full length cDNA coding for bovine lactoperoxidase (bLPO) was amplified by RT-PCR from mRNA extracted from mammary gland cells. The recombinant DNA was introduced into Chinese Hamster Ovary (CHO) cells by the electroporation method. The recombinant bovine lactoperoxidase (rbLPO) expressed in a large- scale production system was purified by a combination of anion- exchange chromatography and cation- exchange chromatography. The purified rbLPO was then characterized in terms of molecular weight, N- terminal amino acid sequence, carbohydrate structure, peroxidase activity, and spectroscopic properties.

The data showed that rbLPO is secreted as a single chain molecule, as two major forms differing in glycosylation. The N- terminal amino acid of rbLPO was Asp101, starting 19 residues upstream from the N- terminus of natural bovine lactoperoxidase (nbLPO). The rbLPO is enzymatically active and its specific absorption spectrum at 413 nm appears to be identical to that of LPO. This indicates that heme is integrated into the recombinant protein. The properties of rbLPO are discussed as compared to the nbLPO expressed in CHO cells obtained previously.

## 1. Introduction

Lactoperoxidase (LPO) is one of the mammalian peroxidases present in milk, saliva and tears. Quantitatively, LPO is one of the prominent enzymes in bovine milk. LPO has the ability to catalyze the oxidation of halides and pseudohalides such as thiocyanate in a reaction involving hydrogen peroxide to form potent oxidants and bactericidal agents. LPO contributes to the antibacterial and bactericidal activities in mammalian exocrine gland secretions which help to protect a variety of mucosal surfaces. LPO consists of a single polypeptide chain of 612 amino acid residues (Cals, 1991) and its molecular mass is approximately 78 kDa. Carbohydrate moieties comprise 10 % of its mass (Carlstrom, 1969). LPO, myeloperoxidase (MPO), eosinophil peroxidase (EPO), and thyroid peroxidase (TPO) are heme- containing enzymes which share 50- 70% identical residues. A high level of homology can be found among the active site- related residues. The structure of the catalytic center of mammalian peroxidases was found to be protoporphyrin IX (Sievers, 1979; Fenna, 1995). The covalent linkages between heme and the protein have been discussed on the basis of the results of spectroscopic studies, chemical studies, and site- directed mutagenesis. Investigation of the heme environment has been conducted focusing on that in MPO since the X- ray crystal structure has been established (Zeng, 1992; Fenna, 1995), however, there are no such data available concerning LPO.

To study the heme group covalently linked to the protein by site- directed mutagenesis, we obtained and characterized the recombinant protein.

## 2. Materials and Methods

### 2.1. CLONING, EXPRESSION, LARGE-SCALE PRODUCTION, AND PURIFICATION OF rbLPO

A bovine LPO cDNA was subcloned into pcDNA3, an expression plasmid vector for use in mammalian cell systems (5.4 kb, Invitrogen). The LPO cDNA fragment was inserted between the *Apa*I and *Eco*RI (Gibco/BRL) sites in the vector.

Recombinant LPO DNA was introduced into CHO DG44 *dhfr-* cells by the electroporation method. The cells were maintained in αMEM+ supplemented with 5% fetal calf serum (FCS), 2 mM L- glutamine, 1% penicillin/streptomycin and nucleotides. Neomycin (geneticin 418) resistant clones were then selected. The expression of rbLPO was detected by ELISA and by measurement of peroxidase activity using O- phenylenediamine dihydrochloride as a chromogen. After subcloning by limiting dilution, large- scale cell culture was conducted in a 6,000 $cm^2$ cell factory unit (Nunc). The cells were cultured in the presence of 3% FCS. Samples of culture supernatant were collected at 3- to 4- day intervals, 8 times in total, over a period of 28 days. Purification of rbLPO was performed using a Q- Sepharose fast-flow column and a CM- Sepharose fast- flow column, each equilibrated with 20 mM potassium phosphate, pH 7.5 as described by Moguilevsky *et al.* (1991). The rbLPO was eluted with a linear gradient of NaCl (0-450 mM) in the same buffer.

### 2.2. CHARACTERIZATION OF rbLPO

Protein concentration was measured by the method of Lowry et al. (Lowry, 1951), with some modifications. The absorption spectrum between 700 nm and 240 nm was measured with a Cary 1E spectrophotometer. The distinct absorbance ratio $A_{413nm}$/ $A_{280nm}$ was then calculated to assess the purity of the LPO.

SDS- PAGE was performed by the method of Laemmli (1970). The rbLPO expressed was analyzed by ELISA and Western- blotting using anti- bLPO monoclonal antibody (mAb) prepared in this study. The monoclonal antibody 1C3- A2 and the anti- mouse IgG (H&L) alkaline phosphatase (AP) conjugate were each diluted 5,000- fold for the ELISA, while 1C3- A2 was diluted 10,000- fold and anti- mouse IgG (H&L) Ap conjugate was diluted 7,500-fold for Western-blotting.

The ABTS assay was performed in 0.1 M acetate buffer containing 1.0 mM ABTS and 0.1 mM hydrogen peroxide (Venezie, 1991; Shindler, 1976) and the product concentration was estimated based on the extinction coefficient of $3.24 \times 10^{-2}$ $\mu M^{-1}$ $cm^{-1}$.

The N- glycans of nbLPO and rbLPO were removed enzymatically using N-glycopeptidase F (recombinant, Boehringer) and the high- mannose sugar chains of both were removed using endoglycosidase H (recombinant, Boehringer).

Automated Edman degradation of recombinant LPO was performed in a Perkin Elmer 492 Protein Sequencer. Samples were prepared by electroblotting on a PVDF membrane (MILLIPORE) from an SDS- PAGE gel. After electroblotting, the membrane was stained with Ponceau S to detect the transferred proteins. The bands of interest were cut out of the membrane.

Natural bLPO purified as described by Morrison (1963) was kindly provided by Dr. S. Nakamura, University of Hirosaki, Japan, and commercially available bLPO was obtained from Sigma.

## 3. Results

rbLPO was successfully expressed in the CHO cell line employed. Sixteen liters of culture supernatant was collected after large- scale cell culture and 13 mg of rbLPO was obtained. Fractions containing a major peak (FI) and two minor peaks (FII) were pooled and concentrated. After concentration of these fractions, the purified rbLPO was green in color. The absorption spectra of nbLPO and rbLPO were identical, with the Soret peak at 413 nm. The enzymatic activity of rbLPO (FI) measured using ABTS as the substrate was weaker than that of nbLPO (184 U/mg *vs* 337 U/mg) and the purity index was lower, 0.54 *vs* 0.88, respectively. The rbLPO (FII) showed lower activity and lower purity than rbLPO (FI). The SDS- PAGE profiles of rbLPO (FI, FII) and nbLPO showed two main bands in each instance, 88 kDa and 82 kDa for FI, 80 kDa and 74 kDa for nbLPO. FII contained mainly the component corresponding to the lower band in FI. They were all immunoreactive with anti- bLPO mAb. The partial N- terminal amino acid sequence of the constituents of the FI and FII fractions was DTTLT and that in the case of nbLPO was EVGCG.

The 88 kDa and 82 kDa constituents of rbLPO (FI) when treated with N- glycosidase F shifted in size to 73 kDa and 69.5 kDa, respectively, while the 80 kDa and 74 kDa constituents of nbLPO both shifted in size to 71 kDa. The molecular weight of nbLPO as indicated by its mobility displayed a significant change after treatment with endoglycosidase H and the difference was about 7 kDa, whereas rbLPO decreased only 4 kDa in size. The N- terminal amino acid sequence of the constituents of rbLPO corresponding to the two bands obtained after deglycosylation with N- glycosidase F showed the same result, i.e., the sequence started with the residue Asp101.

## 4. Discussion

The major difference between nbLPO and rbLPO was in the carbohydrate moiety and its structure (Table 1). Analysis by SDS- PAGE after removal of the N- glycans showed that the carbohydrate moiety of rbLPO (FI) and nbLPO comprises approximately 17% and 11%, respectively, of the total molecular weight. Furthermore, the mannose content in the case of nbLPO is higher and reaches 78%, whereas in rbLPO (FI) mannose structure represents only 22% of the total amount of N- glycans. From the results of N- terminal amino acid sequence and carbohydrate analysis, it was evident that the difference in the electrophoretic mobility between rbLPO (FI) and rbLPO (FII), and the two distinct bands obtained in the case of rbLPO, were due to differences in the carbohydrate moiety. The heterogeneity of nbLPO has been reported by Calstrom (1969), too. Although two bands were observed after digestion of rbLPO with N- glycosidase F, the N- terminal amino acid sequence of the constituent corresponding to the 73 kDa and 69.5 kDa bands showed the same signal peptide cleavage site. The molecular mass of nbLPO and rbLPO should be different, theoretically, by about 2 kDa, because rbLPO is cleaved 19 residues upstream from the cleavage site of nbLPO. Therefore we presume that both nbLPO and rbLPO also contain O- glycans. The cleavage of rbLPO occurs between Arg100 and Asp101, and cleavage of nbLPO occurs between Trp119 and Glu120 as shown in Figure 1. Moreover, we also found that cleavage occurs between Thr115 and Ala116 in the case of a commercial preparation of nLPO. This inconsistency with respect to the N-terminal amino acid residue of nbLPO has been found by other researchers, such as Asp101 and Ser118 reported by Cals *et al.* (1991), Leu129 reported by Sievers (1980). The signal peptide cleavage site in bLPO may be Asp101 and the reason for this inconsistency could be the deletion of some amino acid sequences as assumed by Cals *et al.* (1991).

LPO and MPO are 51% homologous in terms of primary structure. The cleavage site of rhMPO expressed in CHO cell lines was between Gly22 and Ala23 and it was found to consist of a single polypeptide chain, whereas natural human MPO is a tetramer composed of two dimers of light and heavy chains. The light chain of natural MPO starts at Thr140 and the heavy chain starts at Val253 (the position number is counted from Met 1 in rhMPO (Moguilevsky, 1991)).

The ratio of distinct absorption ($A_{413nm}/A_{280nm}$) was 0.54 for the recombinant LPO. This was lower than that of purified natural LPO (0.88). This lower value could be the result of a noncovalent state of the prosthetic heme in recombinant LPO, as was previously shown for the rLPO produced in insect cells (DePillis, 1997).

### References

Cals, M., P. Mailliart, G. Brignon, P. Anglade, and B. R. Dumas (1991) Primary structure of bovine lacto-peroxidase, a fourth member of a mammalian heme peroxidase family. *Eur. J. Biochem.* **198**, 733-739.

Carlstrom, A. (1969a) Lactoperoxidase : Identification of multiple molecular forms and their interrelationships. *Acta Chem. Scand.* **23**, 171-184.

Carlstrom, A. (1969b) Physical and compositional investigations of the subfractions of lactoperoxidase. *Acta Chem. Scand.* **23**, 185-202.

De Wit, J., and A. Van Hooydonk. (1996) Structure, function and applications of lactoperoxidase in natural antimicrobial systems. *Netherlands Milk & Dairy J.* **50**, 227-244.

DePillis, G., S. Ozaki, J. Kuo, D. Maltby, and P. Montellano. (1997) Autocatalytic processing of heme by lactoperoxidase produces the native protein-bound prosthetic group. *J. Biol. Chem.* **272**, 8857-8860.

Dull, T., C. Uyeda, A. Strosberg, G. Nedwin, and J. Seilhamer. (1990) Molecular cloning of cDNAs encoding bovine and human lactoperoxidase. *DNA Cell Biol.* **9**, 499-509.

Fenna, R., J. Zeng, and C. Davey. (1995). Structure of the green heme in myeloperoxidase. *Arch. Biochem. Biophys.* **316**, 653-656.

Laemmli, U. (1970) Cleavage of structural proteins during the assembly of the head of bacteriophage T4. *Nature* **227**, 680-686.

Lowry, O.H., N.J. Rosebrough, A.L. Farr, and R.J. Randall. (1951) Protein measurement with the Folin phenol reagent. *J. Biol.Chem.* **193**, 265-275.

Moguilevsky, N., L. Garcia-Quintana, A. Jacquet, C. Tournay, L. Fabry, L. Pierard, and A. Bollen. (1991) Structural and biological properties of human recombinant myeloperoxidase produced by Chinese hamster ovary cell lines. *Eur. J. Biochem.* **197**, 605-614.

Morrison, M., and D. Hultquist. (1963) Lactoperoxidase. J. *Biol. Chem.* **238**, 2847-2849.

Shindler, J., R. Childs, and W. Bardsley. (1976) Peroxidase from human cervical mucus. The isolation and characterization. *Eur. J. Biochem.* **65**, 325-331.

Sievers, G. (1979). The prosthetic group of milk lactoperoxidase is protoheme IX. *Biochim. Biophys. Acta* **579**, 181-190.

Sievers, G. (1980). Structure of milk lactoperoxidase. A study using circular dichroism and difference absorption spectroscopy. *Biochim. Biophys. Acta* **624**, 249-259.

Venezie, R., J. Jenzano, and R. Lundblad. (1991). Differentiation of myeloperoxidase and glandular peroxidase in biological fluids: Application to human saliva. *J. Clin. Lab. Anal.* **5**, 57-59.

Zeng, J., and R. Fenna. (1992). X-ray crystal structure of canine myeloperoxidase at 3Å resolution. *J. Mol. Biol.* **226**, 185-207.

Table 1. Physico-chemical characteristics of bovine lactoperoxidase

| | natural LPO | recombinant LPO | reported values* |
|---|---|---|---|
| molecular weight | 80 kDa | 82, 88 kDa | 78 kDa |
| number of amino acid residues | 594, 597 | 612 | 612 |
| half-Cys | 15 | 15 | 15 |
| carbohydrate content | approx. 11% (80 kDa) | approx. 17% (88 kDa) | 10% |
| peroxidase activity | 337 (U/mg) | 184 (U/mg) | |
| A412nm/A280nm | 0.88 | 0.54 | |

* From De Wit and Van Hooydonk (1996)

rbLPO (FI, FII) bLPO (A) bLPO (B)

↓ ↓ ↓

**RR DTTLTNVTDPSLDLT ALSW EVGCGAPVPLV**

101 116 120

Figure 1. N-terminal amino acid sequence of bovine lactoperoxidase
The N-terminal amino acid sequence of purified recombinant LPO and that of natural bLPO were determined by the automated Edman degradation method. The position was counted from Met 1 reported by Dull *et al.* (1990). The cleavage site in nbLPO from Sigma (A), and in another nbLPO (B).

# EXPRESSION OF A SOLUBLE αβ HETERODIMERIC T CELL RECEPTOR IN ANIMAL CELLS

Maho Amano, Mamoru Totsuka, Kiyotaka Fujine, Shinya Yamada, and Shuichi Kaminogawa
*Department of Applied Biological Chemistry, The University of Tokyo, 1-1-1 Yayoi, Bunkyo-ku, Tokyo 113-8657, Japan*

**ABSTRACT** We have constructed a c-myc tagged αβ heterodimeric soluble type of T cell receptor (sTCR). The sTCR has a 'leucine zipper' structure composed of basic and acidic peptides connected to the α- and β- chain constant region, respectively, via a cleavable flexible linker. SV40-based expression plasmids for either sTCR α- or β- chain were constructed, and COS-7 cells co-transfected with these two plasmids transiently expressed the sTCR. The sTCR was efficiently secreted with the expected molecular size as estimated by SDS-PAGE analysis. It is anticipated that structural and physico-chemical analyses of the sTCR will help to clarify the detailed molecular mechanism of the T cell response.

## 1. Introduction

The primary event in the cellular immune response is recognition of antigens by T cells. T cells recognize antigens that are presented in the context of major histocompatibility complex (MHC) molecules through the T cell receptor (TCR), and exhibit various responses such as proliferation, cytokine production, help for antibody production, etc.. Although various examples of altered T cell responses caused by antigen analogs have been reported [1] and the three-dimensional structure of TCR has been clarified [2], the detailed molecular mechanism of the T cell response triggered by interaction of TCR and its ligand has not been made clear yet. Comparison of the physico-chemical properties and structures of TCR interacting with an analog peptide or wild-type antigen would help to clarify how recognition of two similar ligands results in different T cell responses. However, TCR is a membrane-bound protein, so it is difficult to apply these analyses to intact TCR molecules. Therefore, a soluble type of TCR (sTCR) is considered to be necessary for this purpose.

We have analyzed the altered responses of a $CD4^+$ T cell clone, G1.19, specific for residues 119-133 (p119-133) of bovine β-lactoglobulin (β-LG), upon stimulation with single amino-acid substituted analogs of p119-133. Here, we have constructed a c-myc-tagged αβ heterodimeric sTCR derived from G1.19, which has a 'leucine zipper' structure composed of basic and acidic peptides connected to the α- and β- chain constant region, respectively, via a cleavable flexible linker, and a c-myc epitope sequence only for the α-chain construct. It is thought that the electrostatic properties of the 'leucine zipper' moiety facilitate dimerization of the α- and β- chains of the sTCR [3]. In the present study, it was our intention to prepare the sTCR by being expressed in mammalian cells in order to later accomplish the crystallographic analysis of the TCR/peptide/MHC complex and measurement of the affinity between TCR and the peptide/MHC complex.

## 2. Materials and methods

### 2.1 CELLS

Murine Th1 clone G1.19 specific for p119-133 of bovine β-LG was established from lymph node cells of β-LG-immunized C57BL/6 mice in our laboratory [4]. The medium used for cell culture was RPMI-1640 (Nissui, Tokyo, Japan) supplemented with 2 mM glutamine, 100 μg/ml streptomycin, 100 U/ml penicillin, 50 mM 2-mercaptoethanol, 10% fetal calf serum (FCS), and 10% T cell growth factors.

COS-7 cells were cultured in Dulbecco's modified Eagle's medium (DMEM) (Life Technologies, Rockville, MD, USA) supplemented with piperazine-N'-2-ethanesulfonic acid, 100 μg/ml streptomycin, 100 U/ml penicillin, 50 mM 2-mercaptoethanol, and 10% FCS.

### 2.2 CONSTRUCTION OF sTCR α- AND β- CHAIN EXPRESSION PLASMIDS AND TRANSFECTION OF CELLS

cDNAs encoding TCR α- or β- chains of G1.19 were synthesized using the 5'-RACE system (Life Technologies). Briefly, first-strand cDNAs were synthesized from total RNA by reverse transcription using an antisense primer complementary to the Cα or Cβ region. A poly C-tail was synthesized at the 3'-end of the first-strand cDNA using terminal deoxy-nucleotidyl transferase. Poly C-tailed cDNAs were amplified by PCR using an anchor primer and a TCR Cα- or Cβ- specific primer. The resulting cDNAs encoding TCR α- and β- chains of G1.19 were cloned and each was ligated with a DNA sequence encoding a cleavable flexible linker, a basic or acidic peptide constituting a 'leucine zipper', and a c-myc epitope sequence only for the α- chain construct, which was prepared using PCR. Sequence analysis of the resulting DNAs was performed by means of a ThermoSequenase Cycle Sequencing Kit (Shimadzu, Kyoto, Japan) and a DNA sequencer DSQ-1000L (Shimadzu). After confirmation of the nucleotide sequence, each construct was subcloned into the expression vector, pME18S, which has a SRα promoter and the origin of SV40 (Fig.1). The resulting expression plasmids were introduced into COS-7 cells by the lipofection method using Lipofect AMINE PLUS reagent (Life Technologies), according to the manufacturer's instructions.

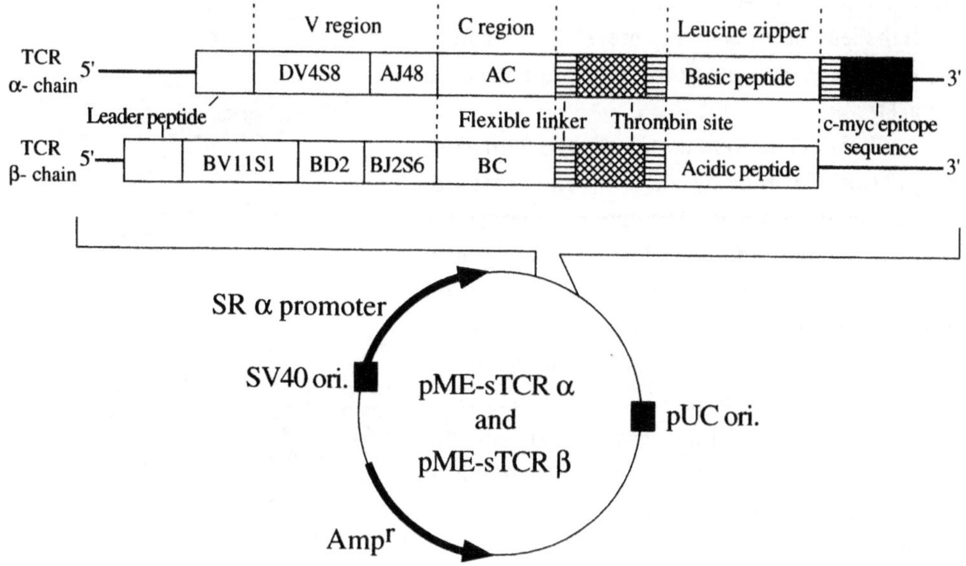


Figure 1. Schematic representation of the structure of soluble type of T cell receptor (sTCR) derived from a β-lactoglobulin-specific murine CD4⁺ T cell clone G1.19, and construction of sTCR expression plasmids, pME-sTCR α and pME-sTCR β.

## 2.3 PURIFICATION AND DETECTION OF sTCR

The transfected cells were harvested 4 days after transfection to obtain the culture supernatant, and 1 mM phenyl methyl sulfonyl fluoride, 5 mM ethylenediamine-N, N, N', N'-tetra acetic acid and 5 mM N-ethylmaleimide, and 0.02% sodium azide were added.

For affinity purification, an immunoaffinity column was prepared by covalently coupling hamster anti-Cβ monoclonal antibody (mAb) (H-57) to NHS-activated Sepharose HP (Pharmacia Biotech, Uppsala, Sweden) according to the manufacturer s instructions. The samples of culture supernatant were applied to the immunoaffinity column equilibrated with DMEM , and then washed with phosphate buffered saline, and eluted with 0.1 M glycine-HCl buffer, pH 3.0. The eluted samples were immediately neutralized by collection in tubes containing 1.5 M Tris-HCl , pH 8.9.

SDS-PAGE analysis was performed as described by Laemmli [5], using a separating gel of 12 % acrylamide under non-reducing conditions, and protein bands were detected by silver staining.

# 3. Results

Analysis of TCR cDNA sequences of G1.19 revealed that the α chain is encoded by

Vδ4.8 joined with Jα48 and Cα DNA, while the β chain is encoded by Vβ11 joined with Dβ2, Jβ2.6, and Cβ DNA. In order to obtain the sTCR with the leucine zipper structure, sTCR α- and β-chain expression plasmids were constructed (Fig.1).

COS-7 cells were co-transfected with these expression plasmids by the lipofection method, and cultured for 4 days. The culture supernatants were collected and applied to immunoaffinity chromatography. The immunoaffinity-purified fraction obtained was examined by SDS-PAGE analysis under non-reducing conditions. As shown in Fig.2, two bands at 62 kDa and 55 kDa were detected in lane 2 which was loaded with the sample, whereas no band with the same mobility as those above was not observed in lane 1 loaded with the negative control.

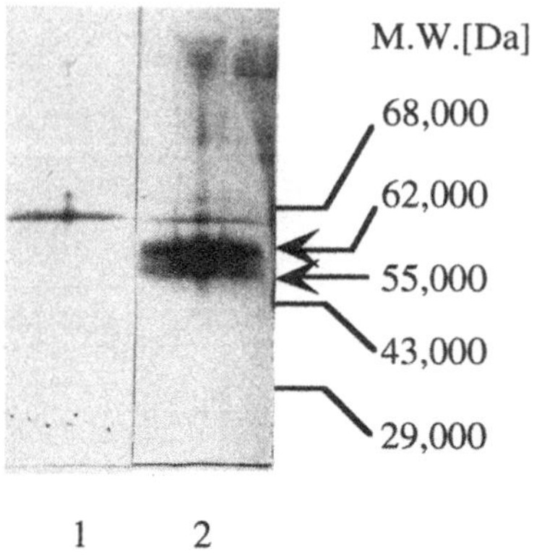


Figure 2. SDS-PAGE analysis of immunoaffinity-purified samples. Lane 1, purified sample from the supernatant of COS-7 cells transfected with pME18S as negative control; lane 2, purified sample from the supernatant of COS-7 cells co-transfected with pME-sTCR α and pME-sTCR β.

**4. Discussion**

In the present study, αβ heterodimeric sTCR derived from a β-LG-specific murine $CD4^+$ T cell clone was successfully expressed and secreted by COS-7 cells which were co-transfected with the α- chain and β- chain expression plasmids.

SDS-PAGE analysis of the immunoaffinity purified fraction from culture supernatants of COS-7 cells transfected with sTCR α- and β- chain expression plasmids showed two bands at 62 kDa and 55 kDa. The band at 62 kDa was considered to contain sTCR, because the expected molecular weight of sTCR is about 60 kDa. Moreover, upon Western blot analysis this band was detected using mouse anti-c-myc mAb (9E10) (data not shown). The band at 55 kDa might be a degradation product of the sTCR.

The yield of sTCR was approximately 50 ng per ml of culture supernatant as estimated by comparing the density of the band with that of the standard. Since a large quantity of purified sTCR is required to characterize its biochemical properties, we are presently trying to establish a baculovirus expression system for large-scale production of the sTCR. With this system, more efficient production of recombinant proteins can be expected.

When a large-scale production of sTCR is achieved, it will be easier to perform crystallographic analysis and measurement of the affinity between TCR and the peptide-MHC complex. Comparison of these properties of TCR interacting with wild-type ligand or altered ligand will help to clarify the detailed molecular mechanism underlying the different responses of T cells depending on the slight structural differences in their specific TCR ligands. From a clinical viewpoint, sTCR might be applicable to treatment of diseases caused by pathogenic T cell responses specific for a certain antigen, such as autoimmune diseases and allergy. This is because sTCR is expected to block the activation of the pathogenic T cells by binding specifically to the relevant peptide-MHC complex.

## 5. References


1. J. Sloan-Lancaster, and P. M. Allen. (1996) Altered peptide ligand-induced partial T cell activation: Molecular mechanisms and role in T cell biology, *Annu. Rev. Immunol.*, **14**, 1-27.
2. P. J. Bjorkman. (1997) MHC restriction in three dimensions: A view of T cell receptor/ligand interactions, *Cell*, **89**, 167- 170.
3. H-C. Chang, Z-Z. Bao, Y. Yao, A. G. D. Tse, E. C. Goyarts, M. Madsen, E. Kawasaki, P. P. Brauer, J. C. Sacchettini, S. G. Nathenson, and E.L. Reinherz. (1994) A general method for facilitating heterodimeric pairing between two proteins: Application to expression of α and β T-cell receptor extracellular segments, *Proc. Natl. Acad. Sci. USA.*, **91**, 11408-11412.
4. M. Totsuka, S. Furukawa, E. Sato, A. Ametani, and S. Kaminogawa. (1997) Antigen-specific inhibition of $CD4^+$ T-cell responses to β-lactoglobulin by its substituted mutant through T-cell receptor antagonism, *Cytotechnology*, **25**, 115-126.
5. K.U. Laemmli. (1970) Cleavage of structural proteins during the assembly of the head of bacteriophage T4, *Nature*, **227**, 681-685.

# POSTTRANSLATIONAL MODIFICATIONS OF RECOMBINANT VON WILLEBRAND FACTOR: LIMITATIONS AND EXPERIMENTAL IMPROVEMENT AT HIGH YIELD EXPRESSION

B. PLAIMAUER, U. SCHLOKAT *, M. HIMMELSPACH, P.L. TURECEK, H.P. SCHWARZ, F.G. FALKNER, AND F. DORNER
*Biomedical Research Center, Hyland/IMMUNO AG (Division of BAXTER, Inc.), Uferstr. 15, 2304 Orth/Donau, AUSTRIA; * for correspondence: fax +43-20100-4000; phone +43-1-20100-4144; e-mail schloku@baxter.com*

## 1. Abstract

Von Willebrand factor (vWF) is a multimeric plasma glycoprotein that promotes platelet aggregation, mediates platelet adhesion to the subendothelium, and stabilizes coagulation factor VIII (FVIII). Recombinant vWF (rvWF) was constitutively expressed at high yield in stable CHO cell clones (CHO-rvWF). Carbohydrate analysis of rvWF and plasma derived (pd) vWF revealed common and divergent structures. The absence of terminal high mannose residues indicated intact and complete glycosylation. Alteration of terminal carbohydrate structures by $\alpha$ (2,6) sialyltransferase coexpression did not influence rvWF mediated platelet aggregation and collagen binding, both of which require appropriate glycosylation and are sensitive to glycosylation changes. Upon increasing rvWF expression by amplification, from 100 ng to 20 $\mu$g rvWF/$10^6$ cells x day, proteolytic propeptide removal had become incomplete resulting in impaired interaction with FVIII. Complete propeptide cleavage could be accomplished by employing recombinant Furin, a ubiquitous endoprotease, and derivatives thereof, either by coexpression *in vivo* or by treatment *in vitro*. Multimerization, also crucial to vWF function, could be significantly improved by cell culture medium modification.

## 2. Introduction

Von Willebrand factor, a large multimeric and multivalent glycoprotein, is involved in the adhesion of platelets to the subendothelium at the site of vascular injury, mediates platelet-platelet interactions, and functions as a carrier for coagulation factor VIII in the circulating blood[1]. Individuals with quantitative or qualitative vWF deficiencies suffer from von Willebrand disease (vWD), which manifests in mild to severe bleeding clinically. vWD patients are currently treated either with the vasopressin analogue DDAVP, where applicable, or with FVIII/vWF complex concentrates from plasma.

vWF is synthesized by megakaryocytes and vascular endothelial cells as a large precursor polypeptide (pro-vWF). Prior to secretion, vWF undergoes complex posttranslational modifications including signal peptide cleavage, carbohydrate processing, sulfation, multimerization by intermolecular disulfide bonding, and propeptide removal. Mature multimers are stored in specific granules, from which they are released upon stimulation, in addition to being secreted constitutively.

Human vWF was previously expressed recombinantly in CHO cells as a potential therapeutic agent for the treatment of vWD patients[2]. Expression of vWF resulted in the secretion of biologically active high molecular weight multimers. Expression of recombinant proteins at high yield is often associated with a reduced capability of the host cell to efficiently exert posttranslational modifications critical to the function of the desired protein. Insufficient modification, resulting in decreased bioactivity of the recombinant protein, may be caused by potential saturation of the cellular enzymatic machinery and/or unfavourable environmental conditions during cell cultivation. Increase of rvWF expression level in CHO cells upon amplification resulted in the production of up to 20μg rvWF/$10^6$ cells x day. However, both the degree of multimerization and the proteolytic propeptide removal became severely impaired, resulting in strongly reduced biological activity of rvWF.

In this report, we summarize specific posttranslational modifications of rWF produced in CHO cells, describe their influence on the functional activity of the protein and discuss approaches that could contribute to an increase of rvWF bioactivity.

## 3. Results/Discussion

### 3.1. PROTEOLYTIC PROPEPTIDE REMOVAL

CHO cell clones constitutively expressing human rvWF were established. At 100ng rvWF/$10^6$ cells x day, propeptide removal of rvWF precursors was complete. However, increasing the expression 100 fold to 10μg rvWF/$10^6$ cells x day by amplification resulted in incomplete propeptide removal (figure 1A).

Furin[3], an endoprotease anchored in the trans-Golgi network (TGN) by a transmembrane domain and known to cleave a wide variety of protein precursors (e.g. coagulation factors, viral surface proteins, bacterial toxins, hormone receptors) in the constitutive secretory pathway, was used to improve rvWF propeptide removal.
Full length wild-type human Furin, when additionally expressed in CHO-rvWF cells, yielded permanent CHO-rvWF/rFurin clones. These were shown to be capable of mediating complete propeptide removal at expression of up to 2μg rvWF/$10^6$ cells x day (figure 1B)[4]. Surprisingly, propeptide removal was found to be performed by a naturally truncated form of recombinant Furin ('shed' Furin, lacking the transmembrane domain) in the conditioned medium rather than intracellularly[5]. Amplification attempts to further increase full length rFurin expression failed. Thus, it was not possible to accomplish complete propeptide removal at rvWF expression levels surpassing 2μg/$10^6$ cells x day. Likely, a rFurin level increased any further becomes toxic to its host cell[6].

Expression of experimentally truncated rFurin molecules lacking the C-terminus and the transmembrane domain, resulted in immediate passage through the TGN into the supernatant, yielding approximately 10 fold higher concentrations of functionally active rFurin molecules in the conditioned medium compared to 'shed' Furin[7]. This suggests that (1) limited tolerance of the host cell to intracellular full length rFurin concentrations is the reason for the failure to increase full length rFurin expression by amplification, and (2) coexpression of these experimentally truncated rFurin derivatives could be used to successfully cleave protein precursor molecules at an expression level of up to 20μg rvWF/$10^6$ cells x day. If expression of recombinant protein precursors exceeds even this level, as has been reported for coagulation factors rFIX and rFX, an *in vitro* processing procedure must be established. Truncated rFurin derivatives, readily secreted and modified by linkage to heterologous affinity epitopes (in order to facilitate their purification), can be used as shown in figure 1C.

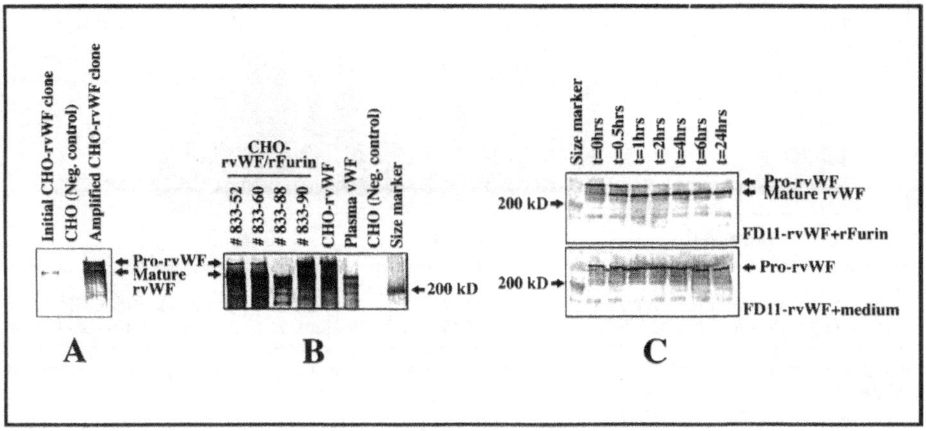


*Figure 1.* Processing of rvWF with and without rFurin *in vivo* and *in vitro* (Western blotting). *A*: Limitation of rvWF precursor processing upon overexpression. Equal amounts of 24 hr conditioned media derived from an initial transfectant expressing 100ng rvWF/$10^6$ cells x day, an amplified CHO-rvWF clone expressing 10µg rvWF/$10^6$ cells x day, and unmanipulated CHO negative control cells were applied. *B*: 24 hr cell culture supernatants obtained from different CHO cell clones stably coexpressing rvWF and rFurin (CHO-rvWF/rFurin). Depending on rFurin expression in individual clones, rvWF is processed to variable degrees. Equal amounts of rvWF were applied. *C*: rvWF propeptide removal *in vitro* by purified C-terminally truncated rFurin. Pro-rvWF in the conditioned medium derived from CHO/FD11-rvWF cells (lacking endogenous Furin) wasincubated with rFurin for 24 hours taking samples at the indicated time points upper panel). In the lower panel, no rFurin was added (negative control).
For Western blotting, reduced samples were applied to SDS-PAGE under denaturing conditions. rvWF was visualized using rabbit anti-human vWF serum and an alkaline phosphatase conjugated second antibody. The blots in *A* and *B* were over-developed to detect even trace amounts of pro-rvWF.

## 3.2. MULTIMERIZATION

The degree of multimerization and the hemostatic activity of vWF strongly correlate[1]. The greater the extent of multimerization, the more collagen and platelet binding sites are available on the repeated subunit structure. Thus, a tight interaction between platelets and the subendothelial matrix upon even high shear stress is ensured. Thus, best ristocetin-induced platelet aggregation (RistoCoF activity) and collagen-binding activity are associated with the largest multimers.

An increase in specific rvWF expression in individual cells resulted in significantly reduced multimerization. Decreased multimerization could be compensated for by modification of the cell culture medium. Supplementation of crucial components therein significantly improved the degree of multimerization (figure 2).

## 3.3. GLYCOSYLATION

Appropriate glycosylation may be critical for functional activity, *in vivo* stability and the absence of immunogenicity of pharmaceutical proteins.The glycosylation pattern

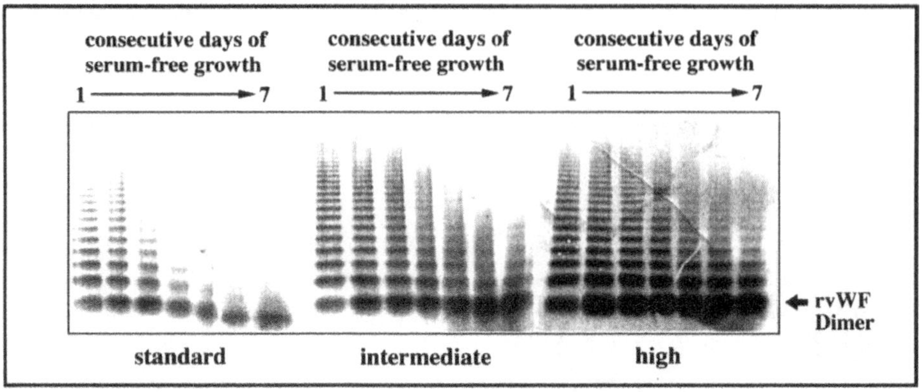


*Figure 2*: Improvement of rvWF multimerization by modification of the cell culture medium. CHO-rvWF cells were grown for 7 days in standard DMEM:HAM'S (1:1) medium, as well as in medium supplemented with intermediate and high concentrations of multimerization promoting components. Samples of the conditioned media were taken daily and the multimer pattern was analyzed applying equal amounts of rvWF.

of recombinant proteins may be determined by the specific carbohydrate processing ability of the host cell, the expression level and the environmental culturing conditions.

Carbohydrate analysis of rvWF demonstrated the presence of N- and O-glycosylation (table 1)[8]. Differential deglycosylation was carried out employing individual combinations of N-glycosidase F, O-glycosidase and sialidase, verified by specific lectin binding, and resulted in impairment or even loss of functional activity (RistoCoF mediated platelet aggregation and collagen binding). The absence of terminal high mannose structures demonstrated complete carbohydrate processing and intact glycosylation.

| Terminal structures | Human vWF | Recombinant vWF |
|---|---|---|
| SA α (2,3) Gal | + | +++ |
| SA α (2,6) Gal | +++ | - |
| Gal β (1,4) GlcNAc | ++ | ++ |
| Gal β (1,3) GalNAc | - | - |
| High Mannose | - | - |

*Table 1*. Analysis of the terminal N- and O-linked carbohydrate structure of plasma derived vWF and CHO cell derived rvWF by specific lectin binding: SA, sialic acid; Gal, galactose; GlcNAc, N-acetylglucosamine; GalNAc, N-acetylgalactosamine. The degree of binding to the corresponding lectin is indicated semiquantitatively.

Typically for hamster cell-derived proteins, only α (2,3)-linked terminal sialic acids were found in rvWF. Functional 2,6-sialyltransferase (2,6-ST), commonly catalyzing α (2,6) sialic acid linkages in human cells, is not present in CHO cells. Coexpression of rat 2,6-ST in CHO-rvWF cells yielded rvWF molecules carrying terminal α (2,3) and α (2,6) sialic acid structures. This alteration of glycosylation, however, did not modulate the functional activity of rvWF in RistoCoF and collagen binding assays.

## 4. Summary

Endogenous Furin in CHO cells was found to mediate rvWF propeptide removal only at an expression of up to 200ng rvWF/$10^6$ cells x day. Complete rvWF precursor cleavage, up to 2μg rvWF/$10^6$ cells x day, was successfully performed by coexpression of full length rFurin. Surprisingly, processing occurred in the conditioned medium by a naturally secreted form of rFurin, rather than intracellularly. Experimental truncation of the transmembrane domain resulted in immediate secretion of Furin and, hence, 10 fold higher concentrations in the cell culture medium. Epitope tagging of Furin molecules, lacking the transmembrane domain which usually anchors the molecule in the TGN, allowed rapid affinity-purification. These molecules were used for processing of rvWF precursor molecules. rvWF multimerization was similarly impaired when specific rvWF productivity in individual cells was increased by amplification. Modification of the cell culture medium by supplementation of specific components improved the degree of rvWF multimerization. Finally, despite heavy glycosylation, carbohydrate structures were shown to be intact and carbohydrate processing complete. Coexpression of α (2,6) sialyltransferase introduced terminal α (2,6) sialic acid structures and, thus, rendered rvWF carbohydrate structure more similar to plasma vWF. This alteration, however, did not influence functional activity.

## 5. Acknowledgments

We are grateful to Alexandra Preininger and Gabriele Mohr for excellent technical assistance, Stephen Leppla and Fabio Dall'Olio for generously providing the CHO/FD11 cells and rat α (2,6) sialyltransferase cDNA, respectively, and Maria O'Rourke for critical reading of the manuscript.

## 6. References

1. Furlan M. (1996). Von Willebrand factor: molecular size and functional activity. Ann. Hematol. 72, 341-348.
2. Fischer, B., Mitterer, A., Schlokat, U., DenBouwmeester, R., and Dorner, F. (1994). Structural analysis of recombinant von Willebrand factor: identification of hetero- and homo-dimers. FEBS Lett. 351, 345-348.
3. Nakayama, K. (1997). Furin: a mammalian subtilisin/Kex2p-like endoprotease involved in processing of a wide variety of precursor proteins. J. Biochem. 327, 625-635.
4. Fischer, B., Schlokat, U., Mitterer, A., Reiter, M., Mundt, W., Turecek, P.L., Schwarz, H. P., and Dorner, F. (1995). Structural analysis of recombinant von Willebrand factor produced at industrial scale fermentation of transformed CHO cells co-expressing recombinant furin. FEBS Lett. 375, 259-262.
5. Schlokat,U., Fischer, B.E., Herlitschka, S., Antoine, G., Preininger, A., Mohr, G., Himmelspach, M., Kistner, O., Falkner, F.G., and Dorner, F. (1996). Production of highly homogeneous and structurally intact recombinant von Willebrand Factor multimers by furin mediated pro-peptide removal in vitro. Biotechnol. Appl. Biochem. 24: 257-267.
6. Creemers, J.W.M. (1994). Structural and Functional Characterization of the Mammalian Proprotein Processing Enzyme Furin. Ph.D. Thesis at the University of Leuven. Belgium.
7. Preininger, A., Schlokat, U., Mohr, G., Himmelspach, M., Stichler, V., Kyd-Rebenburg, A., Plaimauer, B., Turecek, P.L., Schwarz, H.P.,Wernhart, W., Fischer, B.E., and Dorner, F. (1998). Strategies for recombinant furin employment in a biotechnological process: complete target protein precursor cleavage. Cytotechnology, in press.
8. Fischer, B.E., Schlokat, U., Reiter, M., Mundt, W., and Dorner, F. (1997). Biochemical and functional characterization of recombinant von Willebrand factor produced on a large scale. Cell. mol. life. sci. 53, 943-950.

# INCUBATION WITH HEPARIN AS AN IMPROVED HARVEST METHOD FOR HIGH TITRE DISABLED INFECTIOUS SINGLE CYCLE (DISC) HSV-2 FROM MICROCARRIER CULTURES

WRIGHT PA[1], ZECCHINI TA[1], SMITH RJ[1], O'KEEFE, RS[2] and SLATER NK[2]

[1]Cantab Pharmaceuticals Research Ltd.,
184 Cambridge Science Park, Cambridge. U.K. CB4 4GN

[2]Dept. of Chemical Engineering and Applied Chemistry
Aston University, Aston Triangle, Birmingham. U.K. B4 7ET

## 1. Abstract

Disabled Infectious Single Cycle (DISC) HSV-2 is cultured in a complimentary cell line (CR2) in microcarrier cultures up to a 15 L scale. Virus is released from the cells by the addition of a hypotonic saline solution and using low-pressure disruption technique. The resultant material must be further purified to decrease contaminating host cell protein and DNA.

The aim of our study was to demonstrate that incubation with heparin at the point of harvest could be used to release the virus from the cells, providing a favourable alternative to the cell disruption method.

Infected cultures were harvested at approximately 100% CPE by incubation with porcine mucosal heparin.

Virus titres equivalent to those obtained by disruption techniques were consistently obtained using heparin incubation. The intact cells were easily separated from the released virus in the supernatant. The levels of contaminants were reduced for the start of further purification steps.

## 2. Introduction

It has been reported that the incidence of genital herpes world-wide is high and that it is increasing [1]. In trying to combat the disease, a safe and effective vaccine has been sought. Cantab has approached this problem by developing a DISC HSV-2 viable viral vaccine. These are viruses which have an essential gene removed and can only undergo a single round of replication in a vaccinated host. It has been shown that these viruses can be used as an effective vaccines [2]. A complimentary cell line, designated CR2 was prepared from the WHO Vero cell bank approved for use in vaccine manufacture. The cell line was modified to contain the essential gene removed from the virus. In this case CR2 was modified to contain the glycoprotein H (gH) gene under the control of the glycoprotein D (gD) gene promoter [3].
The production process involves the culturing of the CR2 in a microcarrier process [4]. The cells are cultured to almost maximal cell density when the medium is removed the cells washed with medium without FBS and the virus infection medium added. The cells are infected at an MOI of 0.01 with our DISC-HSV. The virus is propagated over a 72 hour period. The virus is harvested from the cells in a number of ways and the aim of this work was to try to demonstrate the differences in the virus harvesting regimes and the improvements we have made to reduce contaminating cellular DNA and proteins.

## 3. Methods

Cells and virus were propagated in microcarrier cultures at a scale of 5 L and $5g.l^{-1}$ of Cytodex 1 microcarriers. The cultures were harvested when approximately 80-100% CPE of the cells was observed. The methods of virus harvest that were compared were either lytic (sonication, hypotonic saline and hypotonic saline/bionebulisation) or non-lytic (heparin release).

### 3.1 LYTIC METHODS

*3.1a Sonication.*
Portions of the culture were sonicated in a cup-horn sonicator (Life Sciences model XL2020) on full power for 1 minute. This disrupts the cells to release virus in to the media. Aliquots of the sonicated suspension were removed and a viable cell count was performed. These counts were used to assess the efficiency of the disruption procedure. The remainder of the suspension was centrifuged to remove cellular debris and the supernatant carefully decanted. A $TCID_{50}$ assay was used to determine the amount of released virus by sonication. Samples were also taken for determination of the levels of cellular DNA and total protein.

*3.1b Hypotonic saline.*
Addition of this reagent causes the cells to swell and become disrupted. This is a method that is not as destructive in comparison to sonication. A portion of the microcarrier culture was drained of culture medium and had hypotonic saline added and incubated at 19°C for 30 minutes. The impeller speed was increased to 300 rpm from the culture speed of 90rpm to ensure cellular detachment from the microcarriers. The resulting cell/virus/microcarrier suspension was filtered through a 76μm stainless steel filter to retain the microcarriers. The cell/virus suspension was centrifuged to pellet the cells and cellular debris to provide a cell free virus suspension. A $TCID_{50}$ assay was used to determine the amount of released virus by the hypotonic saline treatment. Samples were also taken for determination of the levels of cellular DNA and total protein post disruption.

*3.1c Hypotonic saline/bionebulisation.*
A portion of the cell suspension above was treated by bionebulisation to release virus. The bionebulisation was performed at a carrier gas pressure of 50psi [5]. Samples were again centrifuged to remove any cells. The supernatant was sampled for the determination of the levels of BSA, cellular DNA and total protein. A $TCID_{50}$ assay was used to determine the amount of released virus by the hypotonic saline/bionebulisation treatment.

## 3.2 NON-LYTIC METHOD

*3.2a Heparin release.*
For a 5 L culture the culture medium is removed and a solution of porcine mucosal heparin (at a final concentration of 60μg.ml$^{-1}$) in PBS (w/o $Ca^{2+}$and $Mg^{2+}$) was used to release DISC-HSV from the cells in the microcarrier culture. This harvest buffer was incubated at 37°C for 5 hours. The impeller speed was increased to 180 rpm. Samples were taken and centrifuged to remove any cells. A $TCID_{50}$ assay was used to determine the amount of released virus by heparin release. Samples were also taken for determination of the levels of cellular DNA and total protein post virus release.

## 4. Results

The results are shown as a direct comparison of each method for clarity.

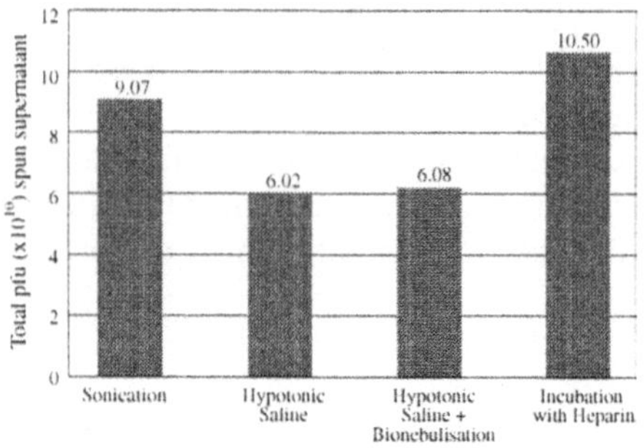


***Figure 1: Comparison of virus titres achieved using the different harvest methods.***
*It can be observed that the method that has released the most virus is the heparin release buffer non-lytic method. This result compares very favourably with the sonicated preparation and the other hypotonic saline lytic methods.*

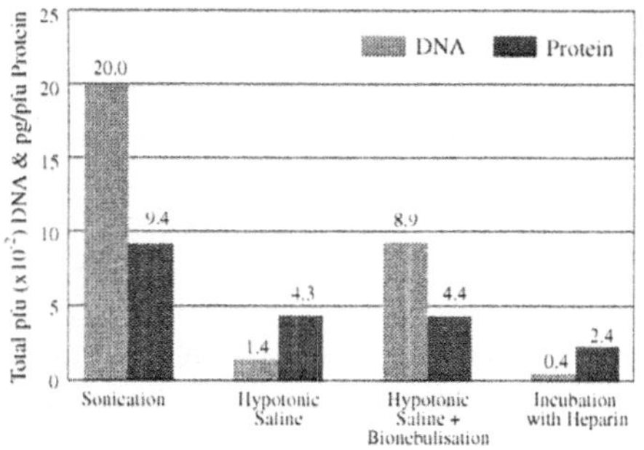


***Figure 2: Comparison of contaminating cellular DNA and protein levels using different harvest methods.***
*The results indicated in this table clearly show the benefit of using a non-lytic method of virus release. The level of DNA in the heparin release buffer viral suspension is almost 50-fold less than that released from the cells when using sonication. The DNA level is even 3-fold better than the value obtained with the hypotonic saline. As well as having the least amount of contaminating DNA the amount of cellular protein is also decreased by 4-fold over that of the sonicated method and it is half of that observed with the other two hypotonic saline disruption methods.*

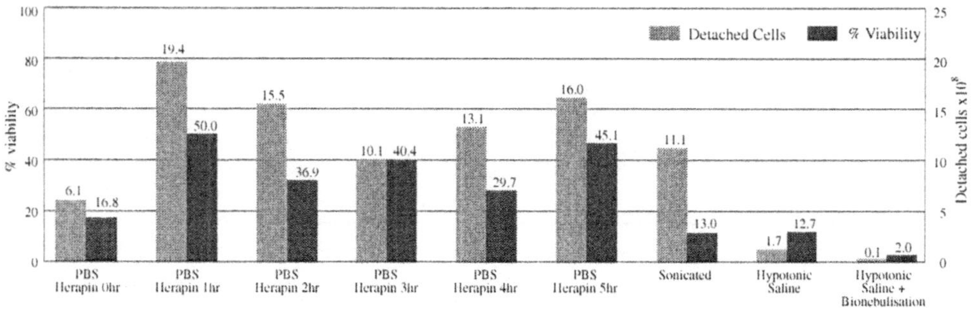


***Figure 3: Comparison of the Total cell counts and % viability of detached cells using the DISC-HSV releasing methods***
*This figure depicts the proportion of viable detached cells from the microcarriers which is highest when using heparin release non-lytic method. This cell viability level probably accounts for the lower levels of DNA and protein observed in Figure 2.*

## 5. Discussion

It has been shown that the non-lytic method of releasing our DISC HSV virus from infected cells has great benefits. The method provides as much cell free virus as the sonication method but has a decreased level of contaminating cellular DNA and protein. This method of heparin release of virus from microcarrier cultures compares very favourably with the results achieved by O'Keeffe *et al* [6] when using heparin to release virus from cell monolayers in culture flasks.

This improved method will be employed on large scale microcarrier cultures to determine if the process can be effectively scaled up. The virus obtained from this harvesting process is an ideal starting point for the further purification that is required to allow the viable viral vaccine to be used in the clinic.

## References

1. Corey L. (1993).
*Herpes Simplex virus infections during the decade since licensure of acyclovir.*
J. Med. Virol.; Suppl.; 1: 7-12.

2. McLean CS, Erturk M, Jennings R, Ni Challanain D, Minson AC, Boursnell MEG and Inglis SC. (1994).
*Protective vaccination against primary and recurrent disease caused by herpes simplex virus type 2 using a genetically disabled by herpes simplex virus type 1 virus.*
J. Inf. Dis.; 170: 1100-1109.

3. Boursnell MEG, Entwistle C, Blakely D, Roberts C, Duncan IA, Chisholm SE, Martin GM, Jennings R, Ni Challanain D, Sobek I and McLean CS. (1997).
*A genetically inactivated Herpes Simplex virus type 2 (HSV) vaccine provides effective protection against primary and recurrent HSV disease.*
J. Inf. Dis.; 175: 16-25.

4. Zecchini TA and Smith RJ. (1997).
*Production of high titre DISC-HSV-2 from a microcarrier culture.*
*In New Developments and New Applications in Animal Cell Technology Proceedings of the 15th ESACT meeting.*
Eds. Merton O-W, Perrin P and Griffiths B. Kluwer Academic Publishers.

5. Degouys V, Montoisy C, Wanderpepen M-P, Togasaki RK, Surzycki SJ, Segault V and Miller AOA. (1997).
*"Bionebulisation": an alternative to Dounce homogenisation for the preparation of mammalian cells nuclei in bulk. pp271-275 in Animal Cell Technology (From Vaccines to Genetic Medicine)*
Eds. Carronda M.J.T., Griffiths B. and Moreira L.P.: Kluwer Academic Publishers Dordrecht, Holland.

6. O'Keeffe R, Johnston MD, Slater NKH. (1998).
*The primary production of an infectious recombinant Herpes Simplex virus vaccine.*
Biotech. Bioeng.; 57: No. 3: 262-271.

# PASSAGING OF MICROCARRIER CULTURES AS AN ALTERNATIVE METHOD FOR THE SEEDING OF LARGE SCALE BIOREACTORS FOR THE PRODUCTION OF HIGH TITRE DISABLED INFECTIOUS SINGLE CYCLE HSV-2 VACCINE (DISC HSV)

ZECCHINI TA, WRIGHT PA AND SMITH RJ.
Cantab Pharmaceuticals Research Ltd.,
184 Cambridge Science Park, Cambridge. U.K. CB4 4GN

## 1. Abstract

Disabled Infectious Single Cycle (DISC) HSV-2 is presently routinely cultured in a complimentary cell line (CR2) in microcarrier cultures up to a scale of 15 L. In small-scale systems, these vessels can be set up from roller bottle cultures. It is obvious that for a production vessel of around 500 L that this is not a feasible option. Therefore, a procedure for passaging the cells from microcarrier culture to microcarrier culture must be defined.
The aim of this study was to provide a suitable scaleable regime for the routine passaging of small-scale microcarrier cultures upto larger production cultures. It was important to ensure that the overall time of the production train was as short as possible and that the virus productivity was maintained post passaging into a production vessel.

## 2. Introduction

Genital herpes is a disease that has a high incidence level and is increasing world-wide [1]. One approach to try to combat the disease has been the use of viable viral vaccines. Cantab is developing a DISC HSV-2 viral vaccine. This is a virus which has an essential gene removed. It is only able to undergo a single round of replication in a vaccinated host. It has been shown that such a virus can be an effective vaccine [2]. The virus is propagated in a complimentary cell line, designated CR2 which was prepared from the WHO Vero cell bank approved for use in vaccine manufacture. The cell line was modified to contain the essential gene removed from the virus. In this case, CR2 was modified to contain the glycoprotein H (gH) gene under the control of the glycoprotein D (gD) gene promoter [3].
The production process involves the culturing of the CR2 in a microcarrier process [4]. The cells are cultured to almost maximal cell density when the medium is removed the cells washed with basal DMEM and the virus infection medium added. The cells are infected at an MOI of 0.01 with the DISC HSV construct. The virus production cycle lasts for approximately 72 hours. The

virus is harvested from the cells and clarified in preparation for further purification procedures. This production process can routinely be performed at small scales from 1 L, 5 L and 15 L culture volumes. Development of a suitable a passaging regime will allow the scale to be increased to a level that is appropriate for large-scale manufacture.

## 3. Scale up issues

For our purposes an ideal manufacturing process must be practical and robust. It should deliver a high cell growth and produce a good level of virus production. The whole system must be cost effective, this cost implication is a very real constraint for full scale manufacture. The process must also be reproducible when scaled up in terms of the total number of cells per unit surface area available for growth and for the total amount of virus produced per unit surface of area.
Our previous laboratory scale production system was in roller bottles. For scaling up the virus production we had chosen a microcarrier system. This worked very well at a small scale (1-5 L) but needed to be scaled further to ensure that this microcarrier system was a realistic option for full-scale manufacture. One of the biggest issues to be considered using this system was the passage of the cultures to achieve a culture large enough to supply sufficient virus for the market demands.
A roller bottle production system could have been used but to supply our projected requirements we would have needed to use over a million roller bottle cultures a year. This same roller bottle equivalents can be achieved from a microcarrier culture system but the scale would need to be in the 100 - 500 L working volume range. Our development cultures were at a scale of 15 L and a process was needed to scale up the cultures to assess the impact of the manufacturing train and the trypsinisation procedures on virus productivity. Small cultures can be set up from roller bottle cultures but for cultures at around 100 L some 800 roller bottle cultures would be needed. Trypsinising and seeding a culture with cells derived from this number of roller bottle cultures is time consuming and brings a large risk from contamination as the number of manipulations is very high. Therefore we have concentrated on a suitable passaging regime to alleviate the need for large numbers of roller bottle cultures to seed large microcarrier cultures. We have currently used a passaging split ratio of 1:5 i.e. a 1 L start culture could be used to inoculate a 5 L culture.

## 4. Methods

### 4.1 Reactors used in passaging regime.

Small scale vessels of total volumes of 2, 7 and 15 L (1, 3 and 10 L working volumes were glass sterilisable vessels from FT Applikon). A stainless in-situ sterlisable 20 L total volume (15 L working volume) vessel was used as well. Finally a 50 L total volume (35 L working volume) *in-situ* sterilisable vessel from Setric was used as the final scale up vessel.

### 4.2 Microcarrier Culture.

The microcarriers used were Cytodex 1 (Pharmacia Ltd.) at a density of $5g.l^{-1}$. CR2 cells were used to inoculate the cultures by addition of enough cells to provide a ratio of 10 cells per microcarrier. The cultures were maintained at 30% DO, 37°C at a pH of 7.2 ± 0.2. The growth medium used was DMEM (Gibco Life Technologies) containing 5% FBS. The impeller was agitated at 90rpm.

4.3 Trypsinisation/Passaging regime.

The cell cultures that were to be passaged to a new larger culture were treated in the following manner. The medium was removed and the culture washed with 4 culture volumes of prewarmed PBS w/o $Ca^{2+}$ and $Mg^{2+}$ (Biowhittaker). To the washed cultures trypsin/EDTA solution (Gibco Life Technologies) was added which was equal to 70% of the original culture volume. The cell/trypsin suspension was incubated at 37°C for 12-15 minutes at an increased agitation rate of 180 rpm. The resulting microcarrier cell suspension was transferred directly into the new vessel which had been primed with microcarriers and complete medium. The new cultures were observed to ensure that the cells addition allowed early plating out of the cells onto the mcirocarriers.

This regime was used for two successive passages before a production culture was infected to determine if the passage regime decreased the potential of the final cultures to produce DISC-HSV virus. Two scaled down production systems were evaluated using the passage regime. Cells were serially passaged from 1 to 3 to 15 L culture volumes. This was a 1:3 passage ratio followed by a 1:5. The ability of the cells to plate out on the fresh microcarriers was observed carefully and compared to our previous data gained during our development. Finally the passage regime was used to passage cells from a 7 L culture upto a 35 L culture. This culture was infected in order to ascertain that the level of virus productivity was not affected by the passaging regime.

4.4 Virus Infection.

Confluent cultures to be infected were treatea as follows:- the culture medium was removed and the cultures washed three times with prewarmed serum free medium. A final volume of serum free DMEM was added and the culture cell density estimated. This density was used to determine the amount of virus to add. The cultures were infected with DISC-HSV at an MOI of 0.01. The infected cultures were harvested when approximately 80-100% CPE of the cells was observed approximately 72 hours post infection. The virus was harvested from the cells by removing the medium and adding a PBS buffer (Biowhittaker) containing $60ug.ml^{-1}$ heparin (CP Pharmaceuticals). The amount of virus harvested was calculated using a $TCID_{50}$ assay. This total amount of virus achieved was used to compare our new system where a passaging regime in microcarriers had been used (in a small scaled down production system) to our previous roller bottle system.

## 5. Results

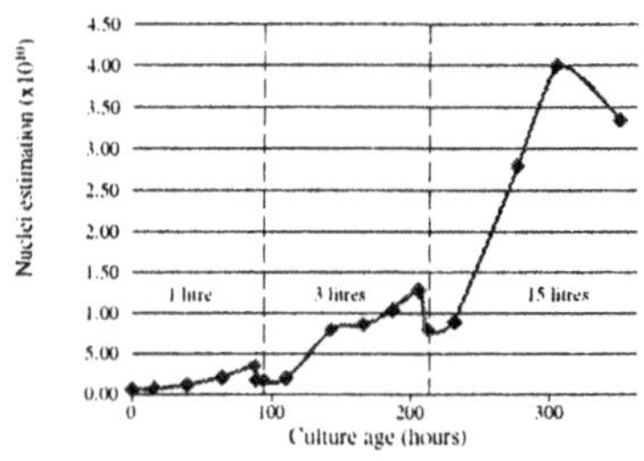


***Figure 1: Serial passage from 1 to 3 to 15 L cultures***

*The graph depicts the cell growth and time for each culture. The passage regime described above was used for the two passages. It can be seen that there is a decrease in the total number of cells at each passage. This may be due to loss of some cells during the washing phase of the passaging regime. It is clear that the regime does not affect the cells ability to be passaged onto new microcarriers nor does it arrest their growth substantially. Upon microscopic observation cells quickly adhered and plated out onto the microcarriers as expected. Figure 1 also indicates that a good cell density per culture is maintained. The density for the 15 L culture that can be seen is approximately 5 times greater than that of the 3 L culture.*

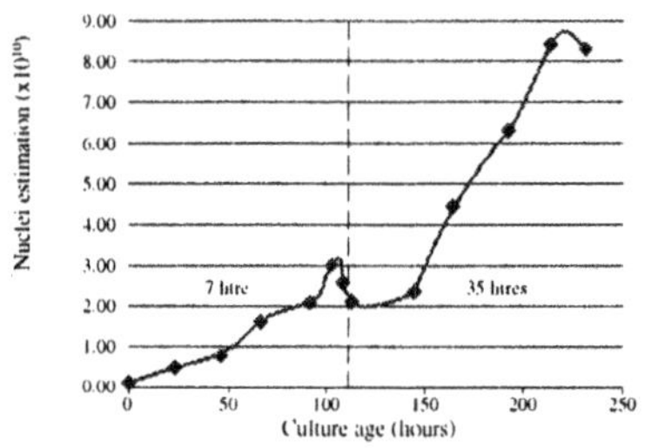


***Figure 2: Serial passage from 7 to 35 L cultures***

*This figure shows that the passage regime is able to be used on larger scale vessels enabling a production system based on microcarrier cultures a real possibility. The 35 L culture was infected and the amount of virus achieved was equivalent to that of smaller cultures and that of the titres achieved in a roller bottle system.*

| Culture size L | Source of cells for innoculation | Final cell density $x10^{10}$ | Total virus titre (total pfu) $x10^{11}$ | Virus titre/roller bottle equivalent $x10^{9}$ pfu/RB |
|---|---|---|---|---|
| roller bottle | roller bottle | 0.08 | 0.01 | 1.0 |
| 1 | roller bottles | 0.15 | 0.1 - 0.2 | 0.8 |
| 15 | roller bottles | 3.7 | 3.9 | 0.99 |
| 15 | microcarriers | 4.0 | 4.0 | 1.0 |
| 35 | microcarriers | 8.5 | 10.0 | 1.1 |

***Table 1: Comparison of the DISC-HSV titres obtained from the relevant production systems.***

*This table shows that the expected virus titre from each of the culture systems is maintained and is equivalent to the virus produced in a roller bottle system.*

| | 15 L seeded from microcarriers | 15 L seeded from 5 L microcarriers | 400 ml roller bottles |
|---|---|---|---|
| Expected final cell density | $4.0 \times 10^{10}$ | | $4.0 \times 10^{10}$ |
| Expected virus titres (total pfu) | $4.0 \times 10^{11}$ | | $4.0 \times 10^{11}$ |
| Media & serum costs | £775 | £720 | £1,700 |
| Microcarriers | £140 | £110 | £0 |
| Roller bottle | £15 | £185 | £1,470 |
| Total costs | £930 | £1,015 | £3,170 |
| Cost per clinical dose @ bulk harvest | 2.3p | 2.5p | 8.2p |

***Table 2: Comparison of the costs of a microcarrier vrs roller bottle production systems.***

*This table shows the economic advantage that a microcarrier system has over that of a roller bottle system. The roller bottle system is labour intensive and has many risks as the cultures all have to undergo the same manipulations. The equivalent surface area can be used in a microcarrier culture reducing the number of manipulations and also producing virus in a true batch system. Even at a comparison on such a small scale the difference in the costs is already highlighted. As the scale of the microcarrier system increases the cost saving is also likely to rise.*

## 6. Discussion

We have demonstrated that the passage regime is able to be used to scale up microcarrier cultures from 1 to 35 L without any observed effects on the cells ability to achieve equivalent cell densities in any production size of vessel. The passaging regime employed is based upon the systems used by Hu *et al*, [5] and that of Tao *et al*, [6]. It is, we believe, that the extensive washing allows the successful transfer of cells to allow good attachment to the new microcarriers in the new culture. Without such a suitable passage regime a scaled up microcarrier system would not be possible. This required maximal cell density appears to be maintained in our current scale. The amount of virus that is also able to be produced from the production scale cultures infected in this study is also maintained. This means that the passage regime can be used to scale up our cultures as a routine standard process and indeed it has been adopted as such a passage regime. In using a microcarrier production system for our DISC-HSV virus the cost of production has also been seen to fall. This has a good economic implication at the large manufacturing scale envisaged, possibly making it a system that will be able to deliver a cost effective dose. We have demonstrated here a suitable manufacturing system for our DISC HSV virus that is robust and practical, that maintains virus production and is likely to be cost effective at a manufacturing scale. We wish to further develop our regime by improving the recovery of cells at the passage and try to increase the passage split ratio from 1:5 to at least 1:8 and try to achieve 1:10. It is our intention to keep developing our passaging and production system by increasing our manufacturing vessel to a working volume of greater than 50 L at the earliest opportunity.

References

1. Corey L. (1993).
*Herpes Simplex virus infections during the decade since licensure of acyclovir.*
J. Med. Virol.; Suppl.; 1: 7-12.

2. McLean CS, Erturk M, Jennings R, Ni Challanain D, Minson AC, Boursnell MEG and Inglis SC. (1994).
*Protective vaccination against primary and recurrent disease caused by herpes simplex virus type 2 using a genetically disabled by herpes simplex virus type 1 virus.*
J. Inf. Dis.; 170: 1100-1109.

3. Boursnell MEG, Entwistle C, Blakely D, Roberts C, Duncan IA, Chisholm SE, Martin GM, Jennings R, Ni Challanain D, Sobek I and McLean CS. (1997).
*A genetically inactivated Herpes Simplex virus type 2 (HSV) vaccine provides effective protection against primary and recurrent HSV disease.*
J. Inf. Dis.; 175: 16-25.

4. Zecchini TA and Smith RJ. (1997).
*Production of high titre DISC-HSV-2 from a microcarrier culture.*
*In New Developments and New Applications in Animal Cell Technology Proceedings of the 15th ESACT meeting.*
Eds. Merton O-W, Perrin P and Griffiths B. Kluwer Academic Publishers.

5. Hu W-S, Giard DJ and Wang DIC. (1985).
*Serial propagation of mammalian cells on microcarriers.*
Biotech. Bioeng.; 27: 1466-1476.

6. Tao T-Y, Ji G-Y, and Hu W-S. (1988).
*Serial propagation of mammalian cells on gelatin-coated microcarriers.*
Biotech. Bioeng.; 32: 1037-1052.

# THE HYBRID SYSTEM USING BOTH PROMOTER ACTIVATION AND GENE AMPLIFICATION FOR ESTABLISHMENT OF EXOGENOUS PROTEIN HYPER-PRODUCING CELL LINES

K. Teruya, X.-Y. Dong*, Y. Daimon, Y. Katakura, Y.-P. Zhang, P. Seto, T. Miura, H. Ohashi** and S. Shirahata

*Graduate School of Genetic Resources Technology, Kyushu University, Fukuoka 812-8581, Japan; *Department of Chemical Engineering, Tianjin University, Tianjin 300072, P. R. China; **Pharmaceutical Laboratory, Kirin Brewery Co. Ltd., Takasaki, Gunma 370-1202, Japan.*

## ABSTRACT

A cell line D-29 easily and rapidly established by the OAP and the GS hybrid system could secrete recombinant human interleukin-6 (hIL-6) at a productivity rate of 39.5 $\mu g/10^6$ cells/day that is one of the highest level in the world. The productivity rate was about 130 times higher than that of a cell line A-7 which was established without both promoter activation and gene amplification. Although D-29 cells had high copy number and high message level of the hIL-6 gene and high secretion rate of hIL-6, it was revealed that large amounts of intracellular hIL-6 proteins were accumulated in D-29 cells compared to A-7 cells. There are two possible signal pathways to decrease the secretion of hIL-6. One is a pathway that accumulation of abnormal hIL-6 induce BiP expression, decreasing the hIL-6 secretion. Another is a pathway that accumulation of intact hIL-6 may activate NF-κB, inducing negative feedback signals to decrease the hIL-6 production. Western blotting analysis of BiP showed no change of expression level of BiP. In contrast, electrophoresis mobility shift assay of NF-κB revealed that NF-κB was activated in D-29 cells. It was suggested that large amounts of hIL-6 mRNA and translated hIL-6 protein resulted in abnormal accumulation of intact hIL-6 in endoplasmic reticulum, then negative feedback signals inhibited secretion of hIL-6 protein. To enhance hIL-6 productivity rate of D-29 cells with release of the negative feedback signals, the effect of pyrrolidinedithiocarbamate, an inhibitor of NF-κB activation, was examined. The suppression of NF-κB activation in D-29 cells brought about 25% augmentation of hIL-6 productivity rate, as expected. In the case of high productive cells like D-29 cells, it was indicated that release of negative feedback signal could increase total amount of recombinant protein secretion.

## 1. Introduction

Recombinant proteins produced by animal cells are favored for clinical use, since animal cells have post-translational machinery which is absent in prokaryotic cells such as *E. coli*. However, poor productivity of animal cells has been an obstacle on mass-production of recombinant proteins using animal cells as hosts. So we have developed the oncogene activated production (OAP) system to enhance recombinant protein production by specific promoter activation [1-3]. Recently the benefit of the gene amplification system with glutamine synthetase gene (GS system) has been reported [4]. Here we developed a new hybrid system which combined the OAP and the GS system on CHO cells, and investigated further enhancement of the cellular productivity.

## 2. Materials and Methods

2.1. Plasmids

Plasmid pEE14-IL-6 was constructed for production of hIL-6 with the GS gene amplification system was described previously report [5]. Human IL-6 cDNA was inserted downstream from the CMV promoter of pEE14 (LONZA Biologic plc, Slough, UK) [4]. Other plasmids used in this study were described previously [1-2].

2.2. Cell culture

Recombinant CHO cells transfected with the GS gene were cultured in G-DME medium. This medium was suited for the GS gene amplification system by adding 30 $\mu$M each of adenosine, guanosine, cytidine and uridine, 10 $\mu$M thymidine, 100 $\mu$M non-essential amino acids (Gibco BRL), 500 $\mu$M glutamic acid and 500 $\mu$M asparagine to Dulbecco's modified Eagles (DME) medium without L-glutamine (Bio WHITTAKER, Walkersville, MD, USA) supplemented with 10% fetal bovine serum (IR VINE SCIENTIFIC, Santa Ana, CA, USA) [5]. CHO cells without GS gene introduction were cultured in 10% FBS/DME. All cell cultures were done in 5% $CO_2$ at 37°C.

2.3. Establishment of recombinant CHO cell lines

Human interleukin-6 (hIL-6) was used as model of bioactive protein. A7 (Conventional) was established by co-transfection with pCMVP-IL-6 and pSV2-*bsr* into CHO $D^-$ cells. To establish *ras* oncogene primed cells, CHO $D^-$ cells were transfected with the plasmid pCMVD-*ras*. The cells were gene-amplified with 50 nM Methotrexate and established clone was named Ras-I. I-13 (OAP system) was obtained by co-transfection with pCMVP-IL-6 and pSV2-*bsr* into Ras-I. Ras I cells were transfected with pEE14-IL-6 and amplified hIL-6 gene using the GS system with 25 $\mu$M of methionine sulfoximine, then cell line D-29 (OAP & GS hybrid system) was established. Y-13 (OAP system) was established by introduction and amplification of hIL-6 gene with GS system.

2.4. Enzyme-linked immunosorbent assay (ELISA)
The amount of hIL-6 secreted into the spent media was measured by the enzyme-linked immunosorbent assay (ELISA) [6]. All assays were done in duplicate.

2.5. Electrophoresis mobility shift assay (EMSA)
Electrophoresis mobility shift assay (EMSA) was performed to investigate activated NF-κB level in various recombinant cell lines producing hIL-6. A probe of NF-κB binding in EMSA was made up of 5'-AGTTGAGGGGACTTTCCCAGGC-3' (sense) and 5'-GCCTGGGAAAGTCCCCTCAACT-3' (antisense).

## 3. Results and Discussion

The cell line D-29 easily and rapidly established by the OAP and the GS hybrid system, could secrete recombinant hIL-6 proteins at a productivity rate of 39.5 $\mu g/10^6$ cells/day, it is one of the highest level in the world (Table 1). Its productivity rate was about 130 times higher than the cell line A-7, which was established using conventional method without both promoter activation and gene amplification.

Table 1. Enhancement of hIL-6 productivity using OAP & GS hybrid system

| Cell lines | System | Productivity ($\mu g/10^6$ cells/day) | Relative productivity |
|---|---|---|---|
| A-7 | Conventional system | 0.3 | 1.0 |
| I-13 | Promoter activation (OAP) | 3.1 | 10.3 |
| Y-13 | Gene amplification (GS) | 7.3 | 24.3 |
| **D-29** | **Hybrid system (OAP & GS)** | **39.5** | **131.2** |

Fig. 1 shows copy number and message level of transfected hIL-6 gene, secreted hIL-6 proteins, and accumulated hIL-6 proteins intracellular, about the conventional, OAP, GS and the OAP & GS hybrid system. In the OAP & GS hybrid system, cells had high copy number and message level, and secretion level of hIL-6 proteins was high. However, large amounts of intracellular hIL-6 proteins were accumulated in this hybrid system, compared to a conventional system. There are two possible signal pathways to decrease the secretion of recombinant proteins. One is a pathway that accumulation of abnormal proteins induce BiP expression, decreasing

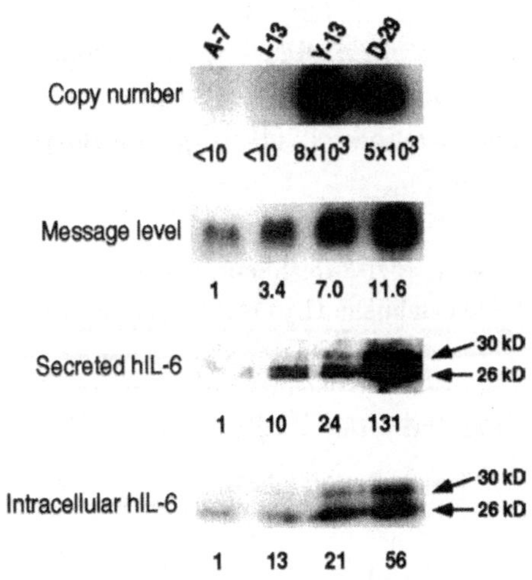


Figure 1. Profiles of various systems to express exogenous hIL-6.

the secretion. Another is a pathway that accumulation of intact proteins may activate NF-κB, inducing negative feedback signals to decrease the protein production [7].

Western blotting analysis of BiP protein showed no change of expression level (Fig. 2A). In contrast, EMSA to detect activated NF-κB in the cells revealed that NF-κB was highly activated in the OAP & GS hybrid system (Fig. 2B). This suggested that large amounts of hIL-6 in endoplasmic reticulum, then the negative feed back signals inhibited secretion of hIL-6 proteins.

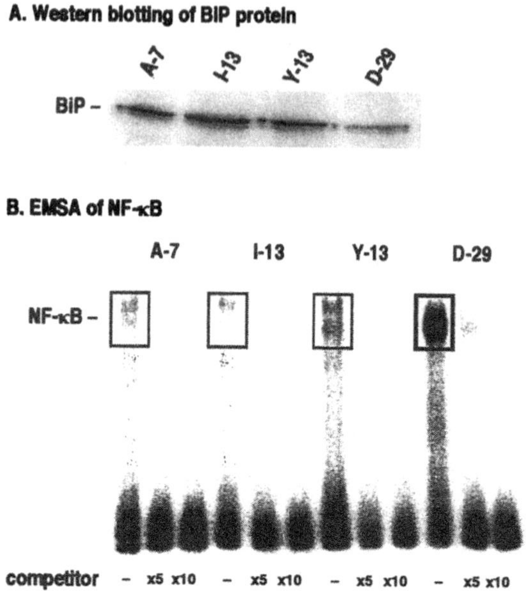


Figure 2. Western blotting of BiP protein (A) and EMSA of NF-κB (B) on various systems to express exogenous hIL-6.

Therefore, it was determined that the effect of pyrrolidinedithiocarbamate, (PDTC) which is an inhibitor of NF-κB activation, to enhance hIL-6 productivity rate of the OAP & GS hybrid system, with release of the negative feedback signals. PDTC treatment occurred inhibition of NF-κB activation in the OAP & GS hybrid system (Fig. 3). The suppression of NF-κB activation in the system brought about 25% augmentation of hIL-6 productivity rate as expected (Table 2).

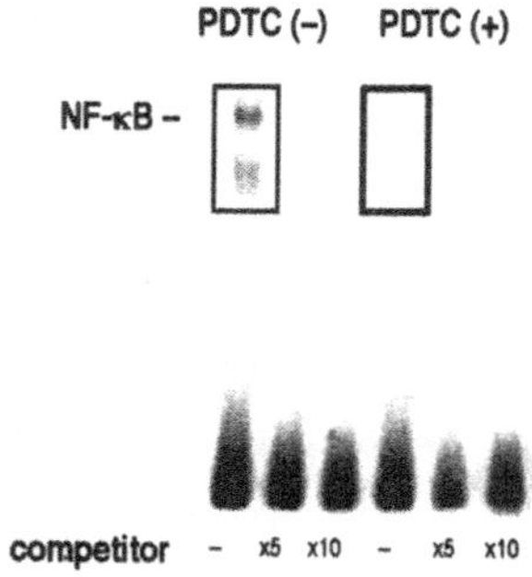


Figure 3. Inhibition of NF-κB activation by PDTC on OAP & GS hybrid system.

Table 2. PDTC-treatment enhanced hIL-6 productivity of D-29 established by OAP & GS hybrid system

| | PDTC (–) | PDTC (+) |
|---|---|---|
| Productivity ($\mu$g/$10^6$ cells/day) | 17.8 | 22.2 |
| Relative productivity | 1 | 1.25 |

In conclusion, the cells had high copy number and high message level and its productivity rate was high in the OAP & GS hybrid system. However, large amounts of hIL-6 proteins were accumulated in the cells, compared to the conventional system. It was suggested that NF-κB activation transduced negative feedback signals to nucleus from endoplasmic reticulum in the cells. Inhibition of NF-κB activation with PDTC showed that the productivity rate augmented in the OAP & GS hybrid system. Hyper-producing cells such as D-29 cells established by the OAP & GS hybrid system indicated that release of negative feedback signals could increase total amount of recombinant protein secretion.

## 5. References


1. Yano, T., Teruya, K., Shirahata, S., Watanabe, J., Osada, K., Tachibana, H., Ohashi, H., Kim, E.H. and Murakami, H. (1995) *Cytotechnology* **16**, 167-178.
2. Teruya, K., Yano, T., Shirahata, S., Watanabe, J., Osada, K., Tachibana, H., Ohashi, H., Kim, E.H. and Murakami, H. (1995) *Biosci. Biotech. Biochem.* **59**, 341-344 16, 167-178.
3. Teruya, K., Shirahata, S., Yano, T., Watanabe, J., Seki, K. Osada, K., Tachibana, H., Ohashi, H. and Murakami, H. (1995) In: Animal cell technology: Developments towards the 21st Century (ed. By Beuvery, E. C. *et al.*), pp 91-95, Kluwer Academic Publishers, Netherlands.
4. Cockett, M. I., Bebbington, C. R. and Yarranton, G. T. (1990) *Bio/Technology* **8**, 662-667.
5. Dong, X.-Y., Katakura, Y., Zhang, Y.-P., Miura, T., Daimon, Y., Ohashi, H. and Shirahata, S. (1998) In: Animal Cell Technology: Basic & Applied Aspects (ed. By Nagai, K. and Wachi, M.), Vol. 9, pp 319-323, Kluwer Academic Publishers, Netherlands.
6. Teruya, K., Shirahata, S., Yano, T., Seki, K., Tachibana, H., Ohashi, H. and Murakami, H. (1993) *Anal. Biochem.* **214**, 468-473.
7. Pahl, H. L. and Baeuerle, P. A. (1995) *EMBO J.* **14**, 2580-2588.

# SYSTEM FOR LARGE SCALE PROTEIN EXPRESSION IN *BOMBYX MORI*

T. ARAKAWA, M. MIYAZAWA, S. TOMITA, N. YONEMURA and S. HAYASAKA
*National Institute of Sericultural and Entomological Science (NISES), Ohwashi, Tsukuba, Ibaraki 305-8634, Japan*

Advances in biotechnology have made it possible to express foreign genes in heterologous organisms. Foreign gene expression systems are widely used to obtain proteins which naturally occur in minute amounts. Numerous human proteins of therapeutic value have been synthesized by expressing their genes in microbes and eukaryotic cells, e.g., peptide hormones and cytokines (Malik, 1989).

In recent years, recombinant nuclear polyhedrosis viruses have been used as vectors to introduce foreign genes into their insect hosts or permissive cell lines. Maeda et al. (1985) developed a vector from a nuclear polyhedrosis virus (NPV) infecting a silkworm, *Bombyx mori.* The recombinant virus replicates in silkworm larvae, and the recombinant protein products accumulate in the larval haemolymph; the product proteins are isolated from the haemolymph. The authors aim at the development of an economically-sound system for large scale protein production using an NPV vector and its host, *B. mori*.

The industrialization of the pharmaceutics production using a baculovirus vector system requires highly productive vectors, a mass rearing system for aseptic insect hosts, an effective vector-inoculation system, and an effective bleeding system which can deal with a large bulk of the hosts. Conventionally, the insect haemolymph has been manually collected by partially punctuating the integument of the living host and squeezing out the haemolymph. Thus, the operation for collectiong a large amount of haemolymph is impractical because it is labor-intensive and time-consuming.

In the present study, the authors developed new effective bleeding method from lepidopteran larvae (under submission). The muscle tissue of the frozen lepidopteran larvae spontaneously contracts after thawing. The haemolymph pressure of the thawed larva builts up due to the body contraction. Piercing the body wall of the larva results in spouting out of the haemolymph. An insect haemolymph collecting system was developed by taking advantage of the spontaneous insect body contraction described above. This method enables a large bulk of insect to be processed. In future, it will support the development of a new drug-production system with insects.

The bleeding process consists of four steps. 1: Larvae are anaesthetized in water in a container at room temperature for about 10 to 30 min. This step stabilizes the larvae and cleans the surface of the larval body. 2: The water immersing the larvae is replaced by 70% ethanol precooled at 4 °C. Then, the container is kept in a -30 °C freezer until the larvae are frozen. The authors leave the container in a freezer overnight routinely. Seventy percent ethanol solution never freezes at -30 °C. The ethanol sterilizes the larval surfaces. 3: Prolegs of the frozen larvae are removed by hand or scissors, which makes holes on the larval body. 4: The larvae with holes on their abdomen are thawed in a physiological saline containing an inhibitor of the haemolymph melanization reaction. Frequently, 150 mM sodium chloride (NaCl) with 0.05% phenylthiourea (PTU), a melanization inhibitor, was used. During the thawing process, the insect body muscles contract and the body size diminishes (Fig. 1). The haemolymph spouts out of the proleg's hole on the abdomen into the surrounding saline. The authors propose this newly developed bleeding method to call freeze-thawing method.

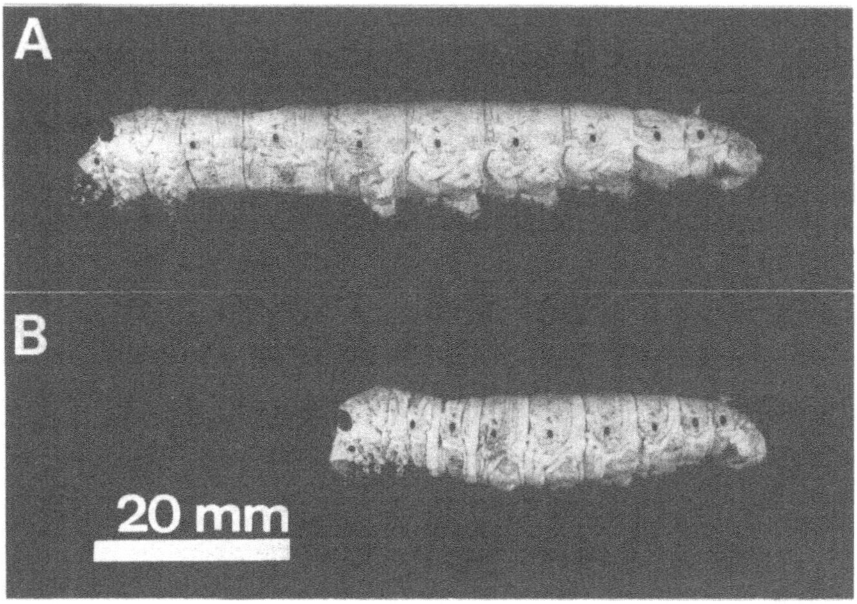


*Figure 1. A 5th-instar* B. mori *larva anaesthetized in water (A) and the same one after bleeding (B).*

Virus-infected and non-infected normal larvae of the silkworm, *B. mori*, were bled with this method. The difference in the body weights before and after the bleeding process was considered to represent the amount of haemolymph discharged.

Three types of bleeding experiments were carried out with non-infected normal larvae. At first, frozen larvae were thawed at 20 °C (step 4 in the above method). Two hundred and forty (4-day old in 5th-instar, 908.9 g) and 537 (3-day old in 5th-instar, 1830.5 g) larvae were bled by this method. One hundred and eighty-six g and 390 g of haemolymph were harvested from the 240 and 527 larvae, respectively. In *B. mori* larvae, the haemolymph constitutes the 25% of the total body weight. In this experiment, 80% or more of the total haemolymph was collected. Next, *B. mori* larvae were thawed at 0 °C and 4 °C, and the protein content in the saline originating from the discharged haemolymph was monitored by a potein assay kit with CBB dye. In general, proteins should be handled at low temperature to avoid their denaturation. At both the temperatures, haemolymph discharge was suggested to complete in 60 min. In this experiment, 73 to 75% of the total haemolymph was collected. Finally, *B. mori* larvae stored in freezers at -14 °C and -80 °C for 39 days were bled. Larvae stored at -80 °C spontaneously contracted during the thawing step, but those stored at -14 °C did not. The muscle tissue should be damaged during the storage at -14 °C. About 85 % of the total haemolymph was collected from the larvae stored at -80 °C, which was two times larger than that from the larvae stored at -14 °C. The larvae stored at -14 °C showed a slightly brownish appearance due to the melanization reaction of the haemolymph which has proceeded during the storage. The bled sample from the larvae was also brownish. Such a coloration was not observed in the larvae store at -80 °C, and also the bled sample was pale-yellow.

This method were also applied to virus-infected *B. mori*. A recombinant nuclear polyhedrosis virus, BmPH-Luc, in which the polyhedrin gene was replaced by a luciferase gene, was proliferated on a cell line, BoMo-15llc. The cultured medium was collected and used as an inoculum for virus-infection (Tomita et al.,1995). The inoculum contained $4.8x10^5$ $TCID_{50}$ units/ml of the virus. It was fifty-times diluted by distlled water and inoculated on 2-day old 5th-instar *B. mori* larvae with a syringe into their haemocoel (50 $\mu$l/larva). Each larva received 480 $TCID_{50}$ units of the virus. Then, larvae were bled every day. Eighty percent or more of the total haemolymph was collected from days 3 to 7. The collected haemolymph showed a clear yellowish color. On day 7, all the larvae exihibited a morphology typical for the virus infection showing a swollen intersegmental membrane. On day 8, i.e. 6 days after the virus inoculation, all the larvae died of the virus infection.

This study presented a simple procedure for bleeding the lepidopteran larvae utilizing a larval spontaneous body-contraction with a freezing-thawing treatment. This method was effective not only on the normal larvae but also on the recombinant virus-infected ones. The freezing treatment terminates the further propagation of viral bodies in larvae at any desired time, which allows us to control the production and accumulation of the target proteins. Stored larvae at -80 °C for longer than 1 month could be bled. This method was successfully done at 0 °C and 4 °C. The texture of the frozen larvae is much harder than that of live ones. Thus, the procedure of the bleeding can be fully automated.

In addition, the development of a method for large-scale virus-infection, which employs a particle gun as the apparatus of virus injection into insect, is underway.

There have been several studies leading to the development of the insect factory. Suzuki et al. (1997) constructed a highly productive NPV vector which had a deletion in the viral cysteine proteinase gene. No proteinase activity corresponding to this proteinase was detected in the haemolymph of the host, *B. mori*, infected with this vector, resulting in the protection of the protein product. Okazaki et al. (1995) developed a peroral virus-inoculation method. It involves feeding the 5th-instar *B. mori* larvae, which had been chilled immediately after ecdysis, with an artificial diet contaminated with the viral vector. The chilling step makes the host permissive to the peroral virus infection. A mass-culture system of aseptic *B. mori* larvae controlled by a personal computer was constructed by Ohura and Peng (1998).

Development of an insect factory is becoming true.

This work was supported in part by Enhancement of Center of Excellence, Special Coordination Funds for Promoting Science and Technology, Science and Technology Agency, Japan.

Maeda,S., Kawai,T., Obinata,M., Fujiwara,H., Horiuchi,T., Saeki,Y., Sato,Y. and Furusawa,M. (1985) Production of human α-interferon in silkworm using a baculovirus vector, *Nature* 315, 592-594.

Maeda,S. (1989) Expression of foreign genes in insects using baculovirus vectors, *Ann. Rev. Entomol.* 34, 351-372.

Malik,V.S. (1989) Biotechnology-the golden age, *Adv. Appl. Microbiol.* 34, 263-306.

Ohura,M. and Peng,Y. (1998) Construction of sulkworm rearing environment automatic control system by personal computer, *J. Seric. Sci. Jpn.* 67, 231-236 (in Japanese with an English abstract).

Okazaki,H., Kanaya,T., Nishimura,S., Ogawa,K. and Watanabe,H. (1995) Peroral inoculation of a baculovirus vector to the silkworm, *Bombyx mori* , treated with a low temperature, *J. Seric. Sci. Jpn.* 64, 504-508 (in Japanese with an English abstract).

Suzuki,T., Kanaya,T., Okazaki,H., Ogawa,K., Usami,A., Watanabe,H., Kadono-Okuda,K., Yamakawa,M., Sato,H., Mori,H., Takahashi,S. and Oda,K. (1997) Efficient protein production using a *Bombyx mori* nuclear polyhedrosis virus lacking the cycteine proteinase gene, *J. Gen. Virol.* 78, 3073-3080.

Tomita,S., Kanaya,T., Kobayashi,J. and Imanishi,S. (1995) Isolotion of p10 gene from *Bombyx mori* nuclear polyhedrosis virus and study of its promotor activity in recombinant baculovirus vector system, *Cytotechnology* 17, 65-70.

# PROTEIN-FREE CULTURE OF *RAS*-AMPLIFIED RECOMBINANT BHK-21 CELLS

YUICHI INOUE[1], SEIJI KAWAMOTO[2], MASAHIRO SHOJI[3], SHUICHI HASHIZUME[3], KIICHIRO TERUYA[4], YOSHINORI KATAKURA[4], AND SANETAKA SHIRAHATA[4]

[1]*Department of Biochemical Science and Technology, Faculty of Agriculture, Kagoshima University, 1-21-24 Korimoto, Kagoshima 890-0065, Japan;* [2]*Department of Molecular Biotechnology, Graduate School of Advanced Sciences of Matter, Hiroshima University, 1-4-1 Kagamiyama, Higashi-Hiroshima 739-8526, Japan;* [3]*Morinaga Institute of Biological Science, 2-2-1 Shimosueyoshi, Tsurumi-ku, Yokohama 230-0012, Japan;* [4]*Graduate School of Genetic Resources Technology, Kyushu University, 6-10-1 Hakozaki, Higashi-ku, Fukuoka 812-0053, Japan*

**Abstract** A *ras*-amplified recombinant BHK-21 cell line (*ras*-rBHK-IgG), which hyperproduces recombinant human monoclonal antibody, was cultured in a protein-free medium. Protein-free medium for *ras*-rBHK-IgG cells was found to be superior to serum containing medium in terms of viability and recombinant antibody production. In particular, antibody production in protein-free culture was shown to be five to ten times higher than that in serum culture. However, when culturing cells at a high density in the hollow fiber bioreactor system, the enhancement of antibody production was not observed. On the other hand, *ras*-rBHK-IgG cells could be maintained for over a month in protein-free culture using the bioreactor system, in contrast with serum culture which only lasted for a half month. The total amount of antibody obtained during cultivation was about two times greater in protein-free culture than in serum culture. Therefore, protein-free culture of *ras*-rBHK-IgG cells was demonstrated to be effective for middle scale production of recombinant human monoclonal antibody.

## 1. Introduction

A *ras*-amplified recombinant BHK-21 cell line (*ras*-rBHK-IgG) has been established, and was shown to be able to hyperproduce the recombinant IgG chimeric human monoclonal antibody (hMAb) AE6F4 using the oncogene activated production system (Yano *et al.*, 1994; Teruya *et al.*, 1995, Shoji *et al.*, 1996). In addition, cells could be easily cultured in ERDF medium supplemented with 80 $\mu$M iron(III) nitrate (Inoue *et al.*, 1996a; 1996b). In this study, we examined the growth and antibody production of *ras*-rBHK-IgG cells in protein-free culture, and compared the data to cells in serum culture to evaluate the effectiveness of protein-free culture of *ras*-rBHK-IgG cells

## 2. Materials and Methods

### 2.1. CELL LINES AND CULTURE CONDITIONS

The *ras*-rBHK-IgG cell line was established as reported previously (Shoji *et al.*, 1996) and is maintained in ERDF medium (Kyokuto Pharmacy Industrial Co.) supplemented with 10% fetal calf serum (FCS).

Cells (ca. 1 X $10^6$ cells) were plated into 35-mm plastic dishes and cultured in ERDF medium supplemented with 80 $\mu$M iron(III) nitrate or with 10% FCS at 37°C in humidified 5 % $CO_2$/95% air for 7 days.

### 2.2. CELL CULTURE USING A HOLLOW FIBER BIOREACTOR SYSTEM

*Ras*-rBHK-IgG cells (ca. 1 X $10^8$ cells) were cultured using the hollow fiber bioreactor system (INTEGRA Bioscience Co., Germany) as described previously (Inoue *et al.*, 1996b). The spent medium in the culture cassette was changed every other day and the antibody concentration in the spent medium was measured using an enzyme-linked immunosorvent assay (ELISA).

## 2.3. MEASUREMENT OF ANTIBODY PRODUCTION

Antibody concentrations were determined by ELISA as reported previously (Shoji *et al.*, 1994), using an anti-human IgG (γ) antibody (4100, TAGO) as the first antibody, and anti-human IgG (γ) peroxidase conjugate antibody (2390, TAGO) as the second antibody.

## 3. Results and Discussion

The growth rate of *ras*-rBHK-IgG cells was considerably slower in the protein-free medium, which results in a longer maintenance period compared with cells cultured in serum containing medium (Fig. 1). *Ras*-transformed cells are known to have low growth factor requirements because they stimulate their own growth via the autocrine mechanism (Sporn *et al.*, 1985; Mckay *et al.*, 1986; Shirahata *et al.*, 1990). In our protein-free medium, cells could grow even from a low initial cell density as 1 x $10^4$ cells (data not shown). Takahashi *et al.* (1994) reported that growth rate suppression of cultured mammalian cells enhanced protein productivities. We also observed that antibody production in protein-free culture was five to ten times higher than in serum culture (Fig. 1).

*Ras*-rBHK-IgG cells were also cultured using the hollow fiber bioreactor system. In the protein-free medium, the enhancement of antibody production was not observed (Fig. 2). This suggests that cells, when cultured in the protein-free medium, may secrete growth-like factor(s) which may enhance the production of antibodies. On the other hand, cells could be also cultured in the protein-free medium for over a month using the bioreactor system, while cells cultured in serum containing medium was viable for only a half month (Fig. 2). This difference is probably due to the slower cell growth rate in the protein-free medium and the availability of surface area for cell attachment. The total amount of antibody produced during cultivation was about two times greater in protein-free culture than in serum culture. Therefore, it was concluded that protein-free culture of *ras*-rBHK-IgG cells was effective for middle scale production of recombinant hMAb.

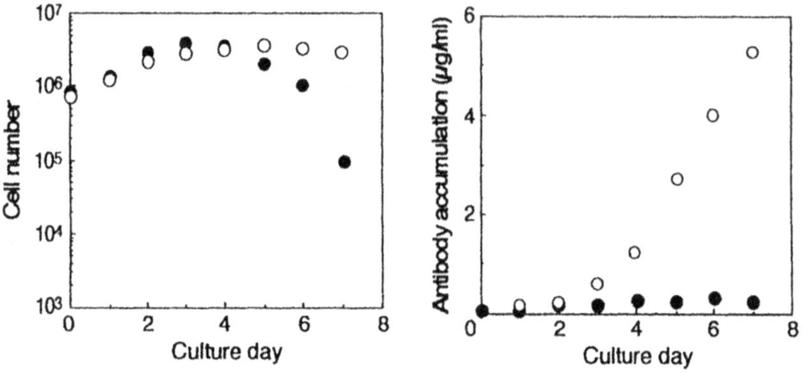


**Fig. 1. Growth and antibody prodction of *ras*-rBHK-IgG cells.** Cells were cultured in 10% FCS-ERDF medium (solid circle) or 80 μM iron(III) nitrate-ERDF medium (open circle).

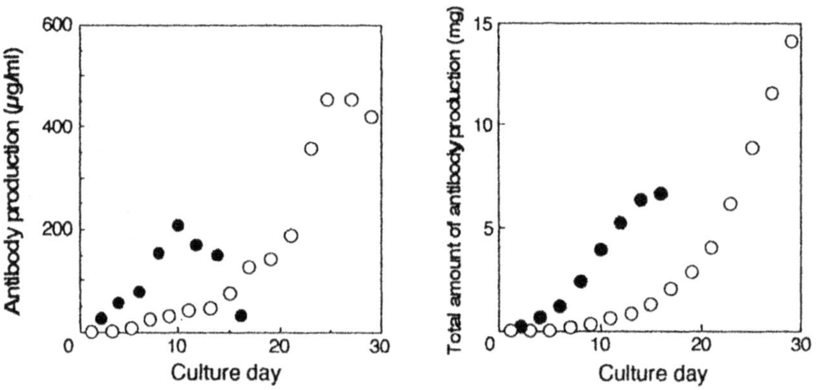


**Fig. 2. Antibody production of *ras*-rBHK-IgG cells cultured using a hollow fiber bioreactor system.** Cells were cultured in 10% FCS-ERDF medium (solid circle) or 80 μM iron(III) nitrate-ERDF medium (open circle).

## 4. References

Inoue, Y., López, L.B., Kawamoto, S., *et al.* (1996a) Production of recombinant human monoclonal antibody using *ras*-amplified BHK-21 cells in a protein-free medium, *Biosci. Biotech. Biochem.* **60**, 811-817.

Inoue, Y., Kawamoto, S., Seki, K., *et al.* (1996b) Production of a recombinant human monoclonal antibody using a novel hollow fiber bioreactor system, *J. Ferment. Bioeng.* **81**, 466-469.

McKay, I.A., Malone, P., Marshall, C.J., and Hall, A. (1986) Malignant transformation of murine fibroblasts by a human c-Ha-ras-1 oncogene does not require a functional epidermal growth factor receptor, *Mol. Cell Biol.* **6**, 3382-3387.

Shirahata, S., Rawson, C., *et al.* (1990) Ras and neu oncogenes reverse serum inhibition and epidermal growth factor dependence of serum-free mouse embryo cells, *J. Cell Physiol.* **144**, 69-76.

Shoji, M., Kawamoto, S., Seki, K., *et al.* (1996) Lung cancer-reacting human recombinant antibody AE6F4: potential usefulness in the sputum cytodiagnosis, *Hum. Antibod. Hybridoma* **7**, 27-36.

Shoji, M., Kawamoto, S., Setoguchi, Y., *et al.* (1994) The 14-3-3 protein as the antigen for lung cancer-associated human monoclonal antibody AE6F4, *Hum. Antibod. Hybridoma* **5**, 123-130.

Sporn, M.B. and Roberts, A.B. (1985) Autocrine growth factors and cancer, *Nature* **313**, 745-747.

Takahashi, K., Tereda, S., Ueda, H. , *et al.* (1994) Growth rate suppression of cultured mammalian cells enhanced protein productivities, *Cytotechnology* **15**, 57-64.

Teruya, K., Yano, T., Shirahata, S., *et al.* (1995) *Ras* amplification in BHK-21 cells produces a host cell line for further rapid establishment of recombinant protein hyper-producing cell lines, *Biosci. Biotech. Biochem.* **59**, 341-344.

Yano, T., Teruya, K., Shirahata, S., *et al.* (1994) *Ras* oncogene enhances the production of a recombinant protein regulated by the cytomegalovirus promoter in BHK-21 cells, *Cytotechnology* **16**, 167-178.

# EFFECT OF SUGAR COMPOSITION ON THE HETEROGENEITY OF ANTIBODY IN HYBRIDOMA CULTIVATION

**Takeshi Omasa, Yuka Kitamoto, Jun-ya Tanaka, Yoshio Katakura, Michimasa Kishimoto, and Ken-ichi Suga**
***Department of Biotechnology, Graduate School of Engineering, Osaka University, 2-1 Yamada-oka, Suita, 565-0871 JAPAN***
**Telephone: +81-6-6879-7437, Facsimile: +81-6-6879-7439**
**e-mail: omasa@bce.bio.eng.osaka-u.ac.jp**

**ABSTRACT**: In order to investigate the effect of the sugar composition of the medium on IgG glycosylation in mouse hybridoma cultivation, the batch culture of mouse hybridoma 3A21 (RCB 1285) was carried out under various sugar compositions. The macroheterogeneity (relative ratio of glycosylated antibody) was analyzed by an enzyme-linked lectin binding assay during cultivation. The relative ratio of glycosylated antibody decreased as cultivation progressed. However, in mannose-enriched cultivation, the relative ratio was maintained at a high level. The microheterogeneity of each oligosaccharide chain was analyzed by two-dimensional mapping techniques. The high-mannose type oligosaccharide chain content was high in mannose-enriched cultivation. The content of oligosaccharide chains terminating in galactose residues, which plays an important role in the immune system, was shown to be high in galactose-enriched cultivation. Using galactose- and mannose-enriched cultivations, the number of oligosaccharide chains terminating in galactose residues could be increased.

## 1. Introduction

Oligosaccharide processing of asparagine-linked carbohydrate occurs by the combined action of at least 11 discrete enzymes. Therefore, the environmental conditions within cell culture strongly affect glycosylation reactions causing oligosaccharide heterogeneity. IgG antibody molecules are glycoproteins. The oligosaccharide component of an IgG molecule is closely related to its effector function. The effector function is reduced, if terminal galactose residue was removed [1]. In our study, we focused on galactose residues and investigated the effect of the sugar composition of the medium on oligosaccharide structure.

## 2. Materials and Methods

The cell line employed in the experiments was the mouse-mouse hybridoma 3A21 (RCB 1285), which produces an anti-RNase A monoclonal antibody (IgG) [2]. The serum-free culture medium was RDF-ITES with bovine serum albumin (BSA) [3]. A serum medium was prepared by supplementing the serum-free medium with 10% FCS. For the case of the investigation of the effect of sugar composition, appropriate amounts of mannose, galactose and/or glucose were added to the medium. The specific growth rate was calculated as previously reported [3]. Antibody was purified from the supernatant of each culture using RNase A affinity column chromatography and used to investigate N-linked glycosylation. The microheterogeneity was analyzed by an enzyme-linked lectin binding assay using concanavalin A. Oligosaccharide chains were cleaved from the protein by glycopeptidase A treatment and were analyzed by the pyridylamination method using a 2-dimensional HPLC system [4].

## 3. Results and Discussion

### 3.1 EFFECT OF SUGAR COMPOSITION OF THE MEDIUM ON THE MACROHETEROGENEITY OF THE OLIGOSACCHARIDES

In order to investigate the effect of the sugar composition of the medium on the heterogeneity, we carried out 7 kinds of batch cultivation. The sugar composition of each batch culture is shown in Table 1.

TABLE 1 Sugar composition of batch cultivations

| | Glucose ($g\ \ell^{-1}$) | Mannose ($g\ \ell^{-1}$) | Galactose ($g\ \ell^{-1}$) | 10% FBS |
|---|---|---|---|---|
| Run 1 | 2.57 | 0 | 0 | + |
| Run 2 | 2.57 | 0 | 0 | – |
| Run 3 | 0 | 2.57 | 0 | – |
| Run 4 | 2.57 | 2.57 | 0 | – |
| Run 5 | 2.57 | 0 | 2.57 | – |
| Run 6 | 0 | 2.57 | 2.57 | – |
| Run 7 | 2.57 | 1.95 | 1.95 | – |

In the case of Run 7, the mannose and galactose concentrations were decreased because of the increase of osmotic pressure in the medium. The macroheterogeneity was investigated in early, mid and late log phase in Runs 2-7 (Figure 1). In our experiments, the macroheterogeneity was analyzed by an enzyme-linked lectin binding assay using Concanavalin A. Using this method, the absolute binding number could not be determined. Therefore, the relative binding number is reported. Except in Run 7, the relative value decreased as cultivation progressed. In cultivations to which mannose was added, Runs 3, 4, 6 and 7, the relative value was higher than in other cultivations. Runs 4 and 7, to which both glucose and mannose were added, gave highest relative value.

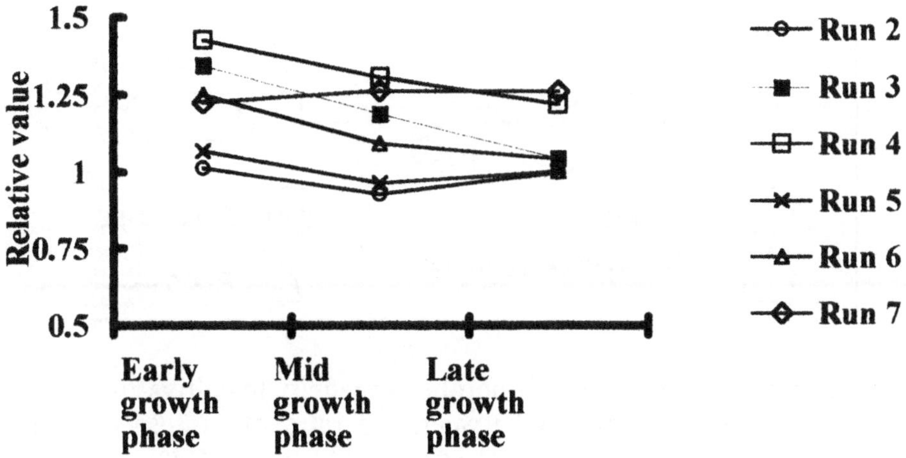


*Figure 1.* Changes of macroheterogeneity during cultivation
(Relative value based on the late growth phase value of Run 2)

## 3.2 EFFECT OF SUGAR COMPOSITION OF THE MEDIUM ON THE MICROHETEROGENEITY OF THE OLIGOSACCHARIDES

At the end of each cultivation, the microheterogeneity, the detailed oligosaccharide structure, was investigated by the pyridylamination method using 2-dimensional HPLC system (Figure 2).

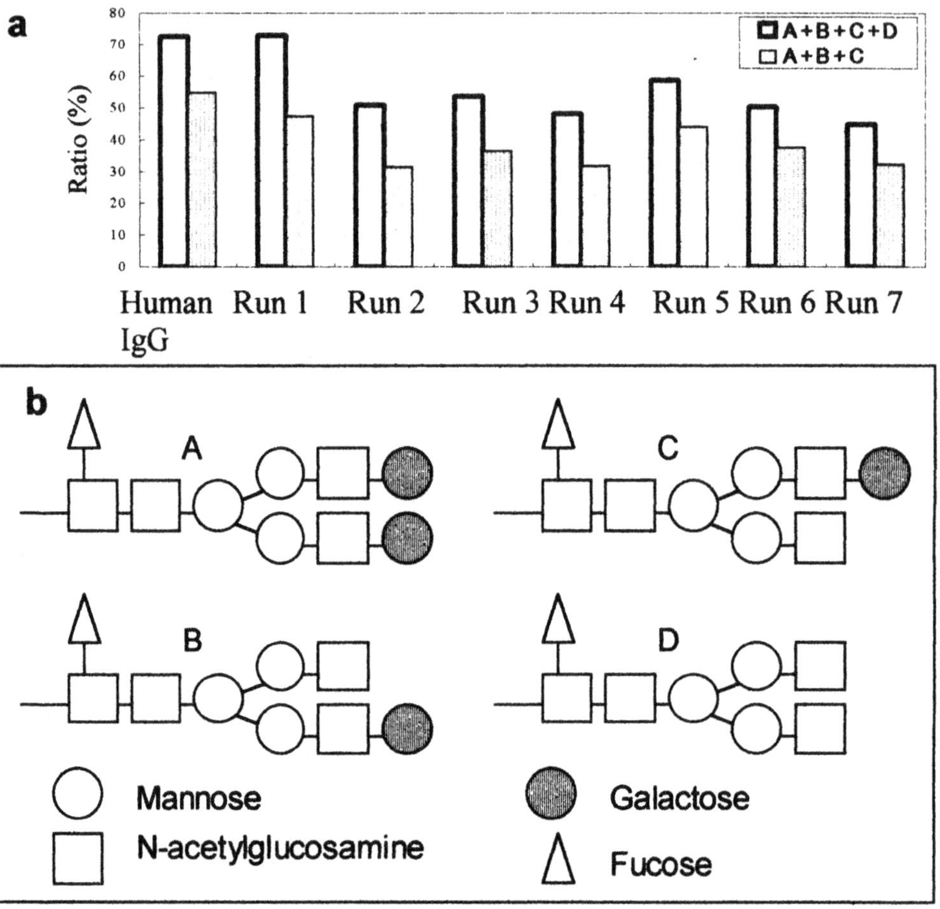


*Figure 2.* The results of microheterogeneity in cultivation.

a: Comparison between the ratio of oligosaccharide chains with structures A+B+C+D to the total oligosaccharide chain content and those with structures A+B+C to the total oligosaccharide chain content.

b: Oligosaccharide structures of A, B, C and D

A, B, C and D represent the structures of the majority of oligosaccharides in human IgG. Therefore, the ratio of oligosaccharide chains with structures A, B, C or D to the total number of oligosaccharide chains (the ABCD ratio) and with structures A,B or C, each of which has one or more galactose residues at the end of the chain, to the total number of oligosaccharide chains (the ABC ratio) was compared between Runs 1 to 7 (Figure 2). In Run 1, the ABCD and ABC ratios were higher than those in other experiments. In serum cultivation, oligosaccharide processing proceeded than in serum-free cultivation. An unknown factor which promotes glycosylation may exist in the serum. Moreover, the ABC ratio increased on addition of galactose to the medium. Based on macro- and microheterogeneity analysis, using galactose- and mannose-enriched cultivation, the number of oligosaccharide chains terminating in galactose could be increased.

## 4. Acknowledgments

We thank Dr. Natsuka and Prof. Hase of Osaka University for directing us in pyridylamination techniques.

## 5. References

1. Tsuchiya N., Endo T., Natsuta K., Yoshinoya S., Aikawa T., Kosuge E., Takeuchi F., Miyamoto T. and Kobata A.: Effects of galactose depletion from oligosaccharide chains on immunological activities of human IgG, *J. Reumatol.* **16**, (1989) 285-290

2. Omasa, T., Higashiyama, K., Shioya, S. and Suga, K.: Effect of lactate concentration on hybridoma culture in lactate-controlled fed-batch operation, *Biotechnol. Bioeng.* **39**, (1992) 556-564

3. Omasa, T., Ishimoto, M., Higashiyama, K., Shioya S. and Suga, K.: The enhancement of specific antibody production rate in glucose- and glutamine-controlled fed-batch culture, *Cytotechnology,* **8**, (1992) 75-84

4. Kobayashi M, Kato, S., Omasa, T., Shioya, S. and Suga K.: Enhancement effects of BSA and linoleic acid on hybridoma cell growth and antibody production, *Cytotechnology*, **15**, (1994) 51-56

5. Hase S.: High-performance liquid chromatography of pyridylaminated saccharides, *Methods Enzymol.*, **230**, (1994) 225-237

# MODULATION OF ANTIGEN BINDING OF HUMAN ANTIBODY VIA GLYCOSYLATION BY HYBRIDOMA CULTURE WITH VARIOUS CARBON SOURCES

Ji-Youn KIM, Yoshinori Katakura, Kiichiro Teruya
and Sanetaka SHIRAHATA
*Graduate School of Genetic Resources Technology,*
*Kyushu University, Hakozaki 6-10-1, Fukuoka, Japan*

**ABSTRACT** The human hybridoma HB4C5 produces a lung adenocarcinoma reactive antibody (mAb–C5) which possesses an unique N–linked carbohydrate on its light chain. We intended to design the culture environment in which the varied glycoforms on the light chain induce the alterd antigen binding affinity. HB4C5 cells were cultured in medium containing diverse carbohydrates such as monosaccharides, oligosaccharides and polysaccharides. Antigen binding ability of mAb–C5 significantly varied depending upon carbon sources. In particular, addition of N–acetylglucosamine and chitosan to medium led to the creation of a certain light chain glycoforms that exhibit increased antigen binding. To clarify the mechanism of carbohydrates arrangement on mAb–C5 light chain, the expression level of a glycogenesis-related enzyme was examined by using specific antibody. The expession level of galactosidase varied depending upon carbon sources, suggesting that the enzyme activity can be regulated by carbon sources.

## 1.INTRODUCTION

Antibodies are glycoproteins that have different amount of carbohydrates attached to the structure portion. Glycosylation of antibodies is thought to play important roles in antibody conformation, solubilization and effector function (1). Some antibodies have glycosylation sites in the variable region which can either increase or decrease antigen binding. It is well known that the glycosylation of glycoproteins is affected by the culture environment. Human monoclonal antibody produced by HB4C5 cells (mAb–C5) is strongly reactive to human lung adenocarcinoma tissues and cross-reactive to Cyt-C and Cpase (2,3). Therefore we designed a culture medium with various carbon sources and demonstrated that the change of glycoforms on the light chain of mAb-C5 resulted in change of the reactivity of mAb-C5 against Cyt-C and Cpase.

## 2. MATERIALS AND METHODS

HB4C5 cells were maintained in ERDF medium supplemented 10 μg/ml of insulin, 20 μg/ml of human transferin, 20 μM of ethanolamine, and 25 nM of selenite at 37°C in a humidified atmosphere of 5% $CO_2$. Antibody content and antigen-binding reactivities were assessed by ELISA. To determine the size of the light chain, antibodies were electrophoresed on a SDS-PAGE gel, then electro-transferred on to a membrane. The antibody light chains were detected with HRP-anti human light chain antibody. Lectin blot assay was performed to detect the structure of carbohydrate chains on the light chain. Galactosidase activity was performed using a simple ELISA-based assay according to the method described by Keusch *et al* (4)

## 3. RESULTS AND DISCUSSION

### 3.1. *Antigen binding activities of antibodies produced by HB4C5 cells in medium containing various monosaccharides*

Glycosylation of glycoproteins in mammalian cells is affected by the culture environment. We detected the effects of varing monosaccharides availability in a glucose-free culture medium on the antigen antibody reaction of the mAb-C5 (Table.1). The antibody produced in medium containing GlcNAc exhibited the highest antigen-binding affinity among all monosaccharides tested. A marked variation in the light chain glycosylation was observed when the produced hybrid

**Table. 1. Antigen - binding activities of antibodies produced by HB4C5cells in medium containing various monosaccharides.**

| | Antigens | |
|---|---|---|
| | Cpase | Cyt C |
| Glucose | 250* | 250 |
| Mannose | 252 | 248 |
| N - acetylglucosamine | 20 | 25 |
| Arabinose | 30 | 50 |
| Ribose | 35 | 50 |
| Xylose | 30 | 50 |
| Ramnose | 45 | 100 |

*Values show the antibody concentration (ng/ml) necessary to obtain an O. D. of 0.5 at 405 nm as determined by ELISA.

immunoglobulin (Ig) molecule was analysed by electrophoresis on sodium dodecyl sulphate (SDS)-polyacrylamide gels (Fig. 1). Furthermore, a ERDF medium containing 1 mM glucose and 20 mM GlcNAc was found to be the optimum to increase the antigen binding affinities of mAb-C5 (data not shown).

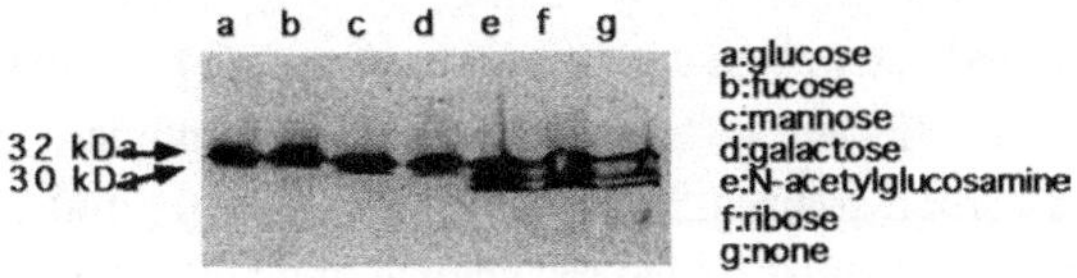


Fig. 1. The effect of monosacchari on glycosylation of light chain in glucose-free medium

### 3.2. *Antigen binding affinities of antibodies produced in oligosaccharides containing medium*

When various concentrations of sodium alginate as an oligosaccharide were added to medium instead of glucose, antigen binding activity was inhibited as shown in Fig. 2. These results demonstrate that the extent of inhibition or enhancement of antigen binding was affected the alternated carbohydrate chains *via* culturing environment.

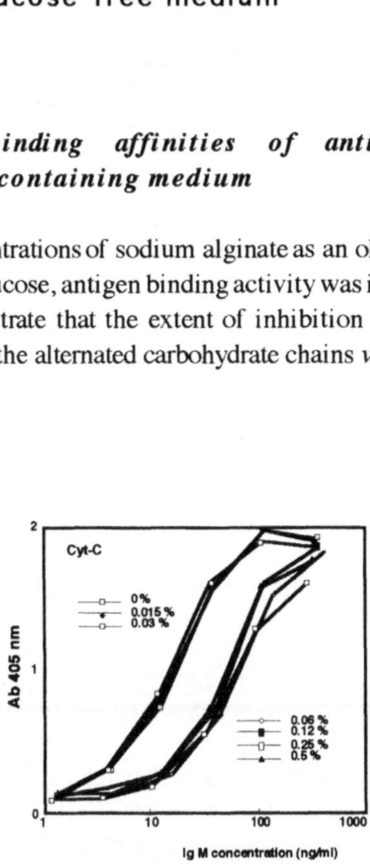


**Fig. 2. Reactivities of the mAb-C5 produced by hybridomas cultured in various concentrations of sodium alginate containing medium**

### 3.3. *Reactivties of galactosidase*

Corresponding to the previous results, we investigated the effective factors in glycosylation. Galactosidase is known to be involved in carbohydrate synthesis. The expression level of galactosidase in HB4C5 hybridomas cultured in medium containing GlcNAc was detected by using specific antibody. Galactosidase expression level differed in GlcNAc containing medium (Fig.3). These results show that glycosylation is affected by the culturing environment, particularly its composition of the carbon sources. It would be possible to modulate the antigen binding ability of a human antibody *via* the change of glycoforms on light chain by cultivation of hybridoma cells in carbon sources controlled medium.

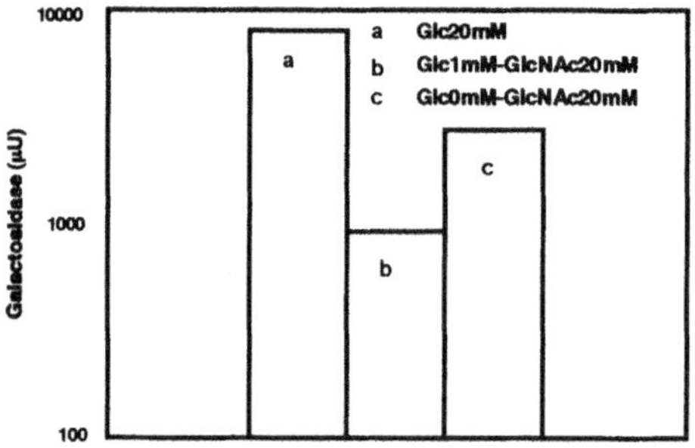


**Fig.3. Galactosidase productivity of mAb-C5 produced in GlcNAc-medium containing various concentrations of glucose**

## 4. REFERENCES


1. Melchers,F. (1973) Biosynthesis, intracellular transport and secretion of immunoglobulins. Effect of 2-deoxy-D-glucose in tumor plasma cells producing and secreting immunoglobulin G1. *Biochemistry*,**12**,1471-1476.
2. Yano T, Yasumoto K, Nagasima A, Hasizumi S, Murakami H and Nomoto K (1988) Immunohistological characterization of human monoclonal antibody against lung cancer. J. Surg. Oncol., **39**,108-113
3. Hasizumi S, Kamei M, Mochizuki K, Sato S, Kuroda K, Kato M, Yasumoto K, Nagahasi H,Hirose H, Tai H, Okano H, Nomoto K and Murakami H(1991) Serodiagnosis of cancer by using Candida cytochrome c recognized by human monoclonal antibody HB4C5. Hum. Antibod. Hybridomas **2**,142-147
4. Jeremy K, Peter M. L, David A. I and Peter J. D( 1995) β-1,4-Galactosyltransferase activity in B cells detected using a simple ELISA-based assay *Glycobiology* , **5**, 365-370

# MANIPULATION OF CYTOKINE PRODUCTION FROM CD8$^+$ T CELLS BY MEANS OF AMINO ACID-SUBSTITUTED ANALOGS OF A PEPTIDE ANTIGEN

Mamoru Totsuka, Masako Kohyama, Masahiro Kakehi, Satoshi Hachimura, Tatsuhiro Hisatsune, and Shuichi Kaminogawa.
*Department of Applied Biological Chemistry, The University of Tokyo, 1-1-1 Yayoi, Bunkyo-ku, Tokyo 113-6857, Japan*

**ABSTRACT** Manipulation of CD8$^+$ T-cell responses specific for an exogenous antigen by epitope variants would be advantageous to develop a novel means of antigen-specific immune regulation. We previously established a CD8$^+$ T cell clone named 5F1 which is specific for peptide 142-149 (p142-149) of $\alpha_{s1}$-casein, a major milk allergen, and these cells produce interleukin 10 (IL-10) and interferon $\gamma$ (IFN-$\gamma$) upon antigenic stimulation. Some of the analog peptides derived from p142-149 with single amino-acid substitutions triggered only IL-10 production whereas others induced production of IFN-$\gamma$ alone or both of these cytokines. These results demonstrate that the signaling pathway involved in induction of IL-10 production in CD8$^+$ T cells differs from that for IFN-$\gamma$ production. Our findings illustrate that cytokine production from CD8$^+$ T cells can be manipulated by using single amino-acid substituted analogs of an antigenic peptide.

## 1. Introduction

Most CD8$^+$ T cells recognize endogenously synthesized antigens, such as tumor antigens and viral proteins, presented by major histocompatibility complex (MHC) class I molecules and show cytotoxic activity. However, it has been demonstrated that exogenous antigens also can be presented by MHC class I molecules and generate CD8$^+$ T-cell responses [1, 2]. These CD8$^+$ T cells have been shown to play an important role in regulation of immune responses [3]. Thus, manipulation of CD8$^+$ T-cell responses would be effective in treatment of some kinds of immune diseases.

Recent studies showed that T cell activation is not a simple on-off-type event; rather, qualitative changes in T cell responses can be induced by amino acid substitutions in antigenic peptides. For example, some analog peptides induce cytokine production without proliferation, cytotoxicity without proliferation and cytokine production, alteration of the pattern of cytokine production, or anergy [4].

In the present study, we approached manipulation of cytokine production from the $CD8^+$ T cell clone 5F1 specific for a peptide corresponding to region 142-149 of bovine $\alpha_{s1}$-casein (p142-149) using a panel of 90 analogs of p142-149 each with a single amino acid substitution at a T cell receptor (TCR)-contact residue.

## 2. Materials and Methods

### 2.1 Animals

Female C57BL/6 mice were purchased from Clea Japan Inc. (Tokyo, Japan).

### 2.2 Antigens

p142-149 was synthesized by means of a peptide synthesizer (model 430A: PE Applied Biosystems, Foster City, CA, USA), and purified by reversed-phase HPLC. Single amino acid substituted peptide analogs of p142-149 (Table 1) were obtained from Chiron Mimotopes Pty LTD (Clayton, Victoria, Australia).

### 2.3 $CD8^+$ T cell clone 5F1

An $\alpha_{s1}$-casein specific $CD8^+$ T cell clone, 5F1 [5], was used in this study. This clone was maintained in a conditioned RPMI 1640 medium which was supplemented with 10% FCS, $5 \times 10^{-5}$ M 2-mercaptoethanol, 100 U/ml penicillin, 100 μg/ml streptomycin, and 10% T cell growth factor. These cells were stimulated with irradiated syngeneic spleen cells as antigen-presenting cells in the presence of trypsin-digested $\alpha_{s1}$-casein at 10 μg/ml every ten days.

### 2.4 Enzyme-linked immunosorbent assay (ELISA)

T cells ($4 \times 10^4$ /well) were plated into 96-well plates with irradiated spleen cells ($2 \times 10^5$ / well) and each peptide in 200 μl of medium. The culture supernatants were collected 48 h later. The levels of IFN-γ and IL-10 in the culture supernatants were assessed by two-site ELISA using anti-IFN-γ monoclonal antibodies (mAbs) R4-6A2 and XMG1.2 and anti-IL-10 mAbs JES5-2A5 and SXC-1 as described previously [5].

### 2.5 Cytotoxicity assay

Cytotoxicity was assayed by measuring the release of lactate dehydrogenase (LDH) into supernatants using an enzymatic assay kit (Cyto Tox 96 Assay kit: Promega, Madison, WI, USA) according to the manufacturer's instructions. Data are expressed as the percentage of the maximal release observed, after subtraction of the release induced by targets pulsed with no peptide. Ten thousand EL-4 cells (American Type Culture Collection, Rockville, MD, USA) were used as the target cells at an Effector : Target ratio of 10.

2.6 T cell proliferation assay

The proliferation of T cells was determined by measuring the uptake of [$^3$H] thymidine. T cells ($4x10^4$ /well) were stimulated by irradiated spleen cells ($2x10^5$ /well) and each peptide in 200 μl of medium. The cells were cultured for 48 h and pulsed with 18.5 kBq of [$^3$H] thymidine (248 Gbq/nmol: New England Nuclear, Boston, MA, USA) during the last 20 h.

## 3. Results

3.1 Analog peptides selectively induced IL-10 production or IFN-γ production in an $\alpha_{s1}$-casein-specific $CD8^+$ T cell clone

The putative TCR contact residues of p142-149 ($NH_2$-Leu-Ala-Tyr-Phe-Tyr-Pro-Glu-Leu-COOH) have been reported to be $Leu^{142}$, $Ala^{143}$, $Phe^{145}$, $Pro^{147}$, and $Glu^{148}$ [6]. On the basis of this information, we designed 90 analog peptides each with a substitution of either of 18 amino acids other than the original amino acid and cysteine at each of the five TCR contact residues .

To evaluate the effects of these single amino acid substitutions, the amounts of IL-10 and IFN-γ produced by 5F1 in response to stimulation with each of the 90 analog peptides were determined by ELISA. Figure 1 shows representative data for each analog peptide capable of triggering IL-10 and/or IFN-γ production. We can categorize the analog peptides into four groups: (1) peptides triggering production of IL-10 but not IFN-γ such as A143L (Leu substituted for $Ala^{143}$) or P147I, (2) peptides triggering

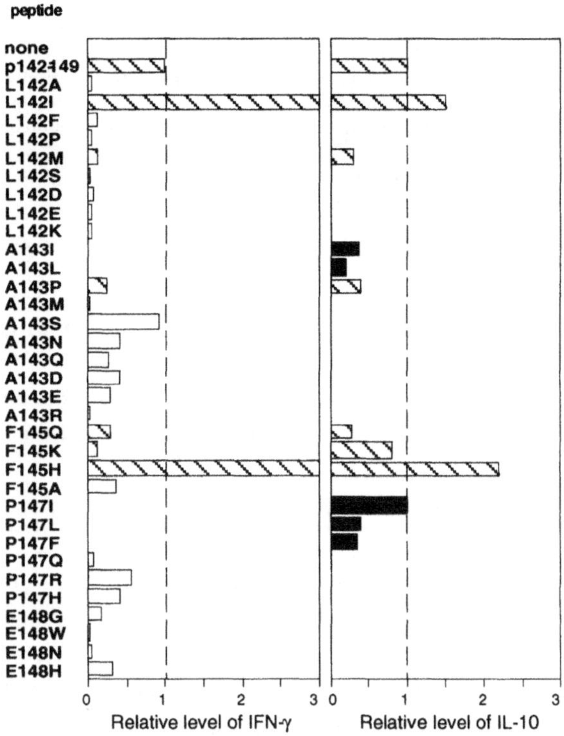


Figure 1. IL-10 and IFN-γ production by 5F1 cells upon stimulation with p142-149 and its single amino acid-substituted analogs. Each peptide was added to the culture at 10 μM. Shown in this figure are representative results for analog peptides which induced IL-10 and/or IFN-γ production; peptides inducing production of both IL-10 and IFN-γ (hatched bars), IL-10 but not IFN-γ (closed bars), or IFN-γ but not IL-10 (open bars). The data are expressed as relative values compared with the level of each cytokine produced upon stimulation with p142-149.

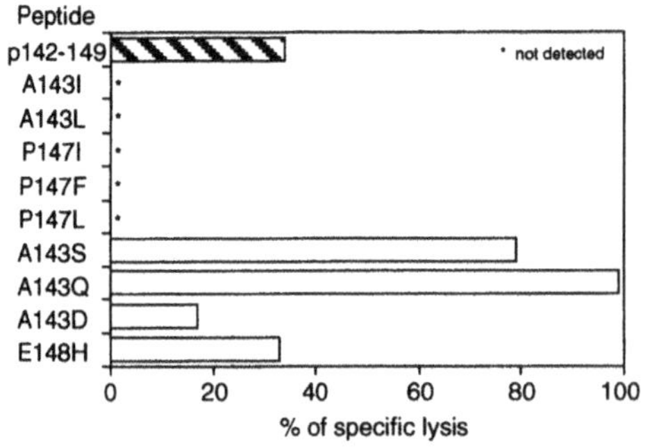


Figure 2. Cytolytic activity of 5F1 cells in response to the analog peptides.
EL-4 cells (1x10[4] /well) were pulsed for 1 h with each peptide at a concentration of 10 μM, and plated with 5F1 cells (1x10[5] /well) in 96-well plates. After a 4-h incubation period, supernatants were recovered and assayed for the presence of LDH released by the target cells. The standard deviation was less than 10% of maximum in each instance.

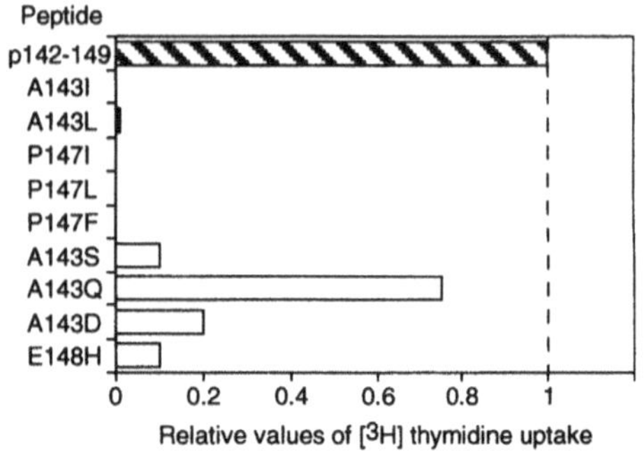


Figure 3. Proliferation of 5F1 cells in response to the analog peptides.
Each peptide was added to the culture at 10 μM. The results are expressed as relative values compared with the level of [$^3$H] thymidine uptake observed upon stimulation with p142-149.

production of IFN-γ but not IL-10 such as A143S or E148H, (3) peptides triggering production of both IL-10 and IFN-γ such as L142I or F145H, and (4) peptides failing to trigger production of either of the cytokines examined. These results demonstrate that the signal requirement for IL-10 production is distinct from that for IFN-γ production in CD8$^+$ T cells. p142-149 at lower concentrations induced weak production of both IL-10 and IFN-γ, and when the concentration of p142-149 was further reduced neither of them was produced (data not shown). Thus, p142-149 could not trigger selective production of either IL-10 or IFN-γ from 5F1 cells.

3.2 Analog peptides which triggered only IL-10 production induced neither cytotoxicity nor a proliferative response

We next assessed the relationship between the putative IL-10 specific cellular signals and the signals for cytotoxicity. 5F1 cells showed significant cytolytic activity for p142-149-pulsed EL4 cells (Figure 2). Each of five peptides, A143I, A143L, P147I, P147L, and P147F, which induced only IL-10 production, was unable to induce cytolytic activity, whereas analog peptides which induced IFN-γ production could induce cytotoxicity (Figure 2).

5F1 cells showed a proliferative response to trypsin-digested $\alpha_{s1}$-casein [7] and to p142-149 (Figure 3). The analog peptides which induced only IL-10 production did not induce proliferation, whereas the analog peptides which triggered IFN-γ production induced a proliferative response (Figure 3). These results indicate that peptides A143I, A143L, P147I, P147L and P147F selectively activated the signaling pathway for IL-10 production, which is independent of that for cytotoxicity and proliferation of the responding CD8$^+$ T cells.

## 4. Discussion

In the present study, we have shown that single amino acid substituted analogs of an antigenic peptide could induce separate production of either IL-10 or IFN-γ from a $CD8^+$ T cell clone. This indicates that a single amino acid substitution in an antigenic peptide can modulate cytokine production by a $CD8^+$ T cell clone in an antigen-specific manner. Furthermore, analog peptides triggering production of IFN-γ induced cytotoxicity and proliferation as well, while those inducing IL-10 production without triggering IFN-γ production did not. Our results demonstrate that the signal transduction pathway for triggering of IL-10 production and the signaling requirements for induction of IFN-γ production, cytotoxicity and proliferation are distinct in $CD8^+$ T cells.

The selective induction of $CD8^+$ T cell responses by altered antigenic peptides as shown here not only facilitates understanding of the relationship between signaling pathways specific for production of a cytokine and antigenic stimulation, but also could be applied as a novel means of controlling the cytokine balance via $CD8^+$ T cells. In certain immune diseases, either the Th1-type or the Th2-type immune response is dominant: for example, multiple sclerosis and rheumatoid arthritis are Th1-dominant diseases, and allergies are Th2-dominant diseases [8]. It has been reported that IFN-γ selectively inhibits Th2-type responses [9], while IL-10 inhibits responses of Th1 cells [10]. Thus, $CD8^+$ T cells secreting IL-10 and/or IFN-γ have the potential to correct the imbalance in Th1-Th2 responses that may be responsible for the deleterious immune responses. If we can choose suitable analog peptides capable of modifying the cytokine production pattern of $CD8^+$ T cells in particular immune diseases, this strategy will lead to clinical application of altered peptide ligands as a $CD8^+$ T cell-mediated immunotherapy.

## 5. References


[1] Raychaudhuri, S., Tonks, M., Carbone, F., Ryskamp, T., Morrow, W. J., and Hanna, N., *Proc. Natl. Acad. Sci. USA* **89**, 8308-8312, 1992.

[2] Hisatsune, T., Nishijima, K., Kohyama, M., Kato, H., and Kaminogawa, S., *J. Immunol.* **154**, 88-96, 1995.

[3] Kemeny, D. M., Noble, A., Holmes, B. J., and Diaz, S. D., *Immunol. Today* **15**, 107-110, 1994.

[4] Sloan-Lancaster, J., and Allen, P. M., *Annu. Rev. Immunol.* **14**, 1-27, 1996.

[5] Nishijima, K., Kohyama, M., Hisatsune, T., Kato, H., Kakehi, M., and Kaminogawa, S., *Biosci. Biotechnol. Biochem.* **61**, 1156-1162, 1997.

[6] Falk, K., Rotzschke, O., Stevanovic, S., Jung, G., and Rammensee, H. G., *Nature* **351**, 290-296, 1991.

[7] Nishijima, K., Hisatsune, T., Kato, H., Kohyama, M., Kakehi, M., Hachimura, S., and Kaminogawa, S., *Cytotechnology* **25**, 89-100, 1997.

[8] Mosmann, T. R., and Sad, S., *Immunol. Today* **17**, 138-146, 1996

[9] Mosmann, T. R., and Coffman, R. L., *Annu. Rev. Immunol.* **7**, 145-173, 1989.

[10] Fiorentino, D.F., Bond, M.W. and Mosmann, T.R., *J. Exp. Med.* **170**, 2081-2095, 1989.

# A NOVEL EFFECT OF NIVALENOL (NIV) ON THE IMMUNE RESPONSE

*NIV, a trichothecene mycotoxin, selectively inhibits antigen-specific Th2 responses.*

CHUNG-YEON CHOI, Ph D[a], HARUYO NAKAJIMA-ADACHI, Ph D[b], SHUICHI KAMINOGAWA, Ph D[b], SUSUMU KUMAGAI, Ph D[b], and YOSHIKO SUGITA-KONISHI, Ph D[a]
[a]*Department of Biomedical Food Research, National Institute of Infectious Diseases*
*1-23-1 Toyama, Shinzyuku-ku, Tokyo, 162-0052, Japan*
[b]*Department of Applied Biological Chemistry, Graduate school of Agricultural Life Science, The University of Tokyo*

**ABSTRACT** In this study, the effect of nivalenol (NIV), a trichothecene mycotoxin produced by *Fusarium nivale*, on the immune response, especially the T helper 2 (Th2) response, was evaluated in ovalbumin (OVA)-specific T cell receptor (TCR)αβ-transgenic (Tg) mice. This mouse develops naive T helper (Th) cells as a Th2 subset and produces OVA-specific serum immunoglobulin E (IgE) upon oral administration of OVA. In *in vitro* experiments using spleen cells from untreated OVA-TCR Tg mice, NIV was found to inhibit of OVA-induced interleukin-4 (IL-4) and interleukin-5 (IL-5) production. Especially, IL-4 production was completely inhibited. In *in vivo* experiments, dietary NIV also suppressed the production of OVA-specific IgE, IgG1 and IgA, as well as total IgE production, although it failed to inhibit total IgG1 and IgA production as determined by monitoring the levels in serum. These results demonstrate that NIV selectively inhibits of antigen-specific Th2 responses, without leading to total loss of Th2 responses. Here we first report a novel function of NIV with respect to the allergic response and this finding has important implications for orally induced antigen-specific modulation of food allergy.

## 1. Introduction

Nivalenol (NIV) is a common environmental mycotoxin[1] and it is a major group B trichothecene mycotoxin.[2] Several studies have shown that administration of this mycotoxin lead to dysregulation of immune responses in mice.[3-4] Recently, elevated serum IgA levels and mesangial IgA nephropathy in mice due to dietary exposure to NIV have been observed.[5] These effects are very similar to IgA nephropathy in humans. The effects of NIV on the immune system, especially with respect to dysregulation of the IgA response, led us to further question whether NIV affects IgE production because the levels of both IgA and IgE are dependent on Th2-type responses.[6] We, therefore, focused on IgE production in this study, in an effort to define the effects of the NIV on immune responses *in vitro* and *in vivo*. It is difficult to study the dietary antigen-induced IgE response in normal mice because serum IgE antibody is scarely produced. In this study, ovalbumin (OVA)-specific T cell receptor (TCR)$\alpha\beta$-transgenic (Tg) mice (OVA TCR-Tg mice) were used. These mice express TCR-$\alpha\beta$ (V$\alpha$3, V$\beta$15) chain genes derived from $OVA_{323-329}$-specific I-$A^d$-restricted $CD4^+CD8^-$ T helper cell clone 7-3-7[7] and show a significant increase in serum IgE levels upon oral administration of OVA. In the present study, we demonstrat for the first time that NIV acts to downregulate the progressive antigen-driven Th2 response. That is, NIV selectively inhibits the production of Th2-type cytokines, IL-4 and IL-5, and the production of Th2-type OVA-specific serum antibodies, IgE, IgG1 and IgA. Moreover, NIV selectively inhibits total IgE production, but not that of other Th2-type antibodies, IgG1 and IgA. Based on these results, we propose the potential efficacy of NIV as a novel therapeutic modality for allergy.

## 2. Materials and methods

### 2.1. MICE AND DIETS

Offspring (male, 8-13 weeks old) of OVA TCR-Tg mice, with a BALB/c genetic background were kindly provided by The University of Tokyo (Tokyo, Japan). In our *in vivo* system, Tg mice were allowed free access to OVA-diet, and drinking water containing 6 ppm NIV (referred to as OVA+NIV-diet in this study) or without NIV (referred to as OVA-diet in this study) for 2, 4 and 8 weeks.Control mice were fed a commercial diet (CE-2) (referred to as CE-2-diet in this study) The other control mice, NIV-control mice, were given CE-2-diet and drinking water containing 6 ppm NIV (referred to as NIV-diet in this study).

### 2.2. CELL CULTURES

In our *in vitro* system, spleen cells ($2.5 \times 10^6$/ml) from OVA TCR-Tg mice were stimulated with OVA (100 µg/ml) in the absence or presence of various concentrations of NIV (30, 60 and 120 ng/ml) and cultured. On day 7, the cells ($2.5 \times 10^5$/ml) were restimulated for 1-3 days with the same antigens as that used for primary stimulation in the presence of antigen-presenting cells (APC). The supernatant was collected in each instance to assess cytokine responses by ELISA, and stored at -70°C until used.

### 2.3. MEASUREMENT OF CYTOKINE LEVELS

A sandwich ELISA was used for determination of cytokine concentrations in the culture supernatants. Purified monoclonal antibodies (mAbs) specific for four murine cytokines (IL-2, IFN-$\gamma$, IL-4, IL-5) (Pharmingen, San Diego, CA.) were used.

2.4. ELISA FOR TOTAL AND OVA-SPECIFIC ANTIBODIES

The levels of OVA-specific antibodies (IgG, IgG1, IgG2a, IgA; Zymed Laboratories, San Francisco, CA) were determined by ELISA. To determine for total Ab, either goat anti-mouse IgG (Fab-specific) or goat anti-mouse IgA (Zymed) was used as the first antibody. Total IgE levels were measured by a sandwich ELISA method. To determine the titer of OVA-specific IgE, biotinylated anti-OVA (5 μg/ml) was used as the secondary antibody. The absorbance at 405 nm of the solutions in each well was recorded using an ELISA plate reader.

## 3. Results

### 3.1. THE EFFECT OF NIV TREATMENT ON CYTOKINE PRODUCTION BY SPLEEN CELLS *IN VITRO*

It is well known that IgE production is dependent on IL-4 released from activated Th2 cells. Therefore, we firstly designed *in vitro* experiments to examine the effects of NIV on cytokine production.

Table 1. The effect of NIV on cytokine responses by OVA-specific TCR-Tg mice splenocyte *in vitro*

| Antigen \ Cytokine | | IL-2 (pg/ml) | IFN-γ (U/ml) | IL-4 (pg/ml) | IL-5 (pg/ml) |
|---|---|---|---|---|---|
| no antigen | | 49.3 | 10.1 | N.D. | 8.6 |
| OVA (100μg/ml) | | 499.2 | 153.7 | 3511.1 | 2632.4 |
| OVA +NIV | 30 ng/ml | 283.6 | 97.1 | N.D. | 565.9 |
| | 60 ng/ml | 345.4 | 99.7 | N.D. | 953.3 |
| | 120 ng/ml | 296.0 | 124.0 | N.D. | 439.7 |

Spleen cells ($2.5\times10^6$/ml) from Tg mice were cultured with OVA (100 μg/ml) in the absence or presence of NIV (30~120 ng/ml) in 24-well plates. On day 7, the cells ($2.5\times10^5$/ml) were restimulated with same antigen as that used for the first stimulation in the presence of irradiated BALB/c spleen cells (3000 rad, $2.5\times10^6$/ml) as antigen-presenting cells (APC). Culture supernatants were collected for IL-2 estimation after 24 hr, for IFN-γ after 72 hr and for IL-4 or IL-5 after 48 hr, and the levels were assessed by ELISA. Results shown are from one of two experiments that gave similar results.

ND = Not detected

As seen in Table 1, the levels of IL-2 and IFN-γ produced by spleen cells cultured with OVA and NIV were only modestly decreased or not affected, as compared to cells cultured with OVA alone. In contrast to Th1-type cytokines, interestingly, the marked increase in levels of IL-4 and IL-5 produced by OVA-restimulated spleen T cells was suppressed by the addition of NIV. Surprisingly, IL-4 production was completely inhibited by the treatment with OVA and NIV at all concetrations tested (30, 60, 120 ng/ml). These data suggest that NIV selectively inhibits the Th2-type cytokine response, and especially, this leads to no production of IL-4 by T cells activated by the antigen.

## 3.2. THE EFFECT OF DIETARY NIV ON ANTIBODY IN SERUM

From the results of our *in vitro* experiments (Table 1), it was expected that dietary exposure to NIV would induce the inhibition of IgE production in the Tg mice because IgE production is regulated by IL-4. To confirm this possibility, we designed an *in vivo* experiment.

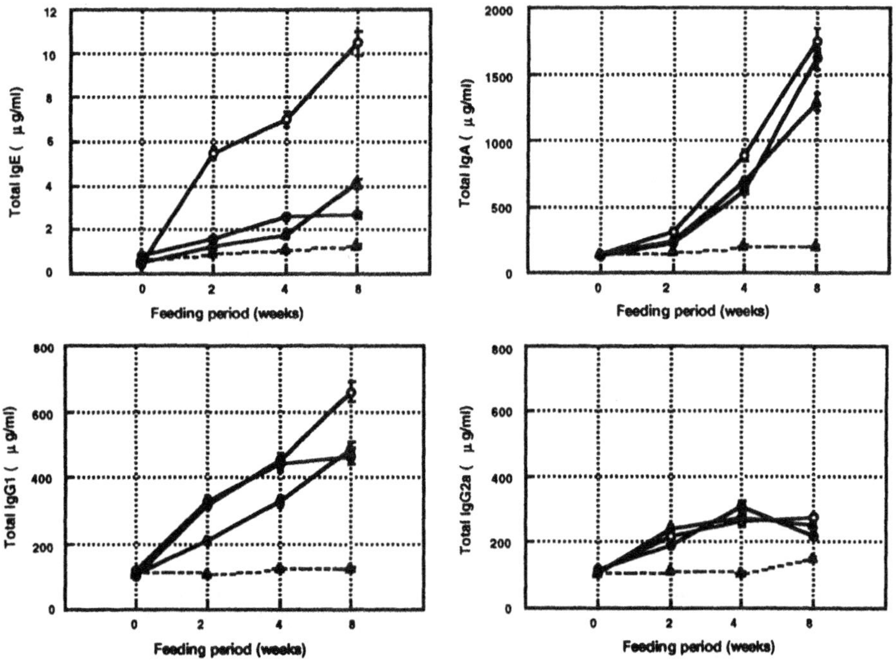


Figure 1. The effect of dietary NIV on total serum levels of immunoglobulins in OVA TCR-Tg mice. Mice were given free accessed to OVA-diet and drinking water with (open circle) or without (closed circle) 6 ppm NIV, or NIV (6 ppm)-diet (open triangle), or control diet (closed triangle). After feeding the mice the assigned diet for 2, 4 or 8 weeks, immuglobulin levels in serum were determined by ELISA.

As shown in Fig. 1, the total amount of each Ig isotype in the serun of OVA-diet mice increased in a time-dependent manner, although the total amount of IgG2a was less than that of other Ig isotypes. Consistent with the data obtained concerning the IL-4 response (Table 1), total serum IgE levels were significantly low in mice fed the OVA+NIV-diet, and this suppression of IgE production was steady for the duration of the experiment. In contrast to the total IgE response, the production of IgA and IgG1 was observed in OVA+NIV-diet mice. From these results, it is evident that dietary NIV selectively inhibited production of total IgE.

The OVA-specific antibody response is shown in Figure 2. Feeding the OVA-diet lead to a significant increase in the levels of all isotypes of OVA-specific serum Ig, above the control level, in a time-dependent manner. In comparison, the OVA+NIV-diet elicited marked decreases in OVA-specific IgE, IgG1 and IgA production. This result suggests that dietary NIV inhibited this established Th2 response in the presence of antigen.

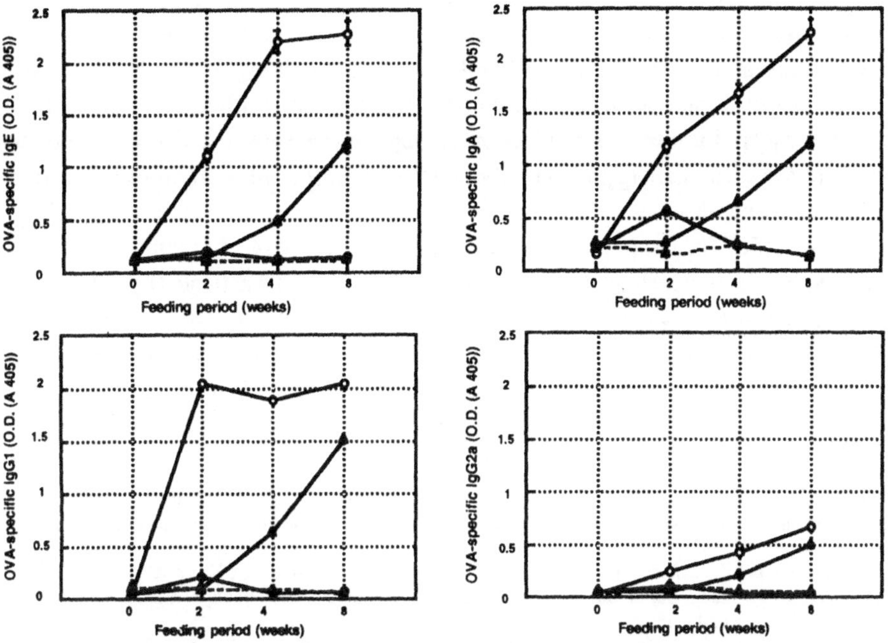


Figure 2. The effects of dietary NIV on serum levels of OVA-specific immunoglobulins in OVA TCR-Tg mice. Mice were given free accessed to OVA-diet and drinking water with (open circle) or without (closed circle) 6 ppm NIV, or NIV (6 ppm)-diet (open triangle), or control diet (closed triangle). After feeding the mice the assigned diet for 2, 4 or 8 weeks, immuglobulin levels in serum were determined by ELISA.

## 4. Discussion

From the *in vitro* experiments conducted in this study, it was found that NIV inhibited the production of IL-4 in spleen cells from untreated OVA-Tg mice when the cells were cultured with the antigen OVA. Especially, IL-4 production as a secondary response was completely suppressed by NIV. From these results, it is suggested that NIV suppresses the production of IL-4 in T cells activated by antigen. Because IL-4 plays a critical role in inducing B-cell Ig-isotype switching leading to IgE synthesis,[6] it was expected that dietary NIV would also suppress IgE production resulting in lower levels in serum. Therefore, OVA-Tg mice were administered OVA orally and simultaneously given NIV in their drinking water. In spite of the hyperinduction of serum IgE observed in the case of mice fed the OVA-diet, production of both OVA-specific and total serum IgE production was dramatically suppressed in mice fed the OVA+NIV-diet. These data suggest that the inhibitory effect of NIV on serum IgE production resulted from a deficiency in IL-4 production in the spleen. In contrast to the total IgE response, there was no difference in total IgA or IgG1 production between the mice fed the OVA-diet and those fed the OVA+NIV-diet. However, significant inhibition of production of OVA-specific serum IgA and IgG1 production was observed in mice fed the OVA+NIV-diet. This suggests that ingested NIV acts as an inhibitor suppressing antigen-specific Th2-type Ig production when an exogenous antigen is administered orally. The most striking finding in our study is that NIV selectively inhibits the antigen-specific Th2 response without loss of the total Th2 response, except for total IgE production. It may not be desirable to tolerize completely the total Th2 response inallergy patients because it is not possible to protect Th2 response-related infectious diseases. Therefore, it is noteworthy that NIV selectively inhibits the antigen-specific Th2 response, but not the total response. In summary, the results presented herein demonstrate that NIV may affect to Th2 cells activated by antigen and their cytokine promoters directly, and consequently induce the selective loss of IL-4 and subsequent hypoproduction of serum IgE. This finding is effective in modulating allergic responses and may provide a novel therapeutic approach for IgE-mediated allergic diseases.

## 5. References


1. Tanaka, T., Hasegawa, A., Yamamoto, S., Lee, US., Sugiura, Y., and Ueno, Y. (1988) Worldwide contamination of cereals by the Fusarium mycotoxins nivalenol, deoxynivalenol, and zearalenone, *J Agric Food Chem* **36**, 979-983
2. Robbana-Barnat, S., Loridon-Rosa, B., Cohen, H., Lafarge-Frayssinet, C., Neish, G.A., and Frayssinet, C. (1987) Protein synthesis inhibition and cardiac lesions associated with deoxynivalenol ingestion in mice. *Food Addit Contam* **4**, 49-55
3. Robbana-Barnat, S, Lafarge-Frayssinet, C., Cohen, H., Neish, G.A., and Frayssinet, C. (1988) Immunosuppressive properties of deoxynivalenol, *Toxicology* **48**, 155-166
4. Hughes, B.J., Hsieh, G.C., Jarvis, B.B., and Sharma, R.P. (1989) Effects of macrocyclic trichothecene mycotoxins on the murine immune system, *Arch Environ Contam Toxicol* **18**, 388-395
5. Hinishita, F., Suzuki, Y., Yokoyama, K., Hara, S., Yamada, A., Ogura, Y., Hashimoto, H., Tomura, S., Marumo, F., and Ueno, Y. (1997) Experimental IgA nephropathy induced by a low-dose environmental mycotoxin, nivalenol, *Nephron* **75**, 469-478
6. Mosman , T.R., and Coffman, R.L. (1989) Th1 and Th2 cells: different patterns of lymphokine secretion lead to different functional properties, *Annu Rev Immunol* **7**, 145-173
7. Sato T, Sasahara T, Nakamura Y *et al.* Naive T cells can mediate delayed-type hypersensitivity response in T cell receptor transgenic mice. Eur J Immunol 1994; 24: 1512-16

# HUMAN-HUMAN HYBRIDOMAS PRODUCING MONOCLONAL ANTIBODIES REACTIVE TO HOUSE DUST MITES ANTIGEN

H.KAWAHARA[1,2], M.MAEDA-YAMAMOTO[2], M.SUZUKI[1,2], K.OSADA[2], K.TSUJI[2]

[1] *Bio-oriented Technology Research Advancement Institution,*
[2]*National Research Institute of Vegetables, Ornamental Plants and Tea, MAFF,*
*2769 Kanaya, Shizuoka 428-8501, JAPAN*

## 1. Introduction

It is known well that human IgE play a important role of type I allergy. For studying the mechanism of allergy or diagnosing, many investigators expected to obtain allergen specific antibodies. However, IgE producing hybridomas were little obtained by fusing human lymphocytes from peripheral blood with a human fusion partner, because the population of IgE producing lymphocytes were much less than that of IgM and/or IgG producers in peripheral blood.

We considered that it was necessary to enhance antibodies production from lymphocytes. Therefore, we tried to develop the method for producing antigen specific antibodies from normal human peripheral blood lymphocytes activated with various immunoactivators.

## 2. Materials and Methods

### 2.1. CELLS AND CELL CULTURE

Human peripheral blood were obtained from four volunteers. The peripheral blood lymphocytes (PBLs) were separated with Lymphocyte Separation Medium (LSM: Organon technica, UT). ICLU-B which was a 6-thioguanine resistant human burkitt lymphoma, was used as a human fusion partner (Kawahara et al, 1998). The fusion partner didn't secrete or produce any antibodies in both the culture supernatant and cytoplasm. Those cells were cultured in 10% fetal bovine serum (FBS: Gibco) - ERDF medium and incubated under humidified 5%$CO_2$ / 95%air atmosphere at 37°C.

2.2. STIMULATION OF HUMAN LYMPHOCYTES WITH IMMUNOACTIVATORS
Human PBLs of $1x10^6$ cells/ml was suspended with 5%FBS-10%human plasama-ERDF medium containing 10$\mu$g/ml muramyl dipeptide (MDP), 100U/ml IL-2, IL-4 and 10U/ml IL-6 as immunoactive reagents and 10$\mu$g/ml house dust mite, *Dermatophagoides farinae* (Mite-Df) as a antigen solution. Human plasma was derived from each original donor of the PBL. The cell suspension was inoculated in 24-well culture plate and set in a incubator quietly. After 10 days, classes of antibodies secreted in culture supernatant and reactivities against the antigen were measured with ELISA. Then, to investigate the mechanism of the lymphocyte stimulation, leucil-leucin methylester (leu-leu-MeO) treated PBLs was also used.

2.3. PRODUCTION OF HUMAN-HUMAN HYBRIDOMAS
ICLU-B of $1x10^7$ was mixed with activated PBLs of $2x10^7$. The mixture was centrifuged, and the supernatant was removed. To the remaining cell pellets, 1ml of 50% polyethylene glycol (PEG; MW.4,000; Merck, Germany) was added. Further, 9ml of ERDF medium were added so that the total volume was adjusted to 10ml. The mixture was recentrifuged, and the resulting cell pellets were suspended in 50ml of a 15% FBS-ERDF medium. The suspension was added to a 96-well culture plate such that 100$\mu$l of the suspension were charged in each well of the plate. On the following day of the fusion, a 15% FBS-ERDF medium containing 400$\mu$M hypoxantine, 0.8$\mu$M amethopterin and 32$\mu$M thymidine was added to a 96-well culture such that 100$\mu$l of the medium were charged in each well of the plate. Then, this medium was changed with a 15% FBS-ERDF medium containing 200$\mu$M hypoxanthine, 0.4$\mu$M amethopterin and 16$\mu$M thymidine by half every 2 or 3 days. After 2 weeks or later, the growing cells were observed by microscopic observation.

## 3. Result and Discussion

It was difficult to obtain antigen specific human monoclonal antibodies (Kawahara et al, 1992). So, we tried to enhance antibodies production of PBLs by using various immunoactive reagents. Table 1 shows the antibodies concentration from activated PBLs with additives. Addition of the mixture of MDP, IL-2, IL-4 and IL-6 to the culture medium (Test 3) enhanced the productivities of IgM, IgG and IgE as compared with the control

Table 1 Effects of various additives on antibody productivities of PBLs treated with or without leu-leu-MeO

| Test | Treatment | Additives | | | | Antibody concentration (μg/ml) | | |
|---|---|---|---|---|---|---|---|---|
| | Leu-leu-MeO | MDP | IL-2 | IL-4 | IL -6 | IgM | IgG | IgE |
| 1 | – | – | – | – | – | 0.54±0.33 | 0.42±0.48 | ND |
| 2 | + | – | – | – | – | 0.43±0.28 | 0.71±0.36 | ND |
| 3 | – | + | + | + | + | 0.81±0.29 | 1.02±0.21 | 0.09±0.03 |
| 4 | + | + | + | + | + | 0.68±0.15 | 1.12±0.14 | 0.01±0.01 |
| 5 | – | – | + | + | + | 0.53±0.24 | 0.55±0.32 | 0.03±0.02 |
| 6 | + | – | + | + | + | 0.47±0.34 | 0.69±0.18 | ND |

+: added, –: not added, ND: not detected
Data were shown as means±SEM in triplicated experiments

(Test 1). Specially, IgE antibody was not detected without these additives (Test 1, 2, 6). All antibodies production was much increased by adding MDP to the IL-2, IL-4 and IL-6 (Test 3, 5). Leu-leu-MeO treatment was known as a enhancer of immunization because of removing CD8 positive cells and cytotoxic T cells from the population of whole PBLs. Therefore, we investigated the effects of leu-leu-MeO treatment on antibodies production of the PBLs. IgM and IgG production of PBLs activated with MDP, IL-2,IL-4 and IL-6 were increased by leu-leu-MeO treatment, but the antibodies production of non-activated PBLs were not always affected by the treatment (Test 1, 2, 4, 6). Whereas, concentration of IgE antibody produced by PBLs without leu-leu-MeO treatment was at least nine times higher than with treatment (Test 3, 4). During adding interferon-γ to the population of lymphocytes treated with leu-leu-MeO, IgE was produced by the cells effectively (data not shown). These results suggested that CD8 positive and cytotoxic T cells in the PBLs were related to IgE production, and interferon-γ was related to initial process of IgE production.

Reactivities of antibodies in the culture supernatant after stimulation of PBLs against Mite-Df were measured with ELISA. Figure 1 shows the reactivities of 3 classes of antibodies produced by activated PBLs. All the obtained antibodies in the culture supernatant reacted with Mite-Df antigen in a dose dependent manner, though control antibodies didn't. These made clear that activated PBLs with MDP, IL-2,IL-4 and IL-6 enhance not only antibody production but antibody specificity.

To obtain human monoclonal IgE antibodies, human-human hybridomas were made by

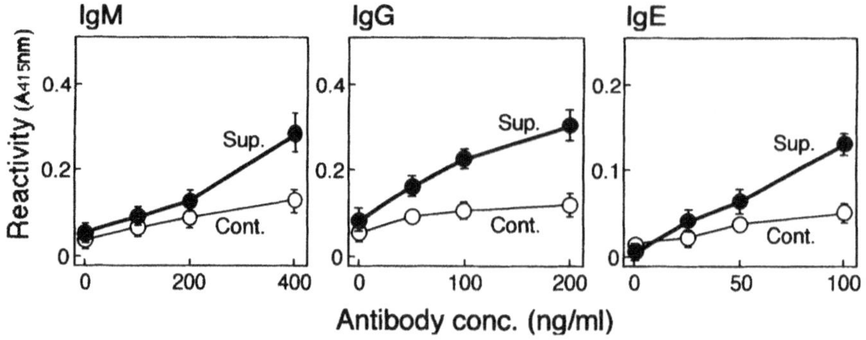


Figure 2 Reactivities of antibodies secreted by PBLs . PBLs were cultured in 5%FBS-10%human plasma-ERDF medium supplemented with MDP, IL-2, IL-4, IL-6. The concentration and reactivities of these antibodies were measured by ELISA. Closed symbol shows the case of supernatant (Sup.) and open ones shows the case of control antibodies (Cont.).

fusing ICLU-B with PBLs stimulated in vitro. As a result, two kinds of human monoclonal IgM antibodies were obtained. Figure 2 shows the reactivity of 1G9 antibody against Mite-Df or other antigens. Human monoclonal IgM antibody, 1G9 were reacted with Mite-Df antigen, though the antibodies were slightly reacted with other allergens from cedar, egg white and ragweed. Recently, various human fusion partners were reported. These cells were almost from B lymphoma, and the cells produced and/or secreted some

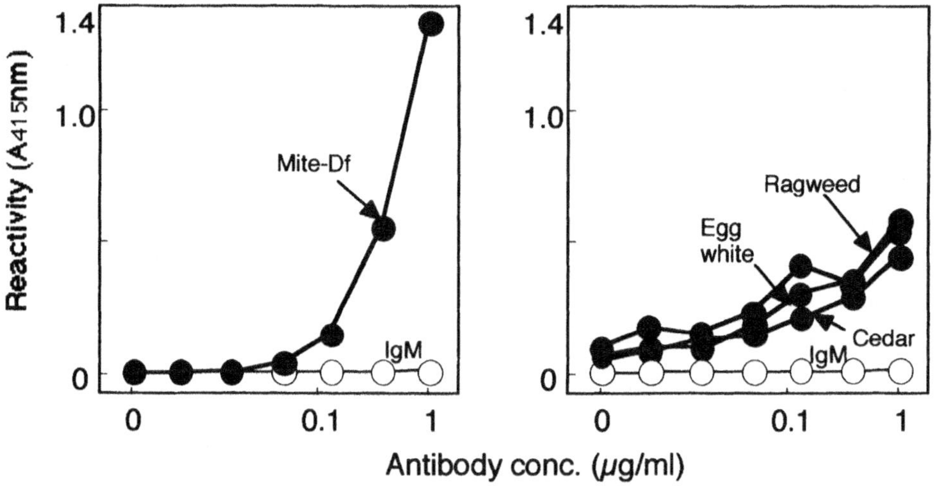


Figure 3 Reactivities of 1G9 monoclonal antibody obtained by hybridomas fused ICLU-B with PBLs activated in vitro. Closed symbol shows the reactivities of 1G9, and open symbols shows conventional IgM.

antibodies into the cytoplasm and/or culture supernatant. It was known that antibody productivity of the fusion partner remarkably effected on that of obtained hybridoma. ICLU-B which was recently established by us, was used as a fusion partner in this study. This cell line was derived from human burkitt lymphoma, and the cell didn't produce or secrete any antibodies in the cytoplasm and/or culture supernatant. Therefore, it was considered that the antibody productivity of the resultant hybridomas was derived from the activated lymphocytes.

In vitro stimulation of lymphocytes was powerful tool to obtain antigen specific antibodies. We developed the effective stimulation of PBLs for antigen specific antibodies. Further, we obtained monoclonal IgM antibodies reactive to mite-Df, although it was not IgE type.

## 4. Acknowledgement

We thank Dr.H.Oda for his helpful technique and Miss N.Matsuda and Mr.N.Tahara for their technical assistance. This work was supported by grants from program for promotion of basic research activities for innovative biosciences.

## 5. References

Kawahara, H., Maeda-Yamamoto, M., Suzuki, M. and Hakamata, K. (1998) New human fusion partners, ICLU-B, -T, -E for making human hybridomas derived from various immunological cells, Hum Antibod, In press.

Kawahara, H., Shirahata, S., Tachbana, H. and Murakami, H. (1992) In vitro immunization of human lymphocytes with human lung cancer cell line A549 Hum Antibod Hybridomas 3, 8-13.

# ESTABLISHMENT OF HUMAN-MOUSE HYBRIDOMAS SECRETING ANTIBODIES TO RICE-ALLERGENS

HIROSHI SHINMOTO[1], KAZUHIKO NAKAHARA[2], RYOICHI ICHIKAWA[3], MASUKO KOBORI[1], and TOJIRO TSUSHIDA[1]

*[1]National Food Research Institute, Tsukuba, Ibaraki 305-8642, Japan, [2]Japan International Research Center for Agricultural Sciences, Tsukuba, Ibaraki 305-8686, Japan, [3] Tokushima Prefectural Industrial Technology Center, Tokushima, Tokushima 770-8021, Japan*

**Summary**

**We obtained 2,202 multi-clone immortalized human B-cell library stocks from seven healthy donors by transforming B-cells with the Epstein-Barr virus. The B-cell library contained various kinds of B-lymphoblastoid cells secreting antibodies to rice, soybeans, milk, and eggs. Finally, we obtained human-mouse hybridomas secreting antibodies to rice allergens. We concluded that the method developed here would be a powerful tool for analyzing food allergens.**

## 1. Introduction

Since it seemed not easy to prepare a standard sera library for allergenicity testing [1], we planned to develop a new system to evaluate the allergenicity of food employing the B-cell immortalizing technique. The specificity of antibody molecules can be defined by V-D-J (heavy chain) and V-J (light chain) rearrangement of the immunoglobulin genes [2]. The possible number of specificities of antibodies was estimated to be more than 100,000,000.

Recently, we reported a method to establish human antibody-secreting cells by transforming B-cells with the Epstein-Barr virus (EBV) followed by cell fusion with murine myeloma cells [3, 4]. In this paper, we outline our approach to the analysis of the allergenicity of food allergens.

## 2. Experimental strategies

### 2. 1. IMMORTALIZATION OF HUMAN B-CELLS

The Epstein-Barr virus (EBV) is a Herpes virus that infects human B-cells (B-lymphocytes) and transforms them into immortalized B-lymphoblastoid cells (BLCs) [5, 6]. BLCs are known to secrete antibodies. By using the EBV transformation method, we obtained antibody-producing BLCs. These were expected to include antibodies against food allergens.

We planned to immortalize more than 200,000 B-cell clones. We estimated that one B-cell out of 1,000 peripheral blood lymphocytes (PBLs) would be immortalized by EBV. Because our one "multi-clone immortalized B-cell library stock" was derived from 100,000 peripheral blood lymphocytes, it would contain 100 clones of immortalized BLCs.

We stored more than 2,000 ampoules of BLCs in a liquid nitrogen container and their supernatants at -80 °C. The supernatants were used for the analysis of specific antibodies to food allergens.

### 2. 2. ANALYSIS OF IMMORTALIZED CELLS SECRETING ANTIBODIES TO FOOD ALLERGENS

Antibody-producing cells were analyzed by an enzyme-linked immunosorbent assay (ELISA). Food allergens such as rice allergens [7, 8], soybean allergens [9], ovomucoid, and milk casein were used as allergens.

We then cloned the BLCs secreting antibodies against food allergens. Cloning of BLCs was very difficult, because the growth of BLCs was cell density-dependent. BLCs could not grow from single cells, even if irradiated feeder cells were added. To solve this problem, we fused BLCs with mouse myeloma cells. The resulting mouse-human hybridomas could easily be cloned [3, 4].

## 3. Results

### 3. 1. IMMORTALIZATION OF HUMAN B-CELLS

PBLs were prepared from seven healthy donors. PBLs infected with EBV were seeded into 96 well micro-culture plates. After 3 to 6 weeks of culture, wells with transformed lymphocytes were transferred to 24 well plates and cultured for 1 week. Cells were further cultured in 6 cm dishes, and cells (BLCs) and the supernatants were frozen. We obtained a total of 2,202 multi-clone stocks of BLCs for the cell library.

### 3. 2. ELISA OF THE CULTURE SUPERNATANTS

Culture supernatants of BLCs were assayed for antibodies to several food allergens. As shown in Table 1, we detected many BLC stocks containing cells secreting food allergen specific antibodies. Antibodies of IgM class were dominant. This result suggested that we obtained a lot of BLCs secreting specific antibodies against many food allergens.

### 3. 3. CELL FUSION OF BLCS WITH MOUSE MYELOMA SP2/ON

The 8-Azaguanine resistant mouse myeloma cell line SP2/O was cultured with ouabain to obtain the ouabain-resistant cell line SP2/ON. BLCs secreting antibodies to rice allergens and SP2/ON cells were fused and the fused cells were cultured in a HAT medium containing ouabain. The cell fusion efficiencies were high with all cells. The clones listed in Table 2 were established after repeated cloning.

All antibodies from established hybridomas reacted with rice 14-16 kDa allergens (data not shown). Additionally, some of them reacted with 25 kDa protein. We are trying to analyze antibody binding sites of rice allergens.

**Table 1. Detection of Antibodies to Food Allergens of BLCs Supernatants**

| Allergen | Positive | Allergen | Positive |
|---|---|---|---|
| Nonspecific | 9 | αs1Casein | 31 |
| Rice | 151 | βLactoglobulin | 32 |
| Soybean | 176 | Ovalbumin | 183 |
| Ovomucoid | 29 | | |

**Table 2. Establishment of Human-Mouse Hybridomas Secreting Antibodies Against Rice Allergens**

| BLC | Hybridoma established |
|---|---|
| 3-207 | 3-207-2-6 |
| 5-39 | 5-39-147-932 |
| 5-133 | 5-133-64 |
| 5-175 | 5-175-7-2 |
| 5-215 | 5-215-24 |

## 4. Discussion

A B-cell library consisting of more than 2,000 stocks of BLCs could be an excellent source of antibodies to a wide spectrum of allergens. It seemed to contain antibodies to various food allergens. The disadvantages of this method lie in the instability of the BLC clones secreting antibodies of interest and, thus, in the difficulty in establishing single clones. In contrast, cell fusion of human lymphocytes (not transformed) with mouse myeloma cells yields high fusion efficiency and enables cloning by limiting dilution. However, it is difficult to establish human-mouse hybridomas secreting antibodies with a certain specificity. The method described in this paper combined the advantages of both the cell library and cell fusion with mouse myeloma, and eliminates disadvantages.

## 5. References

[1] Taylor, S.L. (1995): Evaluation of the allergenicity of foods developed through biotechnology. *IN* : Proceedings of the 3rd international symposium on the biosafety results of field tests of genetically modified plants and microorganisms. ed. by D.D. Jones, The University of California, 185-198.

[2] Okada, A., and Alt, F.W. (1995): The variable region gene assembly mechanism. *IN* : Immunoglobulin genes. ed. by T. Honjo and F.W. Alt, Academic Press, 205-234.

[3] Shinmoto, H., Dosako, S., Tachibana, H., Yamada, K., Shirahata, S., and Murakami, H. (1991): Generation of hybrid hybridomas secreting human IgM class hybrid antiricin and antidiphtheria toxin antibodies. *Hum. Antibod. Hybridomas* **2**, 39-41.

[4] Shinmoto, H., Dosako, S., Tachibana, H., Shirahata, S., and Murakami, H. (1991): Generation of human-mouse hybridoma secreting human IgM class anti-neocarzinostatin antibody and its application to hybrid hybridoma. *Agric. Biol. Chem.* **55**, 2883-2885.

[5] Shinmoto, H., Sato, K., Dosako, S., and Nakajima, I. (1990): Generation of mouse-human hybridomas secreting monoclonal antibodies specific to ricin and diphtheria toxin from peripheral blood lymphocytes without prior immunization. *IN* : Trends in Animal Cell Culture Technology, ed. by H. Murakami, Kodansha, 307-312.

[6] Shinmoto, H., Dosako, S., and Tanaka, S. (1992): Transformation of human colostrum lymphocytes with Epstein-Barr virus. *Tokai J. Exp. Clin. Med.* **17**, 129-132.

[7] Adachi, T., Alvarez, A.M., Aoki, N., Nakamura, R., Garcia, V.V., and Matsuda, T. (1995): Screening of rice strain deficient in 16-kDa allergenic protein. *Biosci. Biotech. Biochem.* **59**, 1377-1378.

[8] Alvarez, A.M., Fukuhara, E., Nakase, M., Adachi, T., Aoki, N., Nakamura, R., and Matsuda, T. (1995): Four rice seed cDNA clones belonging to the α-amilase /tripsin inhibitor gene family encode potential rice allergens. *Biosci. Biotech. Biochem.* **59**, 1304-1308.

[9] Yamanishi, R., Kondo, K., Tsuji, H., and Ogawa, T. (1995): Micro-assay to measure the allergenicity of a Kunitz-type soybean trypsin inhibitor toward Balb/c mice by using RBL-2H3 cells. *Biosci. Biotech. Biochem.* **59**, 1272-1275.

# CELL CULTURE ENGINEERING USING INTELLIGENT BIOMATERIALS

Yoshihiro ITO[1,2]
[1]*Graduate School of Materials Science, NAIST, Ikoma 630-0101, JAPAN*
[2]*PRESTO, JST, Keihanna Plaza, Seika-cho 619-0237, JAPAN*

Growth factor proteins were covalently immobilized on various matrices to manipulate cells. It was shown that insulin and epidermal growth factor stimulated cell growth even after immobilization. Pattern-immobilization of growth factor proteins clearly demonstrated the stimulation by immobilized proteins. In other words, this type of stimulation by non-diffussional growth factors enabled us to regulate tissue formation with artificial biomaterials. These stimulations by the immobilized growth factors may mimick juxtacrine stimulation of membrane-anchored growth factors such as heparin-binding epidermal growth factor, transforming growth factor, and tumor nectrosis factor. The stimulation was enhanced by coimmobilization with adhesion factors. In addition, by coimmobilization with thermosensitive polymers enabled the enhancement of cell growth and easy recovery of cultured cells by lowering temperature.

Keywords: Tissue engineering, Intelligent material, Biomaterial, Immobilization, Growth Factor

## 1. Introduction

Recent progress in biological science has revealed many types of growth factor proteins. The signal proteins regulate various cell functions such as growth, differentiation, secretion, and apoptosis. It is known that the biosignal protein interacts with the cognate receptor on the cell surface to form a complex, and that the complex aggregates on the cell surface before being internalized into the cell. The internalized complex is decomposed in the lysosomes (Figure 1a). However, it has not been known which process was necessary to transduce the signal to the cellular nucleus. If the events on the cell surface are enough for the signal transduction, then biomaterials with immobilized biosignal molecules could have the potential to regulate cellular functions (Figure 1b).

We have been investigating the regulation of cellular functions with immobilized biosignal molecules [1-4], since we found that the immobilized insulin enhanced cell growth. On the other hand, recently some progress has been reported in biomaterial technology, including regulation of geographic location of cells on the biomaterial surface and the control of cell attachment and detachment using thermoresponsive polymers. In this article, the cell manipulation using these newly designed biomaterials are shortly summarized.

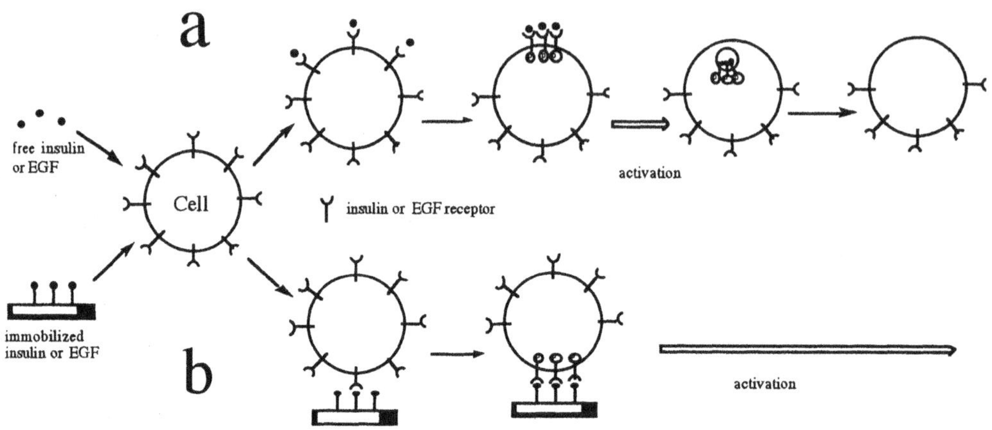


Figure 1 Time course of cell interaction with free (diffusible) growth factors (a) or with immobilized growth factors (b).

## 2. Immobilized growth factors stimulated cells

To visualize the signal transduction and to spatially regulate the cell functions, the growth factors were immobilized on matrices in a prescribed pattern. Insulin or epidermal growth factor (EGF) was micro-pattern immobilized on a poly(ethylene terephtalate) or polystyrene film surface by a photolithographic method [5-7]. Figure 2 illustrated the pattern-immobilization method. The pattern-immobilization was confirmed by immunostaining with an anti-growth factor antibody. Chinese hamster ovary cells overexpressing insulin receptor (CHO-IR) were cultured on the matrices pattern-immobilized with insulin and after culture for 8 hours, the phosphorylation (activation) was investigated by staining with an anti-phosphotyrosine antibody. The signaling proteins were activated only in the cells cultured on the area immobilized with insulin, although the cells adhered independently of the immobilized area. As a result, the cell growth was accelerated only in the insulin-immobilized area. The immobilized insulin did not affect the cell adheison. When nerve growth factor (NGF) was immobilized in a specific pattern on polystyrene film, the surface-dependent differentiation of phaeochromacytoma PC12 cells was observed [8].

CHO cells overexpressing EGF receptors (CHO-ER cells) were cultured on the plate immobilized with EGF in a narrow stripe, of which the size is smalles than that of cell. The contact area (stripes 2 μm in width) between the cells and the immobilized EGF was stained by anti-phosphotyrosine antibody. Since free lateral diffusion and internalization of the bound EGF-EGF receptor complex were prohibited by immobilization of EGF, signal proteins in the interaction regions were only activated. This finding also indicates that the biological signal was transduced only to the cell that intereacted with immobilized EGF.

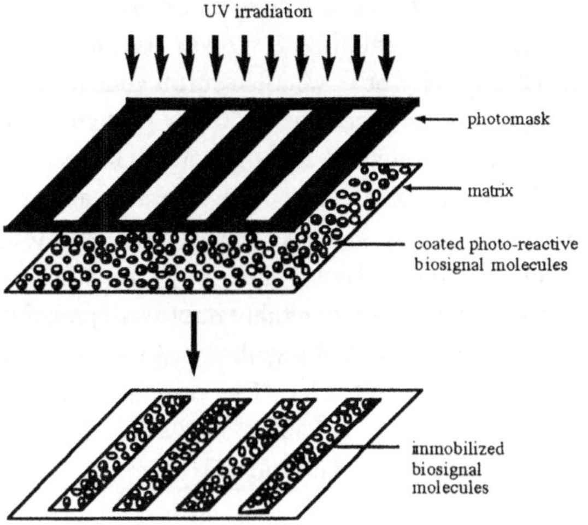


Figure 2 Illustration of pattern-immobilization of growth factors by photo-lithographic method.

The immobilized insulin or EGF was a strong mitogen for anchorage-dependent cells such as STO cells and CHO cells, but a weak mitogen for anchorage-independent cells such as mouse hybridoma cells and human chronic myelogenous leukemia K562 cells. This phenomenon indicated that the signal transduction was triggered by the cell adhesion. In other words, it was demonstrated that the biological signal is transduced by the interaction cells with the immobilized proteins not diffused proteins.

## 3. High activity of immobilized growth factor

The immobilized growth factors not only stimulate the cell growth, but also the cell growth enhancement activities were higher than those of diffusible growth factors. The high activity of immobilized insulin was thought to be due to inhibition of the internalization process as shown in Figure 1b. Generally cells decompose the biosignal molecules in the cells to reduce their stimulation and this phenomenon is called down-regulation. The immobilization was considered to inhibit this down-regulation and as a result the stimulation continued for a long time without reduction [1].

To investigate the above hypothesis, the activation of cellular signaling proteins, insulin receptor subunit in CHO-IR cells, or mitogen-activated protein (MAP) kinase in CHO-ER cells was measured in the presence of immobilized insulin or immobillized EGF, respectively [1,9]. Native diffusible insulin rapidly activated the insulin receptor. However, the activation was transient. On the other hand, the activation by the immobilized insulin continued to increase for up to 12 h. The same time-dependent activation was observed in PI3-kinase, which is downstream of the insulin receptor in the signaling pathway, both in the presence of insulin and the immobilized insulin. Similarly the immobilized EGF activated EGF receptor and the downstream protein, MAPK, for a long time, even though the signaling effect of native EGF was transient. These sustained activations of signaling proteins by the immobilized insulin or EGF should explain the high mitogenic effect of the immobilized growth factors.

**4. Enhancement of signalling effect by immobilized growth factor**

To enhance the cellular regulation ability by immobilized growth factor, other macromolecules were coimmobilized with it. Biologically or physico-chemical enhancement of cell adhesion increased the mitogenic efffect of immobilized growth factor proteins. For biological enhancement of cell adhesion, collagen, fibronectin, and the core RGD peptide were coimmobilized [10]. On these coimmobilized matrices, both cell adhesion and cell growth were remarkably enhanced. For physico-chemical enhancement of cell adhesion, insulin was coimmobilized with cationic polyallylamine and polylysine [11]. The adhesion of STO cells on the polyallylamine-coimmobilized film was remarkably improved, probably by electrostatic interactions with the negatively charged outer cell membrane. As a result, the coimmobilized film markedly enhanced the growth of STO cells. The coimmobilization enhanced even the growth of anchorage-independent cells. The enhancement was considered to due to the increased interactions of immobilized growth factor with the cells.

**5. Applications of immobilized growth factors**

To immobilize growth factors on various matrices with no functional groups on the surface, a photo-immobilization method, which is the same as that for photolithogaphic immobilization, was carried out [13]. photo-reactive insulin and EGF derivatives were synthesized by coupling insulin and EGF with *N*-(4-azidobenzoyloxy)succinimide. An aqueous solution of the insulin or EGF derivative was then cast on the surface of polystyrene film and the water completely evaporated. The film was irradiated with an ultraviolet light and insulin or EGF was coupled to the surface. The amount of photo-immobilized insulin or

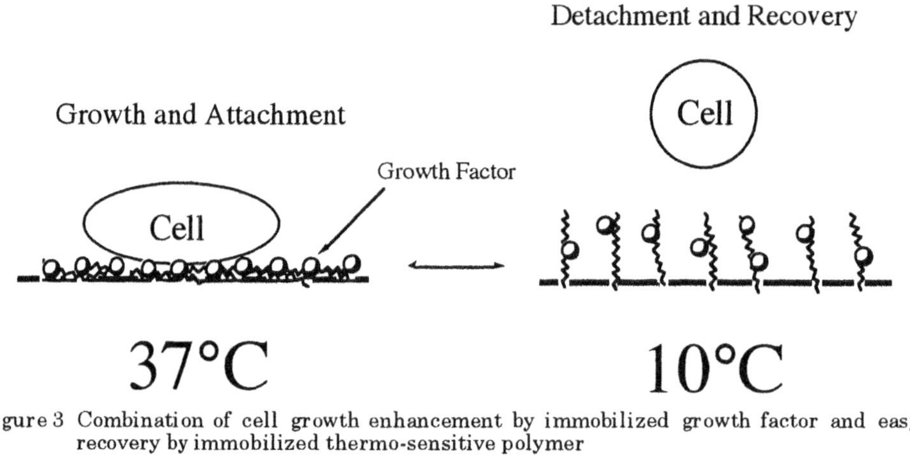


Figure 3 Combination of cell growth enhancement by immobilized growth factor and easy cell recovery by immobilized thermo-sensitive polymer

EGF increased with the amount of insulin or EGF derivative
coated onto the substrate. When mouse fibroblast STO cells and CHO cells cultured on the photo-immobilized insulin or EGF was measured, their growth was remarkably enhanced. The photo-immobilization technique will expand the application of growth factor-immobilized biomaterials.

It was possible to combine growth enhancement and cell adhesion regulation as shown in Figure 3 [13-15]. It is known that cells adhered to a thermo-responsive polymer, poly(*N*-isopropylacrylamide), -immobilized polystyrene plate at 37°C, and then detached from the plate when the temperature was lowered to 10°C. Coimmobilization of insulin with this polymer enhanced cell growth and facilitated harvest of the cells by lowering the temperature.

The immobilization of growth factor enabled serum- or protein-free cell culture engineering. In addition, the immobilized growth factor could be recycled as the enzyme immobilized on bioreactors. Although a few percentage of decrease of the mitogenic activity due to the adsorption secreated from cultured cells was observed, the immobilized growth factor retained its activity even after five times of utilization [1].

## References


1. Ito, Y., Zheng, J., Imanishi, Y., Yonezawa, K., and Kasuga, M. (1996) Protein-free cell culture on artificial substrate with covalently immobilized insulin, *Proc. Natl. Acad. Sci., U.S.A.* **93**, 3598-3601.
2. Ito, Y. (1997) Intelligent biomaterials for tissue engineering, in T. Akaike, T. Okano, M. Akashi, M. Terano, N. Yui (eds.), *Advances in Polymeric Biomaterials Science*, CMC Co., Tokyo, pp. 421-432.
3. Ito, Y., Chen, G., and Imanishi, Y. (1998) Artificial juxtacrine stimulation for tissue engineering, *J. Biomat. Sci. Polym. Eds.*,**9**, 879-890.
4. Chen, G., Ito, Y., and Imanishi, Y. (1997) Mitogenic activities of water-soluble and water-insoluble insulin conjugates, *Bioconj. Chem.* **8**, 106-110.
5. Ito, Y., Kondo, S., Chen, G., and Y. Imanishi, Y. (1997) Patterned artificial juxtacrine stimulation of cells by covalently immobilized insulin", *FEBS Lett.* **403**, 159-162.
6. Ito, Y., Chen, G., and Imanishi, Y. (1998) Micropatterned immobilization of epidermal growth factor to regulate cell function" *Bioconjugate Chem.*, **9**, 277-282.
7. Chen, G., Ito, Y., and Imanishi, Y. (1997) Photo-immobilization of epidermal growth factor enhances its mitogenic effect by artificial juxtacrine signalling, *Biochim. Biophys. Acta.* **1358**, 200-208.
8. Ito, Y. (1998) Regulation of cellular gene expression by artificial materials immobilized with biosignal molecules, *Jpn. J. Artif. Organs* **27**, 541-544.
9. Ito, Y., Li, J.-S., Takahashi, T., Imanishi, Y., Okabayashi, Y., Kido, Y., and Kasuga, M. (1997) Enhancement of the mitogenic effect by artificial juxtacrine stimulation using immobilized EGF,*J. Biochem.* **121**, 514-520.
10. Li, J.-S., Ito, Y., Zheng, J., Takahashi, T., and Imanishi, Y. (1997) Enhancement of artificial juxtacrine stimulation of isnulin by co-immobilization with adhesion factors, *J. Biomed. Mater. Res.* **31**, 190-197.
11. Ito, Y., Zheng, J., and Imanishi, Y. (1997) Enhancement of cell growth on a porous membrane co-immobilized with cell growth and cell-adhesion factors, *Biomaterials* **18**, 197-202.
12. Ito, Y., Chen, G., and Imanishi, Y. (1996) Photoimmobilization of insulin onto polystryene dishes for protein-free cell culture", *Biotechnol. Prog.* **12**, 700-703.
13. Chen, G., Ito, Y., and Imanishi, Y. (1997) Regulation of growth and adhesion of cultured cells by insulin conjugated with thermoresponsive polymers, *Biotechnol. Bioeng.* **53**, 339-344.
14. Ito, Y., Chen, G., Guan, Y., and Imanishi, Y. (1997) Patterned immobilization of thermoresponsive polymer, *Langmuir*, **13**, 2756-2759.
15. Chen, G., Imanishi, Y., and Ito, Y. (1998) Effect of protein and cell behavior on pattern-grafted thermoresponsive polymer, *J. Biomed. Mater. Res.*, in press.

# SPATIAL DEVELOPMENT OF MOUSE BONE MARROW CELLS CULTIVATED IN POROUS CARRIERS

**Y. TOMIMORI, M. TAKAGI, T. SASAKI and T. YOSHIDA**
International Center for Biotechnology, Osaka University, 2-1, Yamada-oka, Suita, Osaka 565, Japan

Key words: Bone marrow cells, Porous carriers, Three dimensional culture, Stromal cells, Progenitor cells

Spatial development of mouse bone marrow cells cultivated in porous carriers was investigated in order to design a reactor with a three-dimensional hematopoietic microenvironment. Because cells from the bone marrow contain anchorage-dependent stromal cells as feeder cells besides hematopoietic cells, three types of porous carriers were used for examining the spatial development of a stromal cell line (SR-4987). Cells could grow to a high density and spread well on a polyester non-woven disc carrier (Fibracell (FC)) and a fibroblast-like shape of the cells was observed under a scanning electric microscope, while cells in porous cellulose beads (Microcube (MC), 500 $\mu$m of pore diameter) spread at a low density; cells on another type of cellulose porous beads (CPB, 100 $\mu$m of pore diameter) were globular. The flatness of the inner surfaces of the carriers could be the reason for the differences in their cell shapes. Mouse bone marrow cells were inoculated to three types of porous carriers in dishes (1.8 $cm^2$). The growth of stromal cells (attached cells) (Fig. 1) and the time-course of total hematopoietic cell (suspension cells) (Fig. 2) concentrations were similar in all cultures. The proportion of progenitor cells (BFU-E, CFU-E, CFU-GM) in the total hematopoietic cell population, after showing an initial decrease, increased after 1 week when cultured in FC while no increase was observed when the cells were cultured on MC or CPB (Fig.

3). The shape of the stromal cells could be the reason for this difference. Comparison of the culture in FC with Dexter culture, which is a standard method for mouse bone-marrow cell cultivation on the bottom surface of a culture dish, revealed that the three-dimensional culture obtained in FC was associated with a marked increase in the proportion of macrophages and erythrocytes among matured cells (Fig. 4).

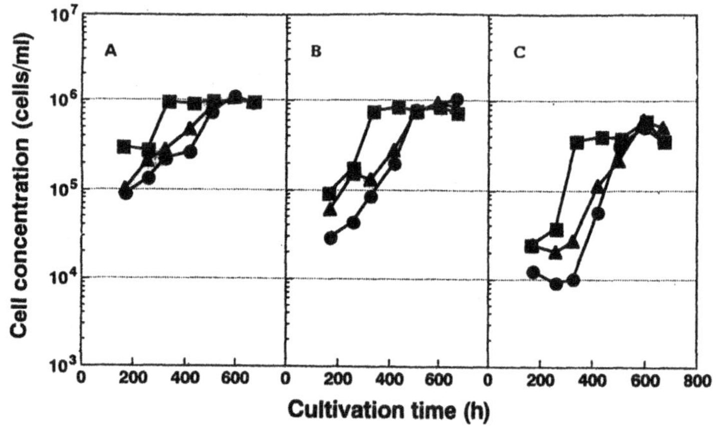


Fig. 1 Growth of stromal cells during the bone marrow cells cultivation in porous carriers. A; CPB, B; Micro cube, C; Fibra cell. Inoculum size of bone marrow cells were 2(●), 5(▲), and $8 \times 10^6$ cells/ml(■), respectively.

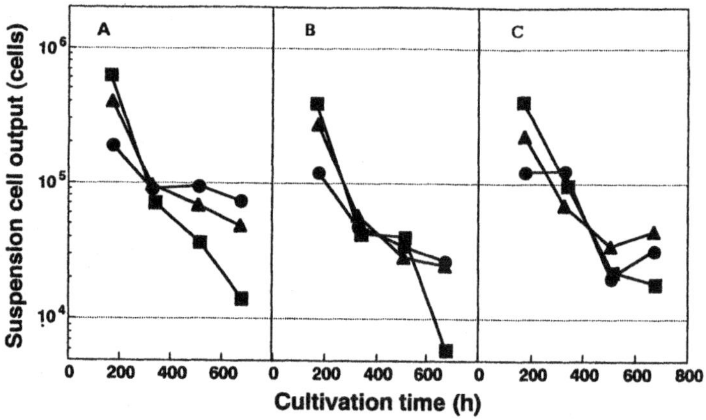


Fig. 2 Time course of suspension cell output during the bone marrow cells cultivation in porous carriers. A; CPB, B; Micro cube, C; Fibra cell. Symbols were same as Fig. 1.

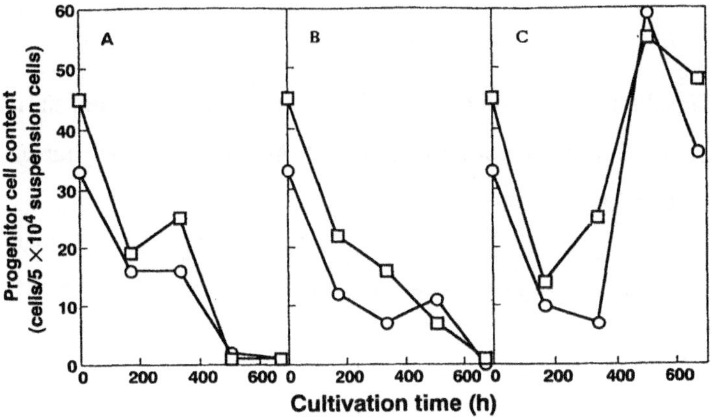


Fig. 3 Progenitor cell content in bone marrow cultures in porous carriers. A; CPB, B; Micro cube, C; Fibra cell. BFU-E(○) and CFU-GM(□) were determined by CFU-assay.

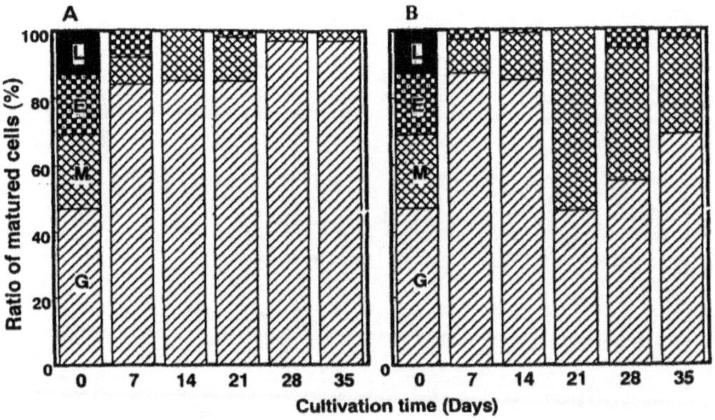


Fig. 4 Matured cell ratio in bone marrow cell cultures. A; Dexter's culture, B; Three dimensional culture in the porous carrier of Fibra cel. Lymphocyte(L), erythrocyte(E), macrophage(M), and granulocyte(G) were analyzed by Wright's staining.

M. Takagi, T. Sasaki, and T. Yoshida (1997) Spatial development of the cultivation of a marrow stromal cell line in porous carriers. 4th Asia-Pacific Biochemical Engineering Conference. 143.

# TUMOR PROMOTING MECHANISM OF BIOMATERIALS:
*No involvement of mutation in cx43 gene in the tumorigenesis induced by polyurethanes in vitro.*


Toshie Tsuchiya, Md. Abu Sayed and Akitada Nakamura.
*Division of Medical Devices, National Institute of Health Sciences,*
*Kamiyoga, 1-18-1, Setagaya-ku, Tokyo 158-8501, Japan.*



Abstract

Polyurethanes (PUs) are widely used biomaterials for medical implants. Previous studies from our laboratory have reported the tumorigenic potentials of these PUs *in vivo* and *in vitro*. PU-components had inhibited the gap junctional intercellular communication (GJIC) in cultures of mouse Balb/c (A31-1-1) and chinese hamster V79 fibroblasts. Since GJIC is crucial for normal differentiation of a cell population, this perturbed gap junction is most likely to play the major role in the PU-induced tumorigenesis. Gap junctions are composed of a multigene family of proteins called connexins. Various tumor promotors have been shown to inhibit GJIC by phosphorylation modification of connexin proteins. Mutations in connexin genes have also been found to lead abnormally regulated cell-cell communication. We are interested to elucidate the mecahnism of tumorigenesis induced by biomaterials including PUs. In the present study, we choose three PU-induced tumor clones, namely, clone-2, 5 and 6, isolated from A31-1-1 cells, for investigation of mutation in their cx43 gene, the major gap junction protein of this cell line. The connexin43 (cx43) gene was amplified by the polymerase chain reaction (PCR). The entire coding region located in the exon 2 of cx 43 gene was sequenced directly from the PCR amplified product. None of the clones showed a difference in their cx43 sequences with the published sequence for mouse cx43 sequence. Thus, probably a post-translational modification or a defect in gene expression, but, not a mutation in the coding sequence of cx43, is involved in the polyurethane-induced inhibition of GJIC during the tumorigenesis *in vitro*.


Introduction

Polyurethane (PU) is a widely used biomaterials for medical implants. Earlier studies from this laboratory have detected the tumorigenic potential of this biomaterials *in vitro* and *in vivo* [1,2]. PU has been implicated to inhibit cell-cell communication in mouse Balb/c

3T3, A31-1-1 and chinese hamster V79 fibroblasts, possibly through alteration in connexin 43 (cx43) protein. Cx43 is a vital protein to form the gap junction channel. Genetic alteration and post-translational modification (*i.e.* phosphorylation) in cx43 has been shown to be involved with the retarded the gap junctional intercellular communication (GJIC) and could be associated in tumorigenesis mechanism [3]. In this study, we aim to examine the possibility of genetic modification in cx43 sequence of the tumor cells.

## Experimental

### *Cell line*

Balb/c 3T3 A31-1-1 cell line was kindly obtained from Dr. T. Kuroki, University of Tokyo.

### *DNA extraction and purification*

PU induced tumor clones 2,5,6 from A31-1-1 cell line were cultured in 10%FCS-MEM. Genomic DNA from clones were isolated as follows: Cells were lysed by the addition of lysis solution and the proteins and lipids were separated from the DNA by precipitation method. The DNA was removed and precipitated with ethanol for purification following the conventional method.

### *Polymerase chain reaction (PCR) for amplication of cx43.*

Cx43 gene which is comprised of two exons, exon 1 and exon 2, separated by a large segment of non-transcribed intron region. The upstream region, exon 1 contains the regulatory messages while the full open reading frame (ORF) lies in the exon 2. A processed cx43 pseudogene, lacking the intronic sequences, has been detected in human cells, as such while analyzing the human cx43 gene for genetic alterations extra steps has to be taken to account this fact. However, such steps are unnecessary for mouse cx43, since they do not have a pseudo copy of cx43 gene.

A 1.3 kb segment of coding region of cx 43 was amplified from the above three clone with specific primers: a sense 5'-ATCACTGAGTTGTTTTCGTGG and an antisense 5'-ACGAAATGAACACCCCAGGTG. A 30 cycle polymerase chain reaction was performed in a Astec Thermal Cycler using pfu DNA polymerase (Stratagene) . The amplified DNA was purified and DNA concentration was measured with Gene Quant II (Pharmacia Biotech).

### *Automated DNA sequencing of cx43*

DNA sequencing was performed with the Thermosequanase kit and in a automated DNA sequencer (Shimadzu). FITC labeled cx43 specific oligomers were prepared and reaction was performed in both directions.

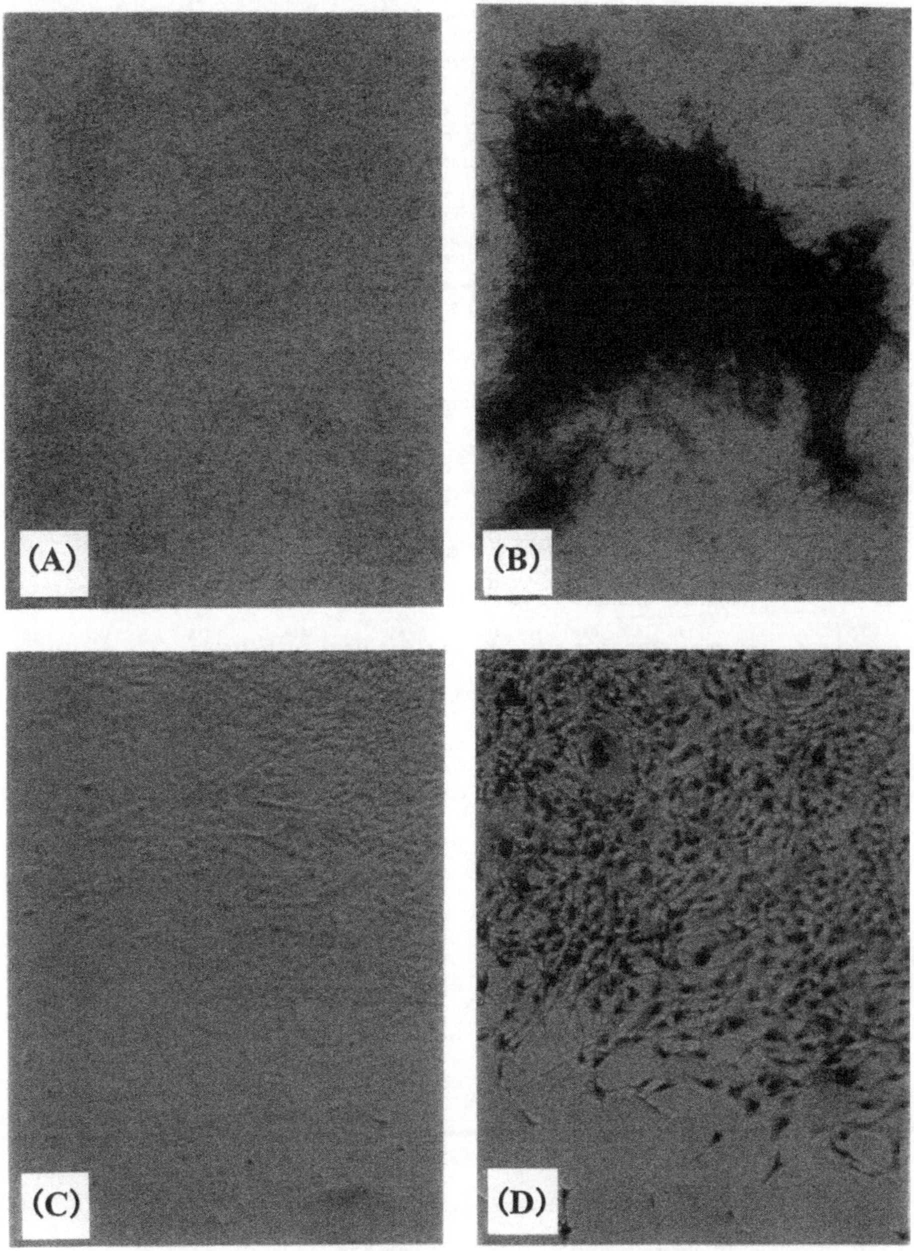


**Figure 1. Appearance of mouse cells cultured on the control glass (A), PU (B,C)** and isolated clone (D). (A) Giemsa stained control cultures of A31-1-1 cells on glass dish (magnification x30), (B) Giemsa stained transformed focus on the PU surface (x30), (C) Photomicrography of transformed cells on the PU surface (x100), (D) Giemsa stained clone isolated from the focus transformed by PU (x100).

```
        10         20         30         40         50         60
AGAAACTATA TATGTGAAAC CATCAATTTA CAGTCTACAA TCACTGAGTT GTTTTCGTGG

        70         80         90        100        110        120
TTTTTGTTTT TGTTTTTCTT AGGAGGTGCC CAGACATGGG TGACTGGAGC GCCTTGGGGA

       130        140        150        160        170        180
AGCTGCTGGA CAAGGTCCAA GCCTACTCCA CGGCCGGAGG GAAGGTGTGG CTGTCGGTGC

       190        200        210        220        230        240
TCTTCATTTT CAGAATCCTG CTCCTGGGGA CAGCGGTTGA GTCAGCTTGG GGTGATGAAC

       250        260        270        280        290        300
AGTCTGCCTT TCGCTGTAAC ACTCAACAAC CCGGTTGTGA AAATGTCTGC TATGACAAGT

       310        320        330        340        350        360
CCTTCCCCAT CTCTCACGTG CGCTTCTGGG TCCTTCAGAT CATATTCGTG TCTGTGCCCA

       370        380        390        400        410        420
CACTCCTGTA CTTGGCTCAC GTGTTCTATG TGATGAGAAA GGAAGAGAAG CTGAACAAGA

       430        440        450        460        470        480
AAGAAGAGGA GCTCAAAGTG GCGCAGACCG_ACGGGGTCAA CGTGGAGATG CACCTGAAGC

       490        500        510        520        530        540
AGATTGAAAT CAAGAAGTTC AAGTATGGGA TTGAAGAACA CGGCAAGGTG AAGATGAGAG

       550        560        570        580        590        600
GTGGCCTGCT GAGAACCTAC ATCATCAGCA TCCTCTTCAA GTCTGTCTTC GAGGTGGCCT

       610        620        630        640        650        660
TCCTGCTGAT CCAGTGGTAC ATCTATGGGT TCAGCCTGAG TGCGGTCTAC ACCTGCAAGA

       670        680        690        700        710        720
GAGATCCCTG CCCCCACCAG GTGGACTGCT TCCTCTCACG TCCCACGGAG AAAACCATCT

       730        740        750        760        770        780
TCATCATCTT CATGCTGGTG GTGTCCTTGG TGTCTCTCGC TCTGAATATC ATTGAGCTCT

       790        800        810        820        830        840
TCTATGTCTT CTTCAAGGGC GTTAAGGATC GCGTGAAGGG AAGAAGCGAT CCTTACCACG

       850        860        870        880        890        900
CCACCACCGG CCCACTGAGC CCATCCAAAG ACTGCGGATC TCCAAAATAT GCTTACTTCA

       910        920        930        940        950        960
ATGGCTGCTC CTCACCAACG GCCCCACTCT CACCTATGTC TCCTCCTGGG TACAAGCTGG

       970        980        990       1000       1010       1020
TCACTGGTGA CAGAAACAAT TCCTCCTGCC GCAATTACAA CAAGCAAGCC AGCGAGCAAA

      1030       1040       1050       1060       1070       1080
ACTGGGCGAA TTACAGCGCA GAGCAAAATC GAATGGGGCA GGCCGGAAGC ACCATCTCCA

      1090       1100       1110       1120       1130       1140
ACTCCCACGC CCAGCCGTTT GATTTCCCTG ACGACAGCCA AAATGCCAAA AAAGTTGCTG

      1150       1160       1170       1180       1190       1200
CTGGACACGA ACTCCAGCCC TTAGCTATCG TGGATCAGCG ACCTTCCAGC AGAGCCAGCA

      1210       1220       1230       1240       1250       1260
GCCGCGCCAG CAGCAGACCT CGGCCTGATG ACCTGGAGAT TTAAACAGGC TTGAACATCA

      1270       1280       1290       1300       1310       1320
AGCTGCCAAT CGATTGTGGA GGAGAAAAAA AAGGGTGCTT GCAGAACGTG CACCTGGGGT

      1330       1340       1350       1360       1370       1380
GTTCATTTCG TTCCCGTGGA GGTGGTACTC AACAACCTCA GTAATGAGGC GTAGAAAACA
```

Figure 2. DNA sequence of mouse connexin 43 in isolated clones. Arrows indicate the full length of coding region of cx43 gene, and DNA sequencing of three isolated clones was done.

## Results and Discussion

Balb 3T3 clone A31-1-1 cells were cultured in 10%FCS-MEM following the transformation assay [1] and transformed foci were observed on the material surface of PU (Figure 1(B) (C)). Three transformed clones were isolated from the transformed foci (Figure 1(D)). Using these isolated clones, DNA sequencing of the full length of coding region of cx43 gene was performed (Figure 2). No base alterations with the published sequence of cx43 gene were observed in cx43 gene from any of the three clones examined (Figure 2). However, the presented study was limited only with *in vitro* tumor clones. Similar examination with *in vivo* tumor clones would be necessary before making a conclusion that PU induced tumorigenesis mechanism does not involve mutation in cx43 coding sequence. Apart from the other possibilities like post-translational modification, we are also interested to examine the pattern of this protein expression in presence of PU. We will also examine the tumor suppression ability of cx43 by overexpressing the protein through genetic manipulation.

## References

1. Tsuchiya, T., Nakaoka, R., Degawa, H., and Nakamura, A.: Studies on the mechanisms of tumorigenesis induced by polyetherurethanes in rats: lechable and biodegradable oligomers involving the diphenyl carbamate structure acted as an initiator on the transformation of Balb 3T3 cells, *J. Biomed. Mater. Res.* **31** (1996), 299-303.
2. Nakamura, A., Kawasaki, Y., Takada, K., Aida, Y., Kurokawa, Y., Kojima, S., Shintani, H., Matsui, M., Nohmi, T., Matsuoka, A., Sofuni, T., Kurihara, M., Miyata, N., Uchima,T., and Fujimaki, M.: Difference in tumor incidence and other tissue responses to polyetherurethanes and polydimethylsiloxane in long-term subcutaneous implantation into rats, *J. Biomed. Mater. Res.* **26** (1992), 631-650.
3. Yamaski, H.: Aberrant expression and function of gap junctions during carcinogenesis, *Environ. Health Perspect.* **93** (1991), 191-197.

# XENOTRANSPLANTATION OF CULTURED NEWBORN PIG THYROID TISSUE FOR THE TREATMENT OF POST-RADIOIODINE HYPOTHYROIDISM IN RATS

I.P. PASTEUR, M.D. TRONKO, I.I. DROZDOVICH,
L.M. VOITENKO, S.F. DONICH

*Institute of Endocrinology and Metabolism, Acad. Med. Sci. Ukraine,
Vyshgorodska Street 69, 254114 Kyiv, Ukraine*

## 1. Introduction

Persistent hypothyroidism resulting from radioiodine or X-rays irradiation of the thyroid gland requires an additional treatment with L-Thyroxine for a long period of time, and even for life [1]. Hormonal therapy using transplantation of thyroid tissue may deliver such patients from a pharmacotherapy [1-3]. Reports already exist on clinical use of autotransplantation of cryopreserved thyroid gland in patients with postoperative hypothyroidism [1-3]. However, all other forms of hypothyroidism exclude the possibility of autotransplantation of thyroid gland.

Moreover, transplantation treatment today faces a big and growing problem: shortage of donor materials [4]. Xenotransplantation (transmit tissues and organs from animals) combined with modern immunosuppressive therapy is a very useful way out of this situation [4]. Porcine tissue is considered as one of the most suitable donor material, and a preliminary organ culturing allows to significantly decrease the immunogenic properties of transplantation material [5].

The aim of the present study was the development of newborn pig thyroid organ culture transplantation technique for reestablishment of thyroid function in Wistar rats with post-radioiodine hypothyroidism.

## 2. Materials and Methods

Thyroid glands were surgically removed from male and female newborn pig under ether anesthesia. Freshly obtained thyroid glands were freed of the fibrous capsule and the connective tissue. Then thyroids were cut into small pieces (<1 $mm^3$), washed and cultured in medium 199 containing a 10% calf serum ("Sigma Chemical Co", USA) and antibiotics (100 IU/ml of penicillin and 100 μg of streptomycin) at 37 C.

Each male Wistar rat (weight of rats: 100-150 g) received the dose of 838 kBq of 131-iodine by intraperitoneal injection. The xenotransplantation of 3-day old newborn pig thyroid organ culture was performed on day 18 after radioactive ablation by injection into the fat tissue of anterior abdominal wall. Total serum thyroxine (TT4) and triiodothyronine (TT3) levels were determined by radioimmunoassay. Radioactivity present in the thyroid grafts and the unoperated fat tissue were counted in a gamma counter. Ratios of >4.0 were considered as positive for the thyroid transplant survival. For light microscopy, thyroid organ culture, thyroid glands and thyroid grafts were examined histologically by the hematoxylin and eosin method.

Statistical processing of experimental results was performed using methods of analysis of variance, and $P<0.05$ was taken as significant.

## 3. Results

The investigations carried out by the authors showed that the serum TT4 and TT3 values in all rats fell to low levels (26.06±2.78 nmol/L and 1.08±0.13 nmol/L, respectively) by 2.5 weeks after radioiodine administration (control rats group: 56.00±3.10 nmol/L and 1.72±0.17 nmol/L, respectively), they remained significantly ($P<0.05$) hypothyroid with respect to TT4 and TT3 for the 3.5-week duration of the study (Figure 1).

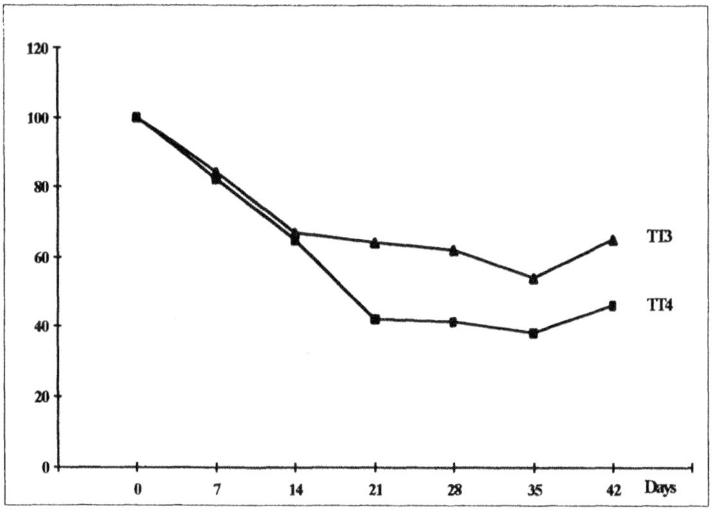


Figure 1. Changes of thyroid function in rats after radioiodine administration

The signs of hypertrophy and sharp vacuolization of thyrocytes expressed by the disintegration of epithelial layer in rat thyroid gland in 20 days following iodine-131 injection were revealed by the histological examination. Spreading of connective tissue was noted too (Figure 2). In 30 days the dystrophic changes make progress and sometimes a paranchimal part of the gland is presented by the distructive zones.

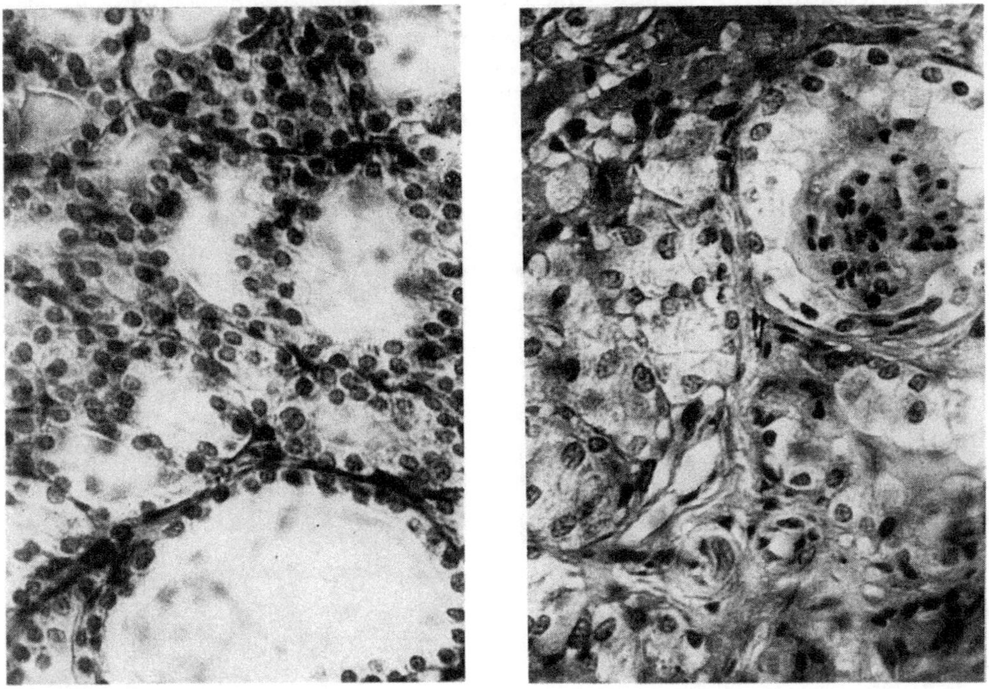

FIGURE 2. Thyroid follicle structure of non-irradiated (left) and irradiated (right) rat thyroid gland. Hematoxylin-eosin stain. × 720

The values of serum TT4 in the rats were generally in the euthyroid range by day 17 after xenotransplantation (Table 1). The serum TT3 values rose rapidly and reached normal levels by day 7 after xenotransplantation.

TABLE 1. Effect of xenotransplantation of newborn pig thyroid organ culture on the restoration of thyroid function in rats with radioiodine-induced hypothyroidism

| Group | Serum TT4 (nmol/L) | Serum TT3 (nmol/L) |
|---|---|---|
| Control | 56.00 ± 3.10 | 1.72 ± 0.17 |
| Radioiodine | 26.06 ± 2.78 * | 1.08 ± 0.13 * |
| xenotransplantation: 7th day | 40.84 ± 9.75 ** | 1.76 ± 0.16 ** |
| xenotransplantation: 17th day | 51.74 ± 3.87 ** | - |

Note: Values shown are the mean ± SD. * $P<0.05$ (compared to control); ** $P<0.05$ (compared to radioiodine).

7 days after xenotransplantation, all the newborn pig thyroid xenografts showed 131-iodine uptake, and by day 17 four of five xenografts were still functional. There is insignificant variation in the degree of radioiodine uptake by functioning newborn pig thyroid transplants.

The epithelial cell swarms with follicular formation were manifested among adipose tissue in 7 day as well as 17 day xenotransplant (Figure 3, left). Follicules and follicle-liked structures formed with impressed thyrocytes are shown at dedline of the examination. Sometimes rather numerous disquamative nuclei are observed in follicular cavity.

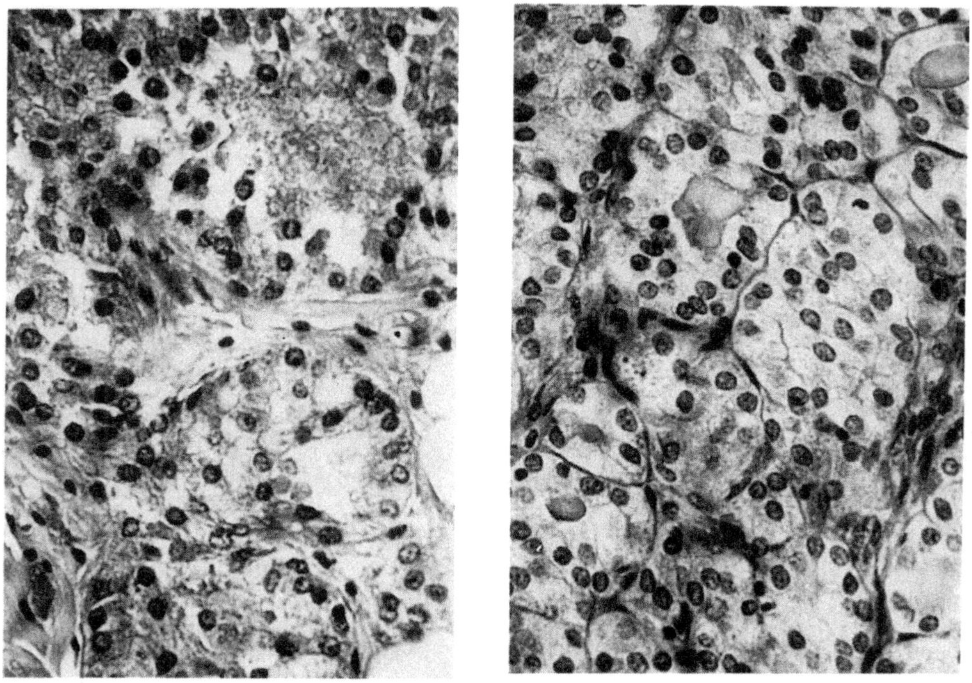

FIGURE 3. Thyroid graft (left) and thyroid gland (right) of irradiated rat after newborn pig thyroid organ culture xenotransplantation. Hematoxylin-eosin stain. × 720

Morphological signs of increased functional activity of thyrocytes that is evidence of the possible reparative processes in consequence of transplantation are observed in thyroid gland of preliminary irradiated rats in the mentioned terms after newborn pig thyroid organ culture transplantation comparing to suitable control. So, the prevalence of follicles with relatively small diameter and rather high epithelial layer (Figure 3, right) was noted in the gland parenchyma. Signs of follicle forming are determined.

## 4. Discussion

Thyroid transplantation requires a functioning thyroid follicle as a minimum unity ensuring thyroid function [3]. For synthesis and secretion of hormones, thyroid tissue of a maximum size of pieces up to 1 $mm^3$ (such a size prevents central necrosis in each tissue sample) is quite sufficient and it guarantees good results in transplantation of thyroid tissue [6,7]. Moreover, thyroid tissue transplantation instead of thyrocyte transplantation is a more simple process from technical point of view [1,3].

When using results of such investigations in transplantology, it should be taken into account that the degree of hormonal compensation of hypothyroidism and real amount of radioiodine absorbed by a thyroid graft depend on its functional activity at the moment of study, on the amount of transplanted tissue and level of thyrotropic hormone which circulates in the blood of the recipient [5]. Though functioning of an additional source of thyroid hormones is recognized as the main mechanism of therapeutic effect of transplantation in case of hypothyroidism [1-3], one may not exclude a possible stimulating effect on the recipient's thyroid gland [2]. So, a study of iodine-accumulating function of a thyroid gland stump in patients with postoperative hypothyroidism showed that, after autoimplantation of cryopreserved thyroid parenchyma, a progressive increase of the capacity for iodine accumulation was observed [2].

Therefore, the newborn pig thyroid organ culture may be used in transplantology for the treatment of persistent hypothyroidism. The major finding in this study was the effect of xenotransplantation on the restoration of thyroid function and morphological features of thyroid gland in rats with postradioiodine hypothyroidism. In recipients of the newborn pig thyroid organ culture euthyroid levels of serum TT4 and TT3 were approached within 7-17 days. The accumulation of 131-iodine at the transplantation site further support the fact that the transplanted newborn pig thyroid tissue had regain its function, suggesting successful experimental application. It should be noted that the thyroid glang of rats reverted to normal in morphofunctional appearance, and resumed a normal thyroid system relationship after xenografting.

**5. References**


1. Shimizu, K., Nagahama, M., Kitamura, Y., et al.: Improvement of thyroid function after auto-transplantation of cryopreserved thyroid tissues in rats: clinical application of the procedure to patients with persistent hypothyroid Graves' disease after thyroidectomy, *Thyroidol. Clin. Exp.*, **8** (1996), 55-62.
2. Puchkar, N.S., Makedonskaya, V.A., Utevsky, A.M., et al.: Autoimplantation of cryopreserved (-196$^{0}$C) thyroid parenchyma as a method of treatment of post-operative hypothyroidism, *Problemy Endokrinologii (Moscow)*, **30** (1984), 42-46.
3. Shimizu, K., Nagahama, M., Kitamura,Y., et al.: Autotransplantation of cryopreserved thyroid tissues for the treatment of irreversible postoperative hypothyroid Graves' disease. Report of the first case. *Thyroidol. Clin. Exp.*, **9** (1997), 23-26.
4. Kemp, E.: Xenotransplantation, *J. Intern. Med.* **239** (1996), 287-297.
5. Lafferty, K.J., Cooley, M.A., Woolnough, J., and Walker, K.Z.: Thyroid allograft immunogenicity is reduced after a period in organ culture, *Science*, **188** (1975), 259-261.
6. Bauer, M.F., and Herzog, V.: Mini organ culture of thyroid tissue: a new technique for maintaining the structural and functional integrity of thyroid tissue in vitro, *Lab. Invest.*, **59** (1988), 281-291.
7. Kitamura, Y., Shimizu, K., Nagahama, M., and Shoji, T.: Cryopreservation of thyroid pieces - Optimal freezing condition and recovery, *J. Jpn. Surg. Soc.*, **95** (1994), 14-20.

# IMPROVED PRODUCTION OF RETROVIRAL VECTORS UNDER SERUM-FREE CONDITIONS FOR THE APPLICATION OF GENE THERAPY

P.A. GERIN, P.F. SEARLE, M. AL-RUBEAI
School of Chemical Engineering & Institute for Cancer Studies
University of Birmingham, Birmingham B15 2TT, England

**Abstract.** The influence of serum on the production of retroviral vectors by the HT1080 human fibrosarcoma-derived packaging cell line FLY-RD18 was investigated. A 5-fold increase in the vector titre was observed under serum-free condition, as compared to medium supplemented with 10 % foetal calf serum. The increase of vector production under serum-free condition was accompanied by a dramatic reduction of the supernatant protein content, leading to much better prospect for further vector concentration and purification. Serum had a negative and dose-dependent effect on virus titre, while it affected neither cell growth nor virus stability and infectivity. An extracellular protease activity was found to enhance the retroviral production. The inhibition of protease activity by serum protease inhibitors can explain the negative effect of serum on vector production.
**Keywords:** retrovirus, gene therapy, serum-free

## 1. Introduction

Gene therapy is a developing therapeutic technology which promises to revolutionise the practice of medicine. Nevertheless, its full-scale application is hampered by the need of large amounts of vectors at reasonable cost. Such vectors are required to ensure the transport, the insertion and the expression of "therapeutic" genetic material into the patient cells, either *in-vivo* or *ex-vivo*. At present, recombinant retroviral vectors are being employed in most clinical trials. Production of retroviral vectors is ensured by specific "packaging" cell lines, i.e. cell lines genetically modified to synthesise replication defective retroviral vectors with the adequate host range. The FLYRD18 packaging cell line was recently developed by introducing the Moloney Murine Leukaemia (MoMLv) virus *gag-pol* gene and the cat endogenous virus RD114 *env* gene into HT1080 human fibrosarcoma cells (Cosset et al., 1995).
Most animal cell lines require supplementation of their culture medium with serum. However, serum is an expensive raw material and it increases the complexity, duration and cost of downstream processing (purification, concentration, recovery of product, etc.). With respect to retroviral vectors, longer purification processes means reduced efficiency due to the short retroviral half-life. Serum also represents a major risk of introducing contaminants (proteins, hormones, viruses, prions, etc.). The safety of gene therapy will require the development of serum-free production processes.

Our aim was to investigate the feasibility of producing retroviral vectors with the HT1080/FLY-RD18 packaging cell line under serum free conditions. Unexpected preliminary results led us to analyse further the influence of serum on the production of retroviral vectors by this cell line.

## 2. Materials & Methods

*Cell lines.* Two packaging cell clones were derived from the HT1080/FLYRD18 cell line (Cosset et al., 1995) by transduction with pLNCX-based vectors (Miller and Rosman, 1989) containing either the human B7/CD80 gene (Linsley and Ledbetter, 1993) or the *E. coli* nitroreductase gene (Green et al., 1997; McNeish et al., 1998). These clones will further be referred to as FLYRD18/LNC-*hB7* and FLYRD18/LNC-*nr*, respectively. A third packaging cell clone, GP+envAM12/LNC-*ntr*, was derived from the murine NIH3T3/GP+envAM12 cell line by transfection with the pLNCX-based vector containing the nitroreductase gene. The A2780cp human ovarian carcinoma cell line was used as target cell line for assessing virus titre.

*Production of retroviral vectors.* Packaging cells were inoculated at a density of $2.0 \times 10^4$ cells/cm$^2$ in 25 cm$^2$ T-flasks and grown for 30 hours at 37°C in 10 ml BM supplemented with 10% FCS (BM+10%FCS). The production phase was then initiated by totally replacing the used medium with fresh medium, the composition of which varied according to the experiment. The production medium was collected after overnight incubation (15-17 hours) at 37°C and the retroviral vectors were separated from cells and cell debris by filtration through 0.45µm pore size filters.

*Infectivity assay.* Titration of retroviral vectors was performed by infecting A2780cp cells inoculated at a density of $1.0 \times 10^4$ cells/cm$^2$ in 6-well plates and incubated overnight in 2 ml BM + 10% FCS. 1.0 ml of virus supernatant diluted 1:10$^2$ to 1:10$^4$ in BM containing 8 µg/ml polybrene was added to each well. After 6 to 10 hours incubation, the medium was diluted with 2 ml BM + 10% FCS and selection operated with 750 µg/ml G418. The number of colonies was counted 11 days post-infection.

## 3. Results & Discussion

### 3.1. INFLUENCE OF SERUM ON VECTOR PRODUCTION

In preliminary experiments, a 3 to 6 fold increase in the vector titre produced by the HT1080/FLYRD18/LNC-*hB7* packaging cell line was consistently observed in basal medium (BM), as compared to BM supplemented with 10% foetal calf serum (FCS) (data not shown). The influence of serum on vector production was further investigated at intermediate serum concentrations. Figure 1a presents the vector titre vs. serum concentration during the vector production phase. Whatever the source of

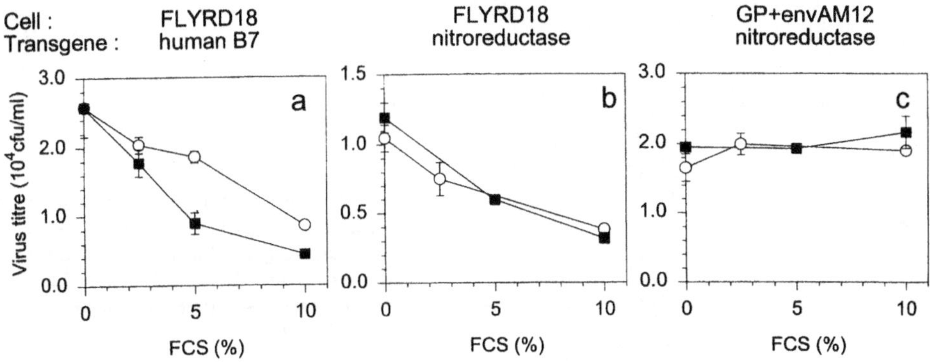


*Figure 1.* Influence of the concentration of foetal bovine serum (FCS) on the retroviral vector production by HT1080/FLYRD18 (a,b) and NIH3T3/GP+envAM12 (c) packaging cell lines producing retroviral vectors containing either the human B7 transgene (a) or the nitroreductase transgene (b,c). Different symbols refer to different FCS sources (a) or duplicate experiments (b,c).

FCS, Figure 1a shows that serum had a negative and dose-dependent effect on vector production, while not on cell growth.

To check for the generality of our result, two other packaging cell lines were tested: a) the FLYRD18/LNC-*nr* line, which is essentially the same as the FLYRD18/LNC-*hB7* line, except than the gene inserted into the vector codes for the nitroreductase enzyme instead of the cd80 surface protein; b) the NIH3T3/GP+envAM12/LNC-*ntr* murine packaging cell line producing vectors containing the same nitroreductase gene as the FLYRD18/LNC-*nr*. Figure 1b show that serum had also a negative and quantitative effect on vector production for the HT1080 derived cell line, while not for the NIH3T3 derived cell line (Figure 1c). There was no negative effect of serum on cell growth. The negative effect of serum on vector production appears to be specific for the HT1080 derived packaging cell line, and does not depend on the gene inserted in the vector. It also affects specifically the virus titre independently of the cell growth.

## 3.2. INFLUENCE OF SERUM ON VECTOR STABILITY

A serum-free vector suspension was supplemented with 10% v/v either serum or BM and incubated at 37°C for increasing periods of time. Figure 2 shows that the vector half-life was longer than 8 hours in the presence as well as in the absence of serum. Figure 2 also indicates that serum does not affect the vector stability in suspension. Vector inactivation by serum can not be regarded as a route to explain the serum negative effect on titre detected in Figure 1.

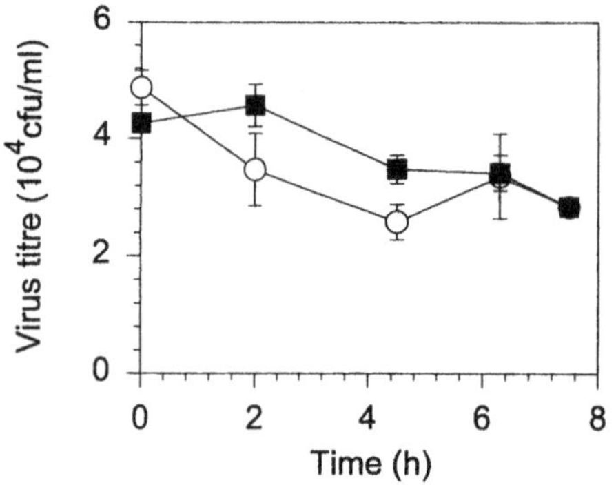


*Figure 2.* Influence of the presence of serum on the inactivation kinetics of FLYRD18-derived rvLNC-*hB7* vectors. Virus particles produced in BM were incubated at 37°C after addition of 10% v/v either BM (open symbols) or FCS (closed symbols).

## 3.3. IMPORTANCE OF EXTRACELLULAR PROTEASE ACTIVITY

To investigate further the nature of serum factors that inhibit vector production, BM+10% FCS was incubated with bovine trypsin before being used as a production medium. Figure 3a shows that trypsin pre-treatment of serum enabled the production of higher vector titre in a dose dependent way. In this experiment, the proteolytic enzyme trypsin was expected to degrade some serum proteins and to inactivate serum protease inhibitors ($\alpha$1-anti-trypsin and $\alpha$2-macroglobulin). The higher vector titre obtained after trypsin treatment suggests that the inhibitor present in serum is a protein or a protease inhibitor.

To refine this result, vectors were produced in BM supplemented with soybean trypsin inhibitor. Figure 3b shows that the vector titre decreased with increasing concentration of soybean trypsin inhibitor. This effect is very similar to the effect of serum (Figure 1). Figure 3 suggests that a protease activity is required for the FLYRD18 to produce infectious retroviral vectors.

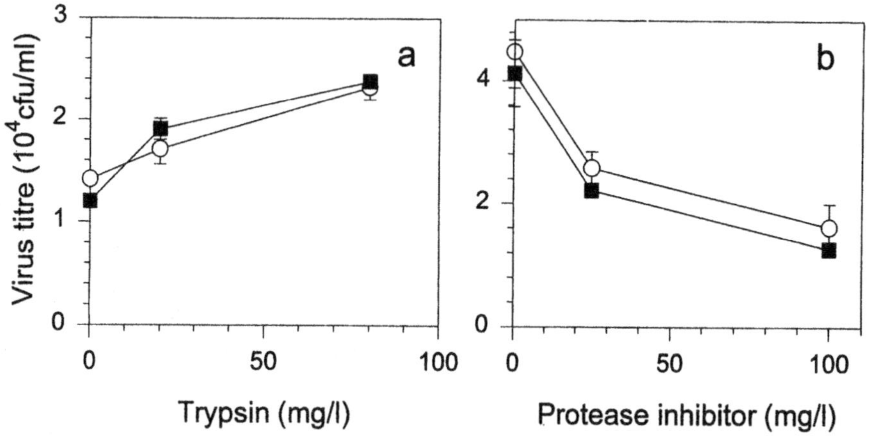


*Figure 3.* Influence of protease inhibition on the production of virus by the FLYRD18/LNC-*hB7* cell line: (a). BM+10% FCS was incubated with trypsin before being used for the virus production; (b) Soybean trypsin inhibitor was added to BM before this medium was used for the virus production.

## 4. Conclusion

The results presented in this study show that serum has a strong negative and dose-dependent influence on the production of retroviral vectors by the HT1080/ FLYRD18 packaging cell line. This negative influence could not be explained by an effect on cell general metabolism. Virus infectivity and stability were not affected by serum. An extracellular protease activity was found to enhance retroviral vector production. The serum negative influence can then be due to inhibition of the protease activity by the serum protease inhibitors. From a technological point of view, the presented results indicate that a vector production process is feasible under serum-free conditions and lead to a five-fold increase in both virus titre and total production. Such process would make further virus concentration and purification (downstream processing) much easier.

## 5. Acknowledgement

P. Gerin is holding a grant funded by the European Community under the BIOTECH programme (Fourth Framework Programme).

## 6. References

Cosset F.L., Takeuchi Y., Battini J.L., Weiss R.A., Collins M.K.L. (1995). High-titer packaging cells producing recombinant retroviruses resistant to human serum. J. Virol. **69**, 7430-7436.

Green, N.K., Youngs, D.J., Neoptolemos, J.P., Friedlos, F., Knox, R.J., Springer, C.J., Anlezark, G.M., Michael, N.P., Melton, R.G., Ford, M.J., Young, L.S., Kerr, D.J., Searle, P.F. (1997). Sensitization of colorectal and pancreatic cancer cell lines to the prodrug 5-(aziridin-1-yl)-2,4-dinitrobenzamide (CB1954) by retroviral transduction and expression of the *E.coli* nitroreductase gene. Cancer Gene Therapy **4**, 229-238.

Linsley, P.S., Ledbetter, J.A. (1993). The role of the CD28 receptor during T cell responses to antigen. Ann. Rev. Immunol. **11**, 191-212.

McNeish, I., Green, N., Gilligan, M., Ford, M., Mautner, V., Young, L., Kerr, D., Searle, P.F. (1998). Virus Directed Enzyme Prodrug Therapy for Ovarian and Pancreatic Cancer using Retrovirally Delivered E.coli Nitroreductase and CB1954. Gene Therapy (in press).

Miller, A.D., Rosman, G.J. (1989). Improved retroviral vectors for gene transfer and expression. Biotechniques 7, 980-988.

# LARGE SCALE PRODUCTION OF RH-ALBUMIN EXPRESSED IN THE MILK OF TRANSGENIC CATTLE - AN ECONOMIC AND TECHNICAL CHALLENGE

Wolfram Eichner[1] and Klaus Sommermeyer[1]

[1]CPFE, Pharma Division, Fresenius AG, 61343 Bad Homburg, Germany

**Albumin from plasma fractionation is generally considered to be a safe and effective product. Nevertheless, a growing number of factors makes recombinant albumin manufacture increasingly attractive, i.e. the perceived risk of viral contamination in serum albumin, likelihood of reduced availability of recovered plasma in the future, and price. As a result of production technology evaluation, the expression of human albumin in the milk of transgenic cows has been identified as the most promising approach. A major challenge for this project is the fact that the production of pharmaceutical proteins on the scale of many tons, essentially free of any impurities at the price level of a commodity has not been achieved in recombinant biotechnology so far. In collaboration with Genzyme Transgenics Corp. (Framingham) project activities were initiated in February 1997.**

The naturally occuring plasma colloid albumin is the most abundant soluble protein in the body of vertebrates and human serum albumin (hSA) is probably the most studied of all proteins. Development of plasma fractionation technology during 1942 was the basis for the production of tons of highly purified albumin for battlefield use. Albumin is used as a plasma volume expander and in albumin deficiency situations, such as trauma, surgery, shock or burns. The function of albumin is not limited to the maintenance of isooncotic pressure (blood vs. tissue). It also serves as a carrier of smaller molecules of many types, such as fatty acids, hormones, and ions as well as an antioxidant (1).

Nowadays the annual worldwide albumin market is approx. 440 t and this market is expected to climb to 550 t/yr. by the year 2000. However, it is believed that a number of different factors will contribute to limited hSA availability in the future:

- General shortages of human blood as a result of the reduced willingness of people to donate blood.
- Changes in transfusion practices due to the availability of new oxygen therapeutics in the future, accompanied by general safety concerns with products derived from blood and/or plasma sources.
- An increasing number of blood factors available from recombinant manufacture.

The last point is particularly important for plasma fractionating companies, because it is expected that the availability of recombinant blood factors will result in significant losses of revenue from plasma products. Production of hSA by means of recombinant technology requires extremely low production and downstream processing costs in order to be price-competitive with the plasma product. Recombinant human albumin was first cloned, expressed (in yeasts) and patented by Genentech in 1981 (2). Human albumin derived from recombinant production will differ from the plasma counterpart in terms of quality (essentially no risk/TSE transmission) and purity (no trace amounts of impurities such as prekallikrein activator). Today, at least two companies (Centeon/Delta Biotechnology Ltd., UK and Green Cross Corp., Japan) have successfully developed large-scale production processes for recombinant hSA from yeasts. These products are in advanced clinical trials. Market launch of the Green Cross product (Albrec) is expected in the year 2000.

Protein-expression in the milk of transgenic animals offers another attractive approach to producing recombinant human albumin at very low costs. Due to their high annual milk yield, cows are considered to be the most suitable 'producers'. An individual dairy cow will produce approximately 8,000 liters of milk per annum, or an estimated 80 Kilograms of albumin per year. Using transgenic cows as naturally occuring 'bioreactors' would enable the production of hSA in the required range of many tons. At Genzyme Transgenics (GTC, Framingham, USA) the general feasibility of transgenic albumin expression was successfully evaluated in mice. Expression levels in mice are known to be predictive of the expression level in larger animals.

Nevertheless, production of proteins using the mammary gland carries several risks. The most obvious technical problem is to efficiently remove the endogenous (bovine) serum albumin, which is fairly similar to the human counterpart (76% homology). During a purification feasibility program initiated by GTC, a proprietary process for the purification of hSA at competitive costs, essentially free of bSA, was successfully developed. However, public acceptance, regulatory requirements and economic aspects (e.g. price development) are considered to be other potential risk factors for transgenic albumin production.

Since the sheep 'Dolly' (3) last year, cloning as a specific aspect of transgenic technology has received considerable public attention. Compared to 'standard' microinjection techniques, this technology opens up new possibilities for more sophisticated genetic manipulation. Therefore, it has been decided to incorporate this technology into the albumin program. A general agreement for the production of human therapeutic proteins in the milk of cloned transgenic animals between GTC and Advanced Cell Technology, Inc. (ACT) was reached in October, 1997. In January 1998, the birth of two calves (George/Charlie), created by combining cloning techniques with genetic engineering was announced by ACT. Birth of cloned bovines transgenic for human serum albumin is expected by late 1998.

Despite these rapid advances, the large scale production of recombinant human albumin in transgenic cattle still lies somewhat in the future.

**Literature:**


1. Peters, 1996, All about Albumin, Academic Press Ltd., ISBN 0-12552110-3
2. Jones, 1996, J. Americ. Blood Res. Assoc. **4:108-110**
3. Wilmut et al., 1996, *Nature* **380:64-66**

# TRANSGENIC RABBITS: A NOVEL MODEL FOR THE STUDY OF ATHEROSCLEROSIS

JIANGLIN FAN*, MASAHIRO ARAKI*, LIHUA WU*, MIREILLE CHALLAH*, HIROAKI SHIMOYAMADA*,
RICHARD M. LAWN†, HIROTOSHI KAKUTA§, HISATAKA SHIKAMA§ AND TERUO WATANABE*.
**Department of Pathology, Institute of Basic Medical Sciences, University of Tsukuba, Tsukuba 305-8575, Japan*
*†Falk Cardiovascular Research Center, Stanford University, Stanford, California 94305*
*§Tsukuba Research Center of Yamanouchi Pharmaceutical Company, Tsukuba, 305, Japan*

## 1. INTRODUCTION

The advent of transgenic technology, in which foreign genes are stably introduced into the mammalian germ line, has dramatically enhanced our basic understanding of many physiological and pathologic processes. Transgenic technology has allowed the rapid elucidation of many fundamental principles of the tissue-specific expression of certain genes and is also particularly valuable in studies of human diseases, such as hypercholesterolemia and atherosclerosis[1]. Most studies of lipoprotein metabolism and atherosclerosis are currently conducted with transgenic mice because mice are small, easily-handled and easy to breed[1]. However, these animals have a major drawback in that the lipoprotein profile of mice (HDL-rich) is different from that of humans. For example, mice lack cholesteryl ester transfer protein (CETP) and have HL that circulates in the plasma[2]. In addition, the rabbit liver does not edit apoB100 mRNA, whereas the mouse liver edits most of the apoB100 mRNA into the B48 form [3]. Unlike mice, in which HDL is the major carrier for cholesterol, rabbits are LDL-rich mammals, as are humans and have high levels of CETP[4]. Therefore, rabbits provide an alternative model for investigating atherosclerosis and lipoprotein metabolism[5]. Additional advantages of using rabbits for the study of atherosclerosis include that rabbits are extraordinarily sensitive to a cholesterol diet, which induces severe hypercholesterolemia associated with the rapid development of atherosclerosis[5]. The availability of rabbit strains with genetic defects in the LDL receptor (WHHL) [6] and the overproduction of LDL (St. Thomas rabbits )[7, 8] has made rabbits a unique model for the study of atherosclerosis. It is known that atherosclerosis is initiated by elevated plasma atherogenic lipoproteins, including low density lipoproteins(LDL), remnant lipoproteins (such as β-VLDL) and lipoprotein (a) [Lp(a)]. To study the functions of lipoprotein (a) [Lp(a)] in lipid and lipoprotein metabolism and the relationship between Lp(a) and atherosclerosis, we generated a transgenic rabbit line expressing human apolipoprotein (a)[apo(a)].

## 2. HUMAN APOLIPOPROTEIN (a) TRANSGENIC RABBITS

Transgenic rabbits were generated by the method as described previously[2, 4, 9, 10]. A total of 1,450 rabbit zygotes were obtained from Japanese white rabbits and microinjected with a 12.8-kb NotI-HpaI fragment of DNA containing human apo(a) cDNA directed by a mouse transferrin promoter[11] and implanted into 41 recipients. Forty-three pups were obtained, of which 5 pups were found to have apo(a) transgene expression by Southern blot analysis. Human apo(a) mRNA distribution in different tissues was investigated by Northern blot analysis. Total RNA was isolated from 15 tissues of the transgenic rabbits including the liver, spleen, stomach, heart, intestine, kidney, adrenal, adipose, muscle, aorta, thymus, testis, lung, brain, and bone marrow using Trizol reagent (Gibco BRL, Life Technology, Frederick, MD). For Northern blot analysis, 10 mg of denatured total RNA was subjected to electrophoresis in a 1.2% agarose gel and transferred to a Nitran nylon membrane and hybridized with the $^{32}P$-labeled human apo(a) cDNA probe as above. We found that human apo(a) mRNA was expressed mainly in the liver, kidney, lung and the brain. This pattern of apo(a) expression in transgenic rabbits was similar to that of transgenic mice[11]. ELISA with a specific antihuman apo(a) monoclonal antibody revealed that one of the transgenic rabbit founders had 0.5 mg/dl plasma level of human apo(a), which is equivalent to low plasma levels of apo(a) in the human population. By agarose gel electrophoresis, we found that apo(a) in transgenic rabbit plasma was located at the pre-β position, which is the same position as in human Lp(a). As nearly all apo(a) binds to low density lipoproteins (LDL) to form Lp(a) in humans, we examined whether human apo(a) in transgenic rabbits could bind rabbit apoB to form human-like Lp(a), or if it simply circulates as free apo(a) in the plasma. By gradient density ultracentrifugation of the plasma and gel electrophoresis, we found that 80% of the apo(a) in the transgenic rabbit plasma was contained in the density fractions 1.02-1.10 g/ml and covalently associated with rabbit apo-B, suggesting the formation of Lp(a) in the transgenic rabbits. To examine the potential interaction between apo(a) and the arterial wall, the rabbits were fed a cholesterol-rich diet for 16 weeks. We found that apo(a) depositions were present along the intima and atherosclerotic lesions in the aorta and the coronary arteries of the transgenic rabbits but not in the control rabbits. Thus, these findings demonstrate the successful expression of human apo(a) and the efficient assembly of Lp(a) in transgenic rabbits.

## 3. DISCUSSION

Atherosclerosis and its complications are the major cause of mortality in Western society. To study its pathogenesis and also to find a therapeutic way to treat this disease depends upon the development of appropriate animal models. Techniques for generating genetically modified animals have increased our understanding of atherosclerosis. In this respect, we attempted to generate novel transgenic rabbits expressing different human genes related with lipid metabolism and atherosclerosis[2, 9, 10]. In this report, we described the development of transgenic rabbits expressing human apolipoprotein (a). Our preliminary studies showed that this transgenic rabbit model will provide a value model for the study of lipoprotein (a) assembly and its relationship with atherosclerosis. Eighty percent of the apo(a) in the transgenic rabbit plasma was associated with rabbit apoB to form human-like Lp(a) particles. The demonstration of apo(a) in atherosclerotic lesions of transgenic rabbits indicates that apo(a) may participate in lesion development. Further studies are necessary to generate higher expressors of transgenic rabbits to investigate the possible role(s) of Lp(a) in the development of atherosclerosis.

## 4. ACKNOWLEDGMENTS

This work was supported by Grants-in-Aid for scientific research from the Ministry of Education, Science, and Culture of Japan, Ono Medical Foundation, Japan, Uehara Memorial Foundation, Japan, Japan Heart Foundation, Japan, Tokyo Biochemical Research Foundation, Ichiro Kanehara Foundation, Takeda Medical Research Foundation and Japan Society for the Promotion of Sciences (JSPS-RFTF96I00202) (JF, TW), and National Institutes of Health grants (RML). M. Challah is the recipient of a postdoctoral fellowship award from the Japan Science Promotion Society.

## 5. References:

1. Breslow, J.L. (1993) Transgenic mouse models of lipoprotein metabolism and atherosclerosis, *Proc Natl Acad Sci USA* **90,** 8314-8318

2. Fan, J., McCormick, S.P., Krauss, R.M., Taylor, S., Quan, R., Taylor, J.M. and Young, S.G. (1995) Overexpression of human apolipoprotein B-100 in transgenic rabbits results in increased levels of LDL and decreased levels of HDL, *Arterioscl Thromb Vasc Biol* **15,** 1889-99

3. Yamanaka, S., Balestra, M.E., Ferrell, L.D., Fan, J., Arnold, K.S., Taylor, S., Taylor, J.M. and Innerarity,

T.L. (1995) Apolipoprotein B mRNA-editing protein induces hepatocellular carcinoma and dysplasia in transgenic animals, *Proc Natl Acad Sci USA* **92,** 8483-7

4. Taylor, M.J. and Fan, J. (1997) Transgenic rabbit models for the study of atherosclerosis, *Front Biosci* **2,** 298-308

5. Overturf, M.L. and Loose-Mitchell, D.S. (1992) In vivo model systems: the choice of the experimental animal model for analysis of lipoproteins and atherosclerosis., *Curr Opin Lipidol* **3,** 179-185

6. Watanabe, Y. (1980) Serial inbreeding of rabbits with hereditary hyperlipidemia (WHHL-rabbit), *Atherosclerosis* **36,** 261-8

7. La Ville, A., Turner, P.R., Pittilo, R.M., Martini, S., Marenah, C.B., Rowles, P.M., Morris, G., Thomson, G.A., Woolf, N. and Lewis, B. (1987) Hereditary hyperlipidemia in the rabbit due to overproduction of lipoproteins. I. Biochemical studies, *Arteriosclerosis* **7,** 105-12

8. Seddon, A.M., Woolf, N., La Ville, A., Pittilo, R.M., Rowles, P.M., Turner, P.R. and Lewis, B. (1987) Hereditary hyperlipidemia and atherosclerosis in the rabbit due to overproduction of lipoproteins. II. Preliminary report of arterial pathology, *Arteriosclerosis* **7,** 113-24

9. Fan, J., Wang, J., Bensadoun, A., Lauer, S.J., Dang, Q., Mahley, R.W. and Taylor, J.M. (1994) Overexpression of hepatic lipase in transgenic rabbits leads to a marked reduction of plasma high density lipoproteins and intermediate density lipoproteins, *Proc Natl Acad Sci USA* **91,** 8724-8

10. Fan, J., Ji, Z.-S., Huang, Y., de Silva, H., Sanan, D., Mahley, R., Innerarity, T. and Taylor, J. (1998) Increased expression of apolioprotein E in transgenic rabbits results in reduced levels of very low density lipoproteins and an accumulation of low density lipoproteins in plasma, *J Clin Invest* **101,** 2151-2164

11. Chiesa, G., Hobbs, H.H., Koschinsky, M.L., Lawn, R.M., Maika, S.D. and Hammer, R.E. (1992) Reconstitution of lipoprotein(a) by infusion of human low density lipoprotein into transgenic mice expressing human apolipoprotein(a), *J Biol Chem* **267,** 24369-74

# *IN VIVO* GENE TRANSFER TO CHICKEN VIA BLASTODERMAL CELLS OF EARLY DEVELOPMENTAL EMBRYOS

S. INADA, M. HATTORI AND N. FUJIHARA
*Department of Animal Science, College of Agriculture, Kyushu University, Fukuoka 812-8581 Japan*

## Introduction

A complete culture sysmtem has been established for the chick embryo from the single cell stage to hatching (Perry, 1988). Sang *et al.* (1989) reported that foreign DNA was introduced into the germinal disc of fertilized ova obtained from the magnum of hen's oviduct. Naito *et al.* (1994) described that the introduction of foreign DNA into somatic and germ cells was successfully achieved by injecting genes into the germinal disc of fertilized ova at the single cell stage, and that most of the chick embryos expressed the DNA in a mosaic (Naito *et al.*, 1991). Petitte *et al.* (1990) developed somatic and germline chicken chimeras by injecting dispersed donor blastodermal cells into the compromised recipient embryos.

On the one hand, the sxpression of DNA (Miw Z as a marker gene) introduced was observed in extraembryonic tissues (Otsuka *et al.*, 1990). Recently, Tajima *et al.* (1993) suggested that primordial germ cells (PGCs) will be noted for manipulations of germlines, sexing and conservation of the genetic materials.

In the present study, we tried to improve the method of exogenous gene transfer to chick embryos by using oviposited fertilized eggs.

## Materials and Methods

### *Preparation of DNA*

The DNA (Suemori *et al.*, 1990; Inada *et al.*, 1997), Miw Z marker gene, was used in this study. The DNA (6.25 μg) was mixed with transfection reagent (TR: 22.5 μl), diluting with 50 μl of Hepes-buffered saline (Inada *et al.*, 1997).

*Microinjection of DNA*

The oviposited fertilized eggs were used as host embryos. For another experimental group, soft shelled eggs were obtained by the injection to hens of oxytocin or arginine vasotocin at 7 h after oviposition of the preceding egg.

Both freshly laid eggs (stage X : Eyal-Gilad and Kochav, 1976) and temporarily stored eggs at around 4 C (in a refrigerator) for 3 h just after oviposition were also used as host embryos. An window of 10-15 mm in diameter was opened at sharp end of the shell, and about 1.5 ml of thick albumen just above the germinal disc was removed. The DNA solution (approximately 0.5 μl) was injected into the central part of the germinal disc and the window was closed with a scotch tape, incubated for additional hours.

In case of the soft shelled eggs, the shell membranes were removedd, followed by the injection of DNA solution. As control groups, Miw Z without TR or TR only were injected into the germinal disc of the freshly oviposited eggs.

*Culture of treated eggs*

he treated eggs were incubated for 50-72 h at 37.5 C with relative humidity of 60 % to the stage 13-15 (Hamburger and Hamilton, 1951), at the stage of which PGCs begin to imgrate from the germinal crescent region to the gonadal anlage via blood stream (Kuwana, 1993).

*Detection of injected DNA*

The expression of MiwZ was detected by histochemical staining of β-galactosidase. Blood samples collected from vitelline artery or dorsal aorta were smeared on a slide glass, fixed with 1 % glutaraldehyde in phosphate-buffered saline (PBS) for 5 min at 4 C. The smears were stained for 1 h at RT (22-25 C) with 0.05% 5-bromo-4-chloro-3-indolyl-β-D-galacto- pyranoside (X-gal), 1 mM $MgCl_2$ , 0.1% Triton X-100, 3 mM potassium ferrocyanide and 3 mM potassium ferricyanide in PBS. The embryos and extraembryonic tissues were fixed with 1% glutaraldehyde for 2 min at 4 C, stained with X-gal for 2 h at 37 C.

*DNA extraction and PCR analysis*

The embryos (stage 13-14) were washed with TNM buffer (20mM Tris-HCl buffer, pH 7.5, containing 100mM NaCl and 1.5mM $MgCl_2$). The total DNA was extracted from homogenized embryos by phenol-chloroform and PCR analysis was carried out. To estimate copy number, PCR reactions of the endogenous gene and glyceraldehyde-3-phosphate dehydrogenase (GAPDH) were also performed on each DNA samples under a standard procedure (Love *et al.*, 1994).

*Viability of manipulated eggs*

Manipulated eggs were cultured from stage 10 to hatching, and viability was checked by candling. Hathced chickes were raised to sexual maturity for progeny test, and PCR analysis was carried out for the detection of the MiwZ in the gonads from chicks.

## Results

*Detection of DNA by X-gal staining*

Survival rates of manipulated embryos were ca. 74% for unincubated eggs and 80% for temporarily stored eggs, respectively. On the one hand, the survival rate for prematurely oviposited eggs was ca. 43%. The expression rates of injected DNA, which were detected by X-gal staining, were ca. 58% for unincubated eggs, 57% for temporarily preserved eggs and 67% for prematurely oviposited eggs, respectively. However, most of them were recognized only in the extraembryonic tissues, but the expression of MiwZ was observed in 6 cases (ca.12%) of aroud 50 embryos from unincubated eggs and 6 cases (ca. 9%) of around 70 embyos from temporarily preserved eggs. No significant ($P>0.05$) difference was observed between two kinds of embryos for the expression of injected DNA.

On the contrary, in the case of prematurely oviposited eggs, much more intensified expression of DNA was observed in the extraembryonic tissues compared with other embryos.

The expression of introduced DNA was recognized mainly in the epidermis (ca. 50%), heart (ca. 34%) and neural tube (ca 17%) in a mosaic manner. In the temporarily stored eggs, the expression of DNA (ca. 17%) was also observed in the primordial germ cells (PGCs) in blood collected from the vitelline artery or dorsal aorta of the embryos.

*Copy number of plasmid DNA*

The treated embryos from unincubated and temporarily stored eggs were classified as single copy (MiwZ present at a level equivalent to one or more copies per cell), mosaic (MiwZ present at a level below one copy per cell), or negative (no MiwZ detectable) at stage 14. The presence of MiwZ was detected in the rates of ca. 69% for unincubated eggs and ca. 74% for temporarily preserved eggs. The MiwZ at a level equivalent to a single copy or more per cell was detected in the rate of ca. 1.7% for unincubated eggs and 4.2% for temporarily stored eggs. Mosaic embryos were also observed in the rate of ca. 67% for unincubated eggs and 69% for shortly stored ones.

*Viability of manipulated eggs*

The hatchability of manipulated eggs were ca. 43% for unincubated eggs and 35% for preserved ones, but the number of dead embryos increased by opening the windows between the stage 10, 15 and just before the hatching. In the present study, the MiwZ was not detected in the gonads of hatched chicks.

## Discussion

Temporary preservation of fertilized egg at 4 C for 3 h did not show any advantages for *in vivo* gene transfection. In this experiment, ca. 4% at a level equivalent to one copy of DNA per cell was observed in a whole embryo examined by PCR analysis.

The PGCs collected from circulating blood of embryos contained injected DNA. In this experiment, DNA was injected just under the blastoderm, at stage of which a single cell layer was formed and PGCs were observed in the central area of germinal disc (Ginsburg and Eyal-Giladi, 1987). The central area of blastodermic layers is a neural presumptive region and destined to form the brain and epidermis of the caput (Rudnik *et al.*, 1948).

In the embryos from the premature oviposition, DNA expression rate was almost the same as fertilized eggs (Naito *et al.*, 1991). The present results suggest that the mosaicism might be inhibited by the injection of DNA into the blastoderm of fertilized eggs. In this study, however, *in vitro* culture and DNA microinjection to the blastoderm led to lower survival rate (ca. 43%).

An advantageous point of this experiment was that an exogenous gene was successfully transferred to embryonic tissues, showing higher hatchability (ca. 39%) than *in vitro* culture method (Naito *et al.*, 1994).

## Acknowlegments

This study was financially supported by the Japanese Society for the Promotion of Sciences and Yuasa International Foundation in Japan. We thank Dr. T. Kuwana for his invaluable helf to develop this skilled technique for manipulating PGCs.

## References

Eyal-Giladi, H. and Kochav , S. (1976) A complementary normal table and a new look at the first stages of the development of the chick. Dev. Biol. **49**, 321-337.

Ginsburg, M. and Giladi, H. E. (1987) Primordial germ cells of the young chick blastoderm originate from the central zone of the area pellucida irrespective of the embryo-forming process. Dev. **101**, 209-219.

Hamburger, V. and Hamilton, H. (1951) A series of normal stages in the development of the chick embryo. J. Morophol. **88**, 49-92.

Inada, S., Hattori, M., Fujihara, N. and K. Morohashi. (1997) *In vivo* gene transfer into the blastoderm of early developmental stage of chicken. Reprod. Nutr. Dev. **37**, 13-20.

Kuwana, T. (1993) Migration of avian primordial germ cells toward the gonadal anlage. Dev. Growth Differ. **35**, 237-243.

Love, J., Gribbin, C.,Mather, C. and Sang, H. (1994) Transgenic birds by DNA microinjection. Bio/Technol. **12**, 60-63.

Naito. M., Agata, K., Otuka, K., Kino, K., Ohta, H., Hirose, K., Perry, M. M. and Eguchi, G. (1991) Embryonic expression of β-actin-*lac* Z hybrid gene injected into the fertilized ovum of the domestic fowl. Int. J. Dev. Biol. **35**, 69-75.

Naito, M., Sasaki, E., Ohtake, M. and Sakurai, M. (1994) Introductionof exogenous DNA into somatic and germ cells of chickens by microinjection into the germinal disc of fertilized ova. Mol. Reprod. Dev. **37**, 167-171.

Otsuka, K., Kino, K., Ohta, M., Hirose, H., Mochii, M. and Agata, K. (1990) Transplantation of blastoderm cells transferred with foreign DNA into chicken blastoderm. Jpn. Poult. Sci. **27**, 221 (abstract).

Perry, M. M. (1988) A complete culture system for the chick embryo. Nature **331**, 70-72.

Petitte, J. N., Clark, M. E. and Liu, G. (1990) Production of somatic and germline chimeras in the chicken by transfer of early blastodermal cells. Dev. **108**, 185-189.

Rudinck, D. (1948) Prospective areas and differentiation protease in the chick blastoderm. Ann. New York Acad. Sci. **49**, 761-772.

Sang, H., Perry, M. M. (1989) Episormal replication of cloned DNA injected into the fertilized ovum of the hen. Mol. Reprod. Dev. **1**, 98-106.

Suemori, H., Kadodawa, Y., Goto, K., Araki, I., Kondoh, H. and Nakatsuji, N. (1990) A mouse embryonic stem cell line showing pluripotency of differentiation in early embryos and ubiqutous β-galactosidase expression. Cell Differ. Dev. **29**, 181-186.

Tajima, A., Naito, M., Yasuda, Y. and Kuwana, T. (1993) Production of germ line chimera by transfer of primordial germ cells in the domestic chicken. Theriogenol. **40**, 509-519.

Thoraval, P. Lasserre, F., Coudert, F. and Dambrine, G. ( 1994) Somatic and germline chicken chimeras obtained from brown and white leghorns by transfer of early blastodermal cells. Poultry Sci. **73**, 1897-1905.

# OVERVIEW OF THE INTERNATIONAL ENDEAVOR TOWARD HARMONIZATION OF TECHNICAL REQUIREMENTS FOR THE CONTROL OF NEW MEDICINES FROM BIOTECHNOLOGY

T. HAYAKAWA
*Division of Biological Chemistry and Biologicals*
*National Institute of Health Sciences*
*1-18-1 Kamiyoga, Setagaya-Ku, Tokyo 158-8501, Japan*

## Introduction

Since 1991, 3 regulatory authorities and 3 pharmaceutical manufacturers' associations from the EU, the USA and Japan, and the Canadian regulatory authority have been positively addressing international harmonization of technical requirements for the control of biotechnology pharmaceuticals. Each party has been making an all-out effort to achieve this objective through the International Conference on Harmonization (ICH) procedure.

## 1. ICH Harmonized Documents on Biotechnology Products

Up to now, 6 topics have been identified and the corresponding harmonized document is being developing by the Expert Working Group (EWG) which is comprised of representatives from the 7 parties. As of the present time, five internationally harmonized documents have been completed. These include: "Derivation and Characterization of Cell Substrate Used for Production of Biotechnological/Biological Products (Cell Substrate)", "Analysis of the Expression Construct in Cells Used for Production of r-DNA Derived Protein Products (Genetic Stability)", "Viral Safety Evaluation of Biotechnology Products Derived from Cell Lines of Human or Animal Origin (Viral Safety)", "Stability Testing of Biotechnological/Biological Products (Product Stability)" and "Preclinical Safety Evaluation of Biotechnology-Derived Pharmaceuticals (Preclinical Safety)". The sixth topic entitled "Specifications: Test Procedures and Acceptance Criteria for Biotechnological/Biological Products" reached consensus within the EWG in February 1998 and was released for internal and external consultation from the regulatory agencies of the EU, the USA and Japan.

It should be noted that the Cell substrate and Genetic Stability documents are related to early stages in the manufacturing process including banked cell substrates (e.g., the master cell bank: MCB and/or the working cell bank : WCB), and cells at the limit, namely production cells cultured up to the proposed limit of drug production runs. Viral safety issues are related to all stages involved in the manufacturing process. Product Stability,

Specifications and Preclinical Safety Testing documents are related to purified bulk and/or final products.

Thus, fundamental issues related to technical requirements for the consistent production of biotechnology pharmaceuticals and for their assessment and control with respect to quality and preclinical safety have been mostly covered.

## 2. Some of the Important Components of Each Specific Document

### 2.1. CELL SUBSTRATE

The objective of the "Cell Substrate" guideline is to provide broad guidance on appropriate standards for the derivation of human and animal cell lines and microbial cells to be used to prepare biotechnological/biological protein products defined in this document, and for the preparation and characterization of cell banks to be used for production.

This guideline covers cell substrates having a cell banking system. In this document, "cell substrate" refers to microbial cells or cell lines derived from human or animal sources that possess the full potential for generation of the desired biotechnological/biological products for human *in vivo* or *ex vivo* use.

Some of the important components of " Cell Substrate" document include: 1) source, history, and generation of the cell substrate, 2) cell banking, and 3) characterization and testing of the cell bank.

It is important to provide supportive documentation which describes the history of the cell substrate that is used in the manufacture of a biotechnological/biological product, as well as any parental cell line from which it was totally or partially derived.

One of the most important advantages of using serially subcultured cells to produce biotechnological/ biological products is the ability to have a characterized common starting source for each production lot, i.e., the preserved bank of cells. The concept of a two-tiered cell bank, in which the MCB is used to generate WCBs, is generally accepted as the most practical approach to providing supply of cell substrate for continued manufacture of the product. With regard to cell banking procedures, it is important to prevent a contaminated cell substrate (or bank) from being used in production and to avoid a loss of product availability or development time.

The characterization and testing of banked cell substrate is a critical component of the control of biotechnological and biological products. For tests of identity, appropriate tests should be performed to determine that the banked cell is what it is represented to be. A critical aspect of cell development and banking is the assessment that the MCB and WCB are biologically pure, i.e., the MCB and WCB are free from adventitious microbial agents and adventitious cellular contaminants. Another dimension to cell characterization is the appropriateness for intended use in production. There are two concerns for cell substrate stability: Consistent production of the intended product and retention of production capacity during storage under defined conditions. Utilization of karyology and tumorigenicity testing for evaluating the safety of a diploid cell line or characterizing a new cell line may be useful depending on the cells, the nature of the product and the manufacturing process.

## 2.2. GENETIC STABILITY

The objective of the "Genetic Stability" guideline is to provide guidance regarding the characterization of the expression construct for the production of recombinant DNA protein products in eukaryotic and prokaryotic cells.

The expression construct is defined as the expression vector containing the coding sequence of the recombinant protein. The purpose of analyzing the expression construct is to establish that the correct coding sequence of the product has been incorporated into the host cell and is maintained during culture to the end of production. The expression construct of the production cells expanded to the proposed *in vitro* cell age or beyond should be analyzed once by the same method that was used for the MCB, except that the protein coding sequence of the expression construct in the production cells could be verified by either nucleic acid testing or analysis of the final protein product.

## 2.3. VIRAL SAFETY

The "Viral Safety" document is concerned with testing and evaluation of the viral safety of biotechnology products derived from characterized cell lines of human or animal origin. The purpose of this document is to provide a general framework for virus testing, experiments for the assessment of viral clearance and a recommended approach for the design of viral tests and viral clearance studies. The main body of the document covers products derived from in vitro cell cultures initiated from characterized cell banks.

Three principal, complementary approaches have evolved to control the potential viral contamination of biotechnology products, i.e., 1) selecting and testing cell lines and other raw materials, including media components, for the absence of undesirable viruses which may be infectious and/or pathogenic for humans; 2) assessing the capacity of the production processes to clear infectious viruses; 3) testing the product at appropriate steps of production for absence of contaminating infectious viruses.

Major areas described in the viral safety document include: 1) cell line qualification, 2) testing for viruses in unprocessed bulk, 3) rationale and action plan for viral clearance studies and virus tests on purified bulk, and 4) evaluation and characterization of viral clearance from unprocessed bulk.

With respect to cell line qualification, the ICH guideline suggests an example of virus tests to be performed once only at various cell levels (MCB, WCB and Cells at the Limit).

At the following unprocessed bulk level, it is recommended that manufacturers develop programs for the assessment of adventitious viruses in production batches.

The ICH document presents an example of the action plan to be adopted, in terms of process evaluation and characterization of viral clearance as well as virus tests on purified bulk, in response to the results of virus tests on cells and/or the unprocessed bulk.

The objective of virus clearance studies is to assess process step(s) that can be considered to be effective in inactivating/removing viruses and to estimate quantitatively the overall level of virus reduction obtained by the process. Two separate approaches will be considered for viral clearance of manufacturing process. The first is viral clearance evaluation studies to determine the ability of the manufacturing process to remove and/or inactivate the viruses that are known or expected to be present in the cell substrates or any other reagents or materials used in the process. The other is viral clearance characterization studies to characterize the capacity of the manufacturing process to remove and/or inactivate other viruses rather than those known or expected to be present.

## 2.4. PRODUCT STABILITY

The purpose of the "Product Stability" document is to give guidance to applicants regarding the type of stability studies which should be provided in support of marketing applications for biotechnological/biological products, in which the active components are typically proteins and/or polypeptides.

Because of the unique chemical and biological properties of these products, special consideration should be given to design a stability testing program as well as to evaluate the stability of a given product.

The product-specific use of appropriate physicochemical, biochemical and immunochemical methods for the analysis of the molecular entity and the quantitative detection of degradation products is critical to confirm its stability during the intended storage period.

Primary data to support a requested storage period for a given product should be based on long-term, real time, real condition stability studies.

The specific areas described in the guideline include: 1) selection of batches, 2) stability-indicating profile that provides assurance that changes in the identity, purity and potency of the product will be detected, 3) storage conditions, 4) testing frequency, 5) specifications, and 6) labeling.

## 2.5. SPECIFICATIONS

Bulk drug substance and final product specifications are key parts of the core documentation for world-wide product license applications. The "Specification" guideline which is under development will provide general principles on the setting and justification, to the extent possible, of a uniform set of international specifications for biotechnological/biological products to support new marketing applications.

The draft guideline identifies a number of general principles for consideration in setting specifications. The principles identified are: 1) characterization of a product, 2) analytical considerations related to reference standards and reference materials as well as validation of analytical procedures, 3) process controls, 4) pharmacopoeial specifications, 5) release limits vs. shelf-life limits, and 6) statistical concepts. The document also provides guidance for justification of the specification and drug substance and drug product specifications.

It should be emphasized that the setting specifications for drug substance and drug product is one part of a total control strategy designed to ensure product quality and consistency. Other parts of this strategy include comprehensive product characterization during development, control of raw materials and excipients, in-process testing, process evaluation/validation, stability testing and evaluation of preclinical/clinical data in relation to quality.

### 2.6. PRECLINICAL SAFETY

It is widely recognized that the special nature of biotechnological/biological protein products requires a unique and product-specific approach to preclinical safety evaluation needed to support clinical development and marketing authorization. Although a flexible, case-by-case, science based approach has been adopted in the EU, Japan and the United States, there has been a need for common understanding and continuing dialogue among the regions. The ICH "Preclinical Safety" guideline is intended to provide general principles for designing scientifically acceptable preclinical safety evaluation programs.

The major general areas considered in the guideline include: 1) specification of the test materials that are utilized in safety assessment studies, 2) considerations related to assessment of biological activity and pharmacodynamics, 3) unique issues related to selection of animal species or models for pharmacology and toxicity studies, 4) guidance related to number/gender of animals, 5) consideration related to designing the route, frequency and method of administration of test materials and dose selection, and 6) specific considerations related to immunogenicity of a product.

The guideline also describes the areas of specific concern. These areas include: 1) safety pharmacology in which the potential for undesirable pharmacological activity is investigated or monitored, 2) exposure assessment utilizing pharmacokinetics and toxicokinetics with relevant assay method(s), and clarification of the definition of metabolism, 3) important considerations related to the design of single and repeated dose toxicity studies, and 4) specific guidance in the design and/or need for assessment of immunotoxicity, reproductive and developmental toxicity, genotoxicity, carcinogenicity and local tolerance.

## 3. Outcome and Future of the ICH Process

After 7 years of endeavor through the ICH process, we have obtained many fruits of our labor. These include: 1) creation of entirely new guidelines, 2) creation of new strategies, concepts and new terminology, 3) creation of new mutual recognition across the three regions, and among their regulators/manufacturers, 4) creation of new scientific data and findings for the evaluation and control of biotechnology pharmaceuticals, 5) description of how to approach a topic, 6) increased efficiency of drug development programs, and 7) resource savings.

Because of the nature of the documents, especially in the biotechnology field, it will be necessary to update them based on new and developing scientific data. The document will always need to be kept in the front line of research.

It is our hope that the ICH will continue to provide an international forum for increased communication, exchange of experiences and practices among the interested parties and also that the outcome of this endeavor will (1) lead to updating the existing guidelines, if necessary and (2) allow the introduction of new guidelines, for example, ones related to gene therapy, cellular therapy or transgenic animal-derived products.

Of course our ultimate goal should be to allow patients increased accessibility to novel medicines awaiting approval.

# REDUCING RISKS OF ANIMAL ORIGIN CONTAMINANTS IN CELL CULTURE.

David W. Jayme, Shawn R. Smith and Mark Z. Plavsic
*Life Technologies, Inc.*
*3175 Staley Road, Grand Island, NY 14072 USA*

**Abstract**

The potential inadvertent introduction of viral or prion agents into the biopharmaceutical production environment has accelerated technical and regulatory actions to ensure biosafety. Primary efforts have focused upon three activities: (a) Validated processing of essential animal-derived medium constituents to eradicate adventitious contaminants; (b) Design and qualification of protein-free, biochemically-defined nutrient formulations without animal origin constituents; and (c) Improved manufacturing process controls to validate sanitization and ensure against cross-contamination.

## Introduction

The specter of introducing etiologic agents via the putative therapeutic drug has haunted biopharmaceutical development and necessitated a critical review of conventional cell culture-based production processes. Three critical elements have emerged for active re-engineering.

***Sourcing and Processing of Animal Origin Constituents:*** A recent conference examining the use of animal sera and derivatives for pharmaceutical manufacture [1] reinforced the reality that a high percentage of commercial bovine sera is contaminated by adventitious viruses, particularly bovine viral diarrhea (BVD) virus and bovine polyoma virus. Although the potential risk for contamination of serum with bovine spongiform encephalopathy (BSE) agent is relatively small due to improved material sourcing and animal husbandry practices, the Strasbourg conference recommended sourcing of sera and derived by-products from secure, traceable global locations documented to be BSE-free.

Where the persistent use of animal sera or other proteinaceous constituents derived from animal tissues or fluids is mandated due to technical constraints, rigorous sourcing criteria and post-collection processing have been recommended to reduce the risk of introducing viral or prion elements into the culture environment. Although various novel approaches for removal or inactivation are under investigation [1-2], gamma irradiation remains the most reliable option to eradicate viruses while retaining post-treatment biological activity.

***Design of Nutrient Formulations Without Animal Origin Constituents:*** While interest in eliminating serum and serum-derived proteins has existed for over a decade, the historical motivations of process economics and technical artifacts have been largely overcome and replaced by regulatory concerns regarding foreign protein antigenicity and introduction of adventitious agents. In re-engineering the composition of our nutrient formulations used for bioproduction applications to mitigate the risk of animal origin materials, we implemented a three-tiered developmental policy involving three "R's": reduction, refinement and replacement. As an initial approach to eliminating serum or another animal-derived medium constituent, the supplemental level of the targeted additive was titered to the lowest level without substantial impact on biological activity. Refinement required analysis of the contributory factors within ill-defined biological supplements and attempted to reconstitute equivalent (or superior) biological activity using biochemically-defined substitutes. The final phase involved a comprehensive qualification of non-animal origin substitutes for traditional nutrient formulations based on demonstrated bioequivalence.

The net result of these three linked processes has facilitated a transition from serum-supplemented cultivation to reduced serum supplementation and generation of serum-free, protein-free and, ultimately, chemically-defined nutrient formulations to support high density bioreactor requirements. To date, we have developed protein-free nutrient formulations which are devoid of animal origin constituents to support hybridoma technology, mammalian cell-based recombinant protein production and human and veterinary vaccine applications. Culture systems adapted to such nutrient formulations frequently exhibit improved biological performance and facilitated downstream recovery while reducing the potential for viral or BSE contamination [1,3-4].

***Manufacturing Process Design and Validation:*** Meticulous surveillance of incoming raw materials must be coupled with facility and process design controls to ensure against inadvertent final product contamination. Facility design issues derive from basic decisions, e.g., Will the manufacturing suite be dedicated to a single task or qualified for multiple applications? Should the design be optimized for a single process or permit flexible changeover to alternate manufacturing products or processes? The response to these questions drives the configurational design of the work area; capacity and situation of air-handling and water delivery systems; equipment location, monitoring and sanitization processes; and training, outfitting, deployment and monitoring of technical personnel.

**Treatment of Animal-derived Proteins to Eradicate Viral Elements**

To validate the commercial inactivation of adventitious agents in biologically-active materials, we spiked scaled-down samples of animal sera and trypsin with high titers of model viruses and mycoplasmas. Candidate viruses and mycoplasmas were selected based upon their biorelevance and broad suitability as representative model agents [2]. Gamma irradiation was monitored using immersed center dosimeters and covered a nominal range of 15-45 kGy. Post-irradiation biological activity of animal sera and trypsin materials was evaluated using appropriate suspension and adherent cell culture proliferative assays (sera) and cell removal and replating assays (trypsin).

Test mycoplasma and acholeplasma species were highly-sensitive to gamma irradiation [M. Plavsic, manuscript in preparation]. Greater than six logs of *M. arginini, M. hyorhinus* and *A. laidlawii* spikes were inactivated by minimal exposures of 25 kGy, while the maximal exposure of 45 kGy had no apparent effect on biological performance in porcine serum samples. Fetal bovine serum lots spiked with five model viruses (infectious bovine rhinotracheitis (IBR), BVD, canine adenovirus, porcine reovirus, and porcine parvovirus) all demonstrated greater than six logs of virus inactivation without significant biological performance reduction when exposed to gamma irradiation over the 25-45 kGy range.

**Design of Animal Origin-free Protein-free Nutrient Media**

Various protein components have been historically associated with serum-free medium, particularly insulin, transferrin and albumin (3). Protein-free medium formulations have been designed which substitute alternative constituents to facilitate iron and lipid delivery and support normal biological proliferation and high density bioproduction. Suspension culture applications have been particularly well-suited to protein-free, chemically-defined nutrient formulations. Adherent cells have been more problematic: technical impediments remain in the effort to replace serum-derived attachment factors (e.g., vitronectin, fibronectin) with substitutes that are stable in solution, available in adequate quantity, and sufficiently economical for large-scale bioreactor requirements.

Improved tracking of sourcing and processing methods for common, non-protein medium constituents also revealed raw materials of animal origin. Many lipids (e.g., sterols, fatty acids), hydrolysates and surfactants used in nutrient medium formulations originate from animal tissues. Even several amino acids used for human intravenous hyperalimentation solutions are obtained from animal sources such as human hair, avian feathers, bovine collagen and bovine or porcine bone gelatin. While the degree to which these raw materials represent a threat of biocontamination remains controversial, many biopharmaceutical manufacturers are attempting to eliminate even the perception of risk by specifying that all raw material components used in nutrient medium formulations must be non-animal sourced, wherever possible.

**Manufacturing Process Controls**

As a Case Study to illustrate requirements for manufacturing systems validation, we reported commissioning studies for our Large Volume Liquid Manufacturing (LVLM) system. This integrated system provides the programmed in-line dilution of multiple concentrated (30-100X) components using a static mixing device to produce large batches (5,000-50,000 liters) of sterile-filtered nutrient medium batches which are dispensed into bulk containers. General processing control was demonstrated through validation of aseptic processing, filter train configuration, automated dispensing, and packaging integrity following compendial guidelines. System performance was correlated with end product integrity, verifying post-mixing formulation homogeneity by monitoring physical parameters, biochemical potency and biological performance over the course of multiple, extended production runs.

Recognizing that the LVLM system would be a multiple use capability, validated sanitization and sterilization were fundamental to ensure against inadvertent end product or culture cross-contamination. Chemical residuals were evaluated with rinsate studies to demonstrate endpoint conductance and endotoxin values and reduction in total organic carbon (TOC) levels. Adhesion and drying of a proteinaceous challenge ("soiling") solution followed by the proposed sanitization cycle demonstrated effective removal of adsorbed residuals in swab studies. Downscale system challenge with a panel of model viruses and prion agent followed by the sanitization cycle provided a quantitative kinetic assurance of agent eradication which was correlated to "worst case" incoming bioburden challenge [5]. System steam sterilization was validated by conventional distribution and cold spot mapping, using multiple calibrated thermocouples and by confirming steam penetration with spore indicators.

**Discussion and Summary**

Terminal product sterilization by gamma irradiation represents the most universally accepted virus inactivation method. Our validation studies demonstrated its effectiveness at eradicating greater than six logs of a panel of model viruses and mycoplasmas in serum, trypsin and other animal origin materials. The process is readily scalable, leaves no adulterative residue, and may be designed to minimize negative impact on biological performance.

Emphasis on elimination of all medium constituents of animal origin remains a controversial issue. Recent association of Creutzfeld-Jacob Disease (CJD) [6] incidence in patients exposed to human blood-derived therapeutics raised anxieties regarding use of albumin, transferrin and other human plasma origin raw materials for biopharmaceutical manufacture [7]. Transferrin requirement for many cell culture systems has been substituted by various inorganic iron carriers or chelates [3-4]. Conversely, one may argue that amino acid and lipid constituents of animal origin have been used for decades in human intravenous solutions without apparent clinical problem. Some of the anxiety may be traceable to the perception that, while most viral infections are rapidly observed and may be clinically managed, rapid qualitative detection and effective therapy for prion-induced encephalopathies remain elusive.

The most effective safeguard against the risk of introducing adventitious contaminants into the final biotherapeutic product appears to be an integrated approach, including all aspects of raw materials sourcing (low risk origination, rigorous vendor qualification, etc.), treatment of suspect materials by validated eradication processes (rather than virus screening which is subject to test sensitivities and statistical limitations), secure storage of intermediate and raw materials, and handling within a controlled, sanitary, validated manufacturing process.

The summary guidance reiterated by the Strasbourg Conference provides clear direction, particularly to those involved at the product and process development phases of therapeutic product development and commercialization: "Manufacturers of biological, medicinal and veterinary products are encouraged to reduce or eliminate the use of animal-derived products in their manufacturing process." [1]

**References**


1. Animal Sera, Animal Sera Derivatives, and Substitutes Used in the Manufacture of Pharmaceuticals: Viral Safety and Regulatory Aspects, Conference Highlights published May 6, 1998, Strasbourg, France.
2. Daley JP, Danner DJ, Weppner DJ and Plavsic MZ, Focus (1998) 20(3): 86-88, *Virus Inactivation by Gamma Irradiation of Fetal Bovine Serum.*
3. Jayme DW and Smith SR "Options to Eliminate Animal-Derived Components of Cell Culture Media" in Animal Cell Technology: Basic & Applied Aspects, volume 10 (Proceedings of JAACT '97, Nagoya, Japan)(in press).
4. Jayme DW "An Animal Origin Perspective of Common Constituents of Serum-free Medium Formulations" in Proceedings of the Strasbourg International Conference on Animal Sera, Animal Sera Derivatives and Substitutes Used in the Manufacture of Pharmaceuticals: Viral Safety and Regulatory Aspects. (in press).
5. Federal Register (1998) 63(185): 51074-51084 *International Conference on Harmonisation; Guidance on Viral Safety Evaluation of Biotechnology Products Derived from Cell Lines of Human or Animal Origin.*
6. Important Drug Withdrawal Information Regarding Buminate 25%, Baxter Corporation, customer letter, dated November 21, 1994.
7. Urgent Withdrawal Notice: PENTEX Human Transferrin Products, Bayer Corporation, letter from Joseph Montalto, dated October 9, 1996.

# CYTOTOXICITY TESTING FOR EVALUATING FOOD SAFETY

**Shiro Yamashoji**[1] **and Kenji Isshiki**[2]
*1 Kobe Gakuin Women's College, 27-1 Hayashiyama-cho, Nagata-ku, Kobe 653-0861, Japan*
*2 National Food Research Institute, 2-1-2 Kannondai, Tsukuba, Ibaraki 305-8642, Japan*

## 1. Abstract

Different cytotoxicity tests are used to determine food safety. We propose rapid cytotoxicity testing based on menadione-catalyzed $H_2O_2$ production by viable cells, which depends on both intracellular NAD(P)H concentration and plasma membrane-bound quinone oxidoreductase (EC1.6.99.2). Damage to either the cytosolic NAD(P)H production system or plasma membrane causes a loss of menadione-catalyzed $H_2O_2$ production resulting from cytotoxic events and is rapidly determined by colorimetrical assay of $H_2O_2$. This assay requires 10 min, and is much faster than MTT reduction or neutral red inclusion assays requiring 4 h. In cytotoxicity testing of food additives such as antioxidant BHA and BHT or p-hydroxybenzoate derivative preservatives, cytotoxic events were observed 4 h after mixing these food additives with animal cells. Natural toxins such as tomatine, solanine, and chaconine contained in tomatoes and potatoes were also detected 4 h after incubation with animal cells. We are now using this technique to test different foods, including whole foods.

## 2. Introduction

Animal experiments are expected to be useful than analytical methods in evaluating toxic food material as a whole. Public concern mounting about the use of animals in experiment has led to a new "3Rs": Reduce animal use, Replace animals with alternatives, and Refine methods to minimize pain.

We propose simple, rapid cytotoxicity testing using animal cells as an alternative to animals. Menadione-catalyzed $H_2O_2$ production by viable animal cells was observed in cell cultures, and determined by chemiluminescent assay as described elsewhere[1,2]. Our colorimetrical assay of menadione-catalyzed $H_2O_2$ production has broad and useful applications in food safety evaluation.

## 3. Materials and Methods

NIH/3T3 cells were cultured in DMEM supplemented with 10% fetal bovine serum. Cells were grown on tissue dishes in a humidified atmosphere of 95% air and 5% $CO_2$ at 37℃. Exponentially growing cells were trypsinized, and single cells were collected by centrifugation and resuspended in a 96-well tissue culture plate. Cells numbered 40,000 per well, and were cultured overnight under the above conditions.

To 100 $\mu$ l of cell culture medium in each well culture plate, 5 $\mu$ l of test solution was added and the mixture incubated under the above conditions for 4 h. Each well was washed twice with 100 $\mu$ l of MEM, and 100 $\mu$ l of MEM containing 0.2 mM menadione was added to each well. After 10 min of incubation in 95% air and 5% $CO_2$, 150 $\mu$ l of DA-64/peroxidase solution or Bis-MAPS-C3/peroxidase solution was added to each well. The absorbance at 725 nm (DA-64/peroxidase) or 649 nm (Bis-MAPS-C3/peroxidase) was measured with a microplate reader 1 min later. DA-64/peroxidase solution contained 0.1% bovine albumin, 10 U/ml peroxidase, and 200 $\mu$ M DA-64 in 1 M acetate buffer (pH 5.0). Bis-MAPS-C3/peroxidase solution contained 200 $\mu$ M Bis-MAPS-C3 in the above solution excluding DA-64. The viability of cells exposed to toxic compounds was determined as a percentage based on the absorbance of wells containing intact cells.

DA-64 [N-(Carboxymethylaminocarbonyl)-4,4'-bis(dimethylamino)-diphenylamine sodium salt] was purchased from Wako Junyaku Kogyo, and Bis-MAPS-C3{Bis[4-(N-propyl-N-sulfopropyl)amino2,6-dimethylphenyl] methane disodium salt} from Dojindo Laboratories.

## 4. Results and Discussion

The $H_2O_2$ concentration produced by intact NIH/3T3 cells under the conditions in Materials and Methods was about 15 $\mu$ M. DA-64 and Bis-MAPS-C3 in the presence of 15 $\mu$ M $H_2O_2$ and confluent cells gave the absorbance of about 0.4 and 0.3. Absorbance was about 0.04 at 725 and 649 nm when cells had died. The change in menadione-catalyzed $H_2O_2$ production by cells exposed to toxic compounds was thus observed at absorbance from about 0.4 or 0.3 to 0.04.

Table 1 shows the 50% inhibition concentration ($IC_{50}$) of antioxidants, preservatives, and alkaloids showing clear cytotoxicity 4 h after incubation

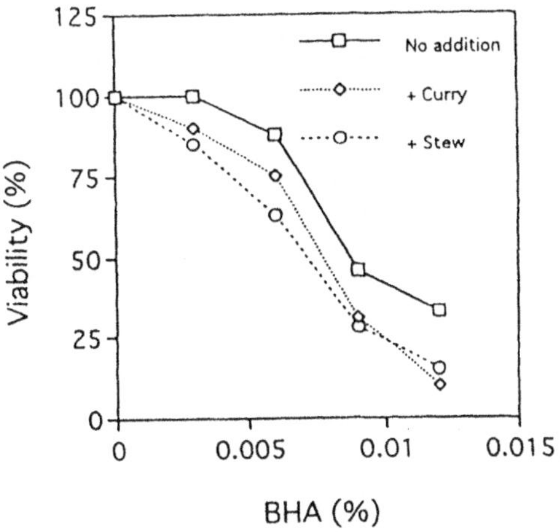


Fig.1 BHA Dose response
BHA was mixed with curry or stew, and 5 μl of the mixture added to 100 μl of cell culture medium. Incubation time was for 1 h.
Viability was determined under the conditions described in Methods.

**Table 1 $IC_{50}$ of alkaloids and food additives**

| | $IC_{50}$(%)* |
|---|---|
| Solanine | 0.004 |
| Tomatine | 0.0027 |
| Chaconine | 0.0014 |
| BHA | 0.012 |
| BHT | 0.01 |
| Ethyl p-hydroxybenzoate | 0.024 |
| n-propyl p-hydroxybenzoate | 0.015 |
| n-butyl p-hydroxybenzoate | 0.013 |
| Isobutyl p-hydroxybenzoate | 0.008 |

*$IC_{50}$ after 4 h incubation

with cells. When $IC_{50}$ is lower than the concentration in food, the cytotoxicity is observed by the direct addition of the food extract to cell culture,e.g., $IC_{50}$ of BHA was 0.12% and lower than the concentration in food, showing the direct determination of BHA in food. The dose response of BHA mixed with retort curry or stew was similar to the standard (Fig.1).

The above results indicate that the rapid assay of menadione-catalyzed $H_2O_2$ production with mammalian cells will prove highly useful as a screening test for food safety evaluation.

## 5. References


1) Yamashoji, S., Nishimoto, F., Usuda, M., Kubota, H. and Isshiki, K. (1992) Application of the chemiluminescent assay to cytotoxicity test : Detection of menadione-catalyzed $H_2O_2$ production by viable cells, *Anal.Biochem.*, 207, 255-260.

2) Isshiki, K., Asano, M. and Yamashoji, S. (1995) Cytotoxicity testing for food safety evaluation, in Beuvery, E.C., Griffiths, J.B. and Zeijlemarker, W.P. (eds.), Animal cell technology: Developments toward the 21st century, Kluwer Academic Publishers, Dordrecht, pp.999-1003.

# FACTORS SPECIFICALLY EXPRESSED IN OSTEOBLASTS

JITSUTARO KAWAGUCHI, KEISUKE HORIUCHI, HISAAKI KUDO, SUNAO TAKESHITA AND AKIRA KUDO

*Department of Life Science, Tokyo Institute of Technology, Yokohama 226-8501, Japan*

Summary

The subtractive screening of cDNA libraries derived from the mouse osteoblastic cell line, MC3T3-E1 has yielded cDNA clones functioned in osteoblast differentiation. One of them, termed OB-cadherin, which is identical to cadherin-11, is involved in classic cadherin family which is generally identified as $Ca^{2+}$ dependent adhesion molecule. Two isoforms of the human OB-cadherin/cadherin-11 transcripts were isolated, and one encodes a typical primary structure of cadherin and the other is a variant form encoding a different cytoplasmic domain because of alternative splicing. Results of immunohistochemical analyses using fetal calvaria indicated that OB-cadherin/cadherin-11 is expressed on osteoblasts and osteocytes. The high expression of the variant form was detected in osteosarcoma by Northern blot and RT-PCR analyses . The anomalous expression of the OB-cadherin/cadherin-11 molecules, the variant form as well as the secretory form found at protein level, modulates cell adhesive properties.

Another gene OSF-2 included in the fasciclin I family is specifically expressed in periosteum in adult mice found by both RNA in-situ hybridization and immunohistochemistry, and termed periostin. Periostin is a secreted molecule as 90 kD, and the purified protein of periostin from the supernatant of MC3T3-E1 cells functioned in cell attachment. The transcription of periostin gene was upregulated by TGF-b and interleukin 6. The recent development of zebra fish in biotechnology led us to clone up a full length of the zebra periostin gene. The homology between mouse and zebra periostin is high (about 65%), suggesting the common function in bone formation. RNA in-situ hybridization revealed that zebra periostin was expressed in somites and fins.

1. OB-cadherin/cadherin-11

Osteoblasts are the skeletal cells responsible for bone formation. They synthesize and regulate the deposition and mineralization of the extracellular matrix of bone. It has been proposed that osteoprogenitor cells arise from mesenchymal stem cells which are multipotential in nature and capable of giving rise to a number of committed and restricted cell lineages including the osteogenic line. Based on the morphological and histological studies, osteoblastic cells are categorized in a presumed linear sequence progressing from osteoprogenitor to preosteoblasts, osteoblasts, and lining cells or osteocytes. During development, extracellular matrix is important to be investigated at molecular level, and specifically, cell adhesion molecules are involved for destiny of cells.

Cadherins are transmembrane $Ca^{2+}$ dependent homophilic adhesion receptors that are well known to play important roles in cell recognition and cell sorting during development (1). One of the most important and ubiquitous types of adhesive interactions required for the maintenance of solid tissues is that mediated by the classic cadherins, subgrouped into type I and type II (2). The classic cadherin family is defined by their highly conserved cytoplasmic domain, which associates with catenins: a-catenin, b-catenin, plakoglobin and p120 (3, 4). There is another subgroup called the type II cadherin classified on the basis of small but significant variations in amino acid sequence. The HAV motif, which is important for $Ca^{2+}$ dependent adhesion, and is common in the type I classic cadherin family but not in the type II cadherin family. Although some of type II molecules are known to be expressed in loosely associated cells and to be responsible for weaker intercellular adhesion, detailed properties of type II cadherin still remain obscure.

A type II molecule, OB-cadherin identical to cadherin-11(5) was isolated by us and others to be expressed on the osteoblastic cells, stromal cells and mesenchymal cells, in which HAV site is replaced by QAV. The mouse OB-cadherin gene was isolated from the mouse osteoblastic cell line, MC3T3-E1, and two types of human OB-cadherin/cadherin-11 genes were also isolated from the human osteosarcoma cDNA library (6). One was the counterpart of the mouse gene and the other was the variant OB-cadherin/cadherin-11 gene having an insertion fragment of 179bp at the position corresponding to the transmembrane region, resulting in a frameshift mutation downstream from the insertion point. It would produce a truncated protein, which lacks one-third of transmembrane domain and the entire cytoplasmic domain. This structure of the variant form will be predicted as behaving like a dominant negative to block the function of the cadherin characteristic.

In this report, we examined the expression and the function of the variant form and investigated that the variant form was the splice variant by the analysis of the genomic DNA structure and it produced a 80kD protein detected in Western blot analysis and may function in co-operation with the intact OB-cadherin/cadherin-11 (7).

1). Three forms of cadherin-11

To investigate cadherin-11 protein, a Western blot analysis using a mouse monoclonal antibody against human cadherin-11 was performed by using cell extracts from mouse L cell transfectants, OBL cells and human calvarial cells (CAL). From three different types of L cell transfectants, the intact, the variant and the double, both the intact and the variant, 120 kD and 85 kD bands, were detected as proteins for the intact and the variant respectively. b-catenin in L cell transfectants became stable in the presence of the intact of cadherin-11 molecule. In variant L cells, no b-catenin was detected, indicating that the variant form was not associated with b-catenin. OBL and CAL cells expressed the intact form as well as b-catenin, but the variant was not detected because of low expression. Since cadherins are associated with b-catenin, the association between cadherin-11 and b-catenin was examined by combination of immunoprecipitation with an anti b-catenin antibody and successive Western blot analyses with antibodies against cadherin-11 and b-catenin. The intact form was co-immunoprecipitaed with b-catenin, and the variant form was not detected. Same results were obtained in OBL.

To investigate whether the variant form is secreted because it lacks a part of the membrane domain, the culture supernatants of each transfectant and OBL were concentrated 20 fold and were tested by Western blotting. The data showed that the variant form is obviously not the secreted form since the secreted form is not present in the conditioned medium of L cell transfectants of the variant form, but interestingly, the secreted form of cadherin-11, 80 kD in size, was found from the intact and the doble L cells, but not from the variant L cells. The result implies that the secreted form is generated from cleavage of the intact form, but not by alternative mRNA splicing. The secreted form was also detected in the supernatant from OBL after long exposure. The three forms are shown in Fig. 1.

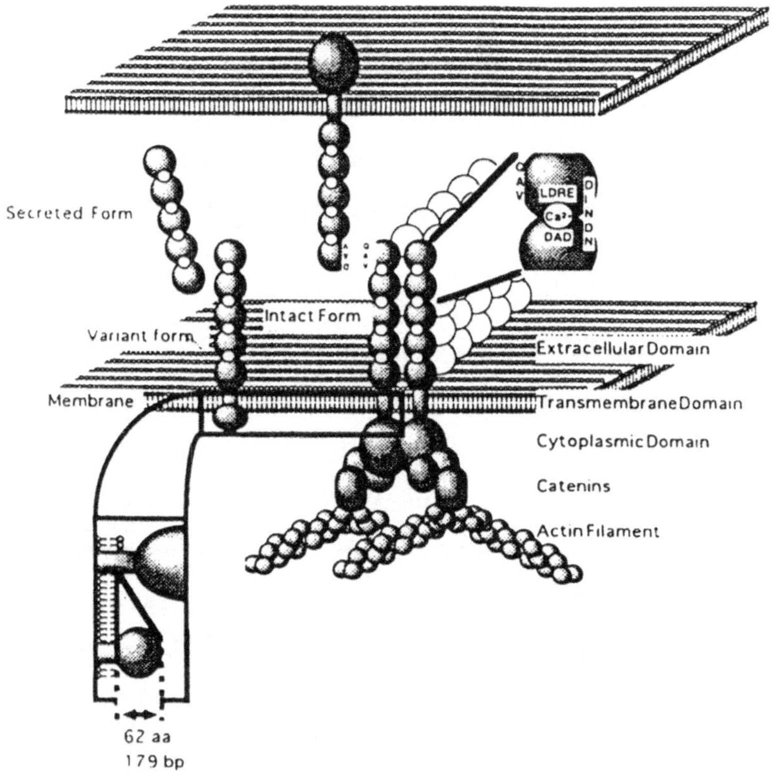


*Figure1.* Three isoforms of OB-cadherin./cadherin-11

Human cadherin-11 has three isoforms. One is the intact form with a typical cadherin structure, and the secreted form is derived from the intact form with proteolysis. The variant form does not have the ordinary cytoplasmic domain because of alternative splicing.

2). Function of the variant form of cadherin-11

To examine Ca $^{2+}$ dependent cell adhesion, a typical characteristic of cadherin molecules, cell aggregation assays were performed (in Fig. 2). Following trypsinization in the presence of $CaCl_2$, the intact L cells slowly reaggregated in the presence of $CaCl_2$. The variant L cells or L cells expressing the vector alone (none) did not show aggregation. Although several transfectants expressing different amounts of the variant form were tested, none of them demonstrated aggregation . To test the effect of the variant form in cell adhesion in the presence of the intact form, double transfectants were newly established by introduction of the variant form of the cadherin-11 gene into the intact L cells. Double transfectants aggregated faster than the intact L cells. These results show that the variant form coupled with the intact form activates the cell adhesion. To generalize this result, we further examined more transfectants with various expression levels of the variant form. All the double transfectants showed faster aggregation kinetics than the intact form alone, although the expression level of the intact form is similar each other. Thus we conclude that the existence of the variant form causes faster cell aggregation in the presence of the intact form, suggesting that the variant form is functional in bone marrow cells or cells in femoral and humeral head for aggregation.

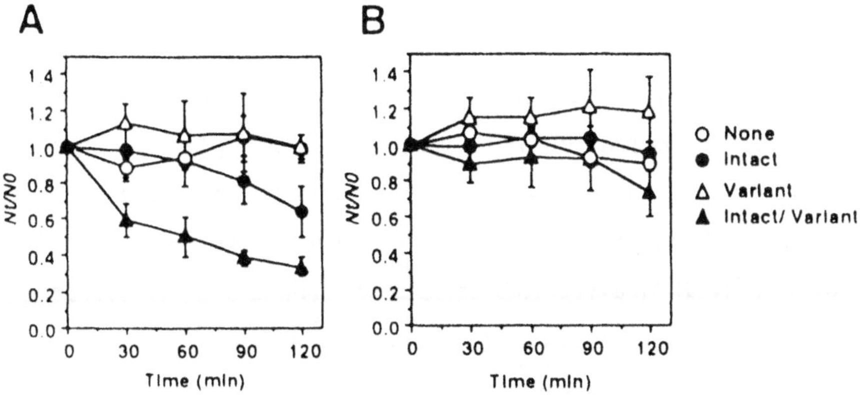


*Figure* 2. Cell aggregation of cadherin-11 L cell transfectants.

Cells were dissociated by TC-treatment, and were allowed to aggregate in the presence (A) or absence (B) of 1mM $CaCl_2$. L cell with the vector alone (none), with the cadherin-11 intact form (Intact), with the cadherin-11 variant form (Variant), with the cadhrin-11 intact and variant form (Intact/Variant).

3). Strong expression of cadherin-11 and b-catenin in normal osteoblasts but faint in osteosarcoma.

To identify localization of cadherin-11 in tissues, fetal calvaria and osteosarcoma were examined by immunohistochemistry using monoclonal antibodies against cadherin-11 and b-catenin. In fetal calvaria, the strong expression of cadherin-11 as well as b-catenin was detected in both resting and formative osteoblasts lined along the surface of the woven bone, thus did mainly around the cell membrane. Interestingly, cadherin-11 was expressed in osteocytes, which have an abundant cytoplasm and a large nucleus, but not in fibroblasts. Adjacent to cadherin-11, b-catenin was also expressed in osteoblasts. On the other hand, in osteosarcoma, we could find only low immunoreactivity of cadherin-11 and b-catenin in the cytoplasm but no obvious signals on the surface of tumor cells.

2. Periostin/OSF-2

Bone is formed by two processes, intramembranous ossification and endochondral ossification. The former contributes to the formation and growth of the flat bones of the skull and addition of bone to the periosteum of long bones, while the latter process is responsible for formation of the rest of the bones in the body. At the histological level, both of these processes have been well studied, but the mechanisms involved at the molecular level are not understood. It is known that the periosteum and the periosteal collar are responsible for bone formation and that this tissue contains mesenchymal stem cells and pre-osteoblasts (8). The periostin/OSF-2 gene was isolated in previous studies aimed at identifying novel genes expressed specifically in the osteoblast by using the techniques of subtraction hybridization and differential screening between cDNA libraries of MC3T3-E1 and NIH3T3 cells (9). Mouse periostin comprises 811 amino acids and computer analysis of the deduced amino acid sequence revealed a complex protein structure with four repeats of a characteristic domain. A similar structure had been reported for fasciclin I, a homophilic cell-cell adhesion molecule expressed in the central nervous system of insects (10) and big-h3, a molecule induced by TGFb that promotes fibroblast attachment and spreading (11).

In recent years several novel proteins with a similar structure to fasciclin I have been identified in a variety of species. These include big-h3 (human and mouse), Algal-CAM (plants)(12), MPB70 (mycobacteria) (13) and midline fasciclin (insects) (14). These genes have a 130-150 amino acid long homologous domain, which is characterized by two conserved stretches of 13 and 14 amino acids, respectively. Fasciclin I, big-h3, midline fasciclin and periostin contain four homologous domains, while Algal-CAM contains two and MPB-70 consists of one, shown in Fig. 3. Outside the two well-conserved stretches, the overall sequence homology is generally low. Although their biological functions and the underlying molecular mechanism of their effects are still poorly characterized, several reports suggest that these proteins act as adhesion molecules. In this report we describe a potential function for periostin as a cell adhesion molecule for pre-osteoblasts based on its localization in tissues and capacity to induce osteoblast attachment and spreading. This novel gene was previously called "osteoblast-specific factor 2" or OSF-2, but a subsequent study also used this acronym to describe a transcription factor also known as Cbfa1 (15). Therefore, based on its tissue specificity, we have renamed this protein "periostin". Herein we describe the characterization of this protein.

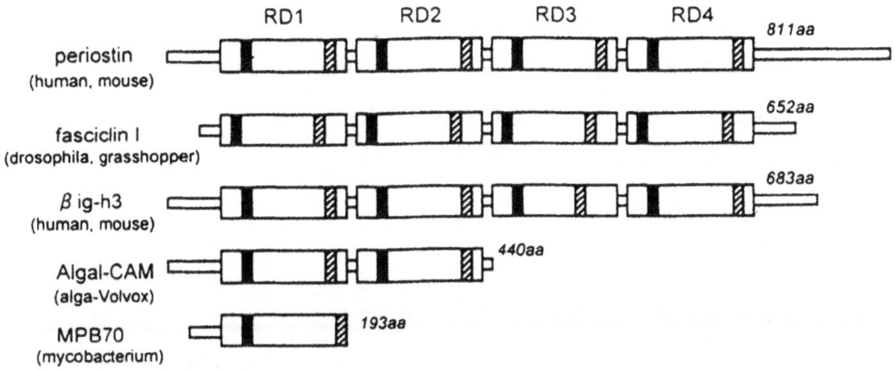


*Figure3.* A diagram of the structure of proteins in the fasciclin I family.

Two highly conserved sequences in each repeated domain (RD) are indicated as black boxes (a) and hatched boxes (b). The proteins shown are; mouse periostin/OSF-2, drosophila fasciclin I, human β ig-h3, Volvox Algal-CAM, and mycobacterium MPB70.

1). High levels of expression of periostin are observed in the periosteum but not in the endosteum and bone matrix.

We had previously investigated the tissue specific expression of periostin by RNA dot-blot analysis and showed that periostin is only expressed in primary calvarial osteoblasts and in MC3T3-E1 cells. Although a weak signal was observed in the lung, in no other tissues including the brain, heart, kidney, liver, muscle, placenta, spleen, testis and thymus, could the transcripts for periostin be detected. To evaluate the expression pattern of periostin in bone tissue in detail, immunohistochemistry was performed using anti-periostin antiserum (16).

Sections of the tibia of a 5-week old mouse revealed strong positive staining in the periosteum but not in endosteum nor bone matrix. At high magnification, a positive reaction was observed in the cambial layer and to a lesser extent in the fibrous layer of the periosteum. The staining pattern indicates that the periostin protein is not anchored to the surface of the cell but secreted into the surrounding extracellular matrix possibly by osteoblasts or osteoblast precursors. Although alkaline phosphatase positive cells, possibly pre- and mature osteoblasts, were present both in the periosteum and on the surface of trabecular bone, expression of periostin was all but restricted to the alkaline phosphatase positive cells in the periosteum. No expression was observed in cartilaginous tissue including the articular surface, growth plate and calcified cartilage. Essentially identical findings were observed using in situ hybridization with the periostin RNA antisense probe. The periosteal osteoblasts were positive for periostin mRNA.

2). Periostin supports cell spreading and attachment in vitro.

Solid-phase binding assays were performed using MC3T3-E1 cells and the purified periostin protein from the conditioned media of MC3T3-E1 cells and the periostin-Fc recombinant protein. MC3T3-E1 cells were trypsinized and incubated in wells coated with either purified periostin (20 mg/ml), recombinant periostin-Fc protein (30 mg/ml), fibronectin (5 mg/ml) or BSA alone. The cells spread on the periostin, periostin-Fc and fibronectin coated wells, but not on the BSA alone coated wells.

On periostin-Fc coated plates, cell spreading occurred in a concentration dependent manner shown in Fig. 4. Periostin also supported cell attachment under reducing conditions, so secondary structure is not essential for this effect. An addition of 5 mM EDTA completely inhibited cell adhesion on both periostin and fibronectin coated plates, while GRGDS peptide (1 mM) had no effect.

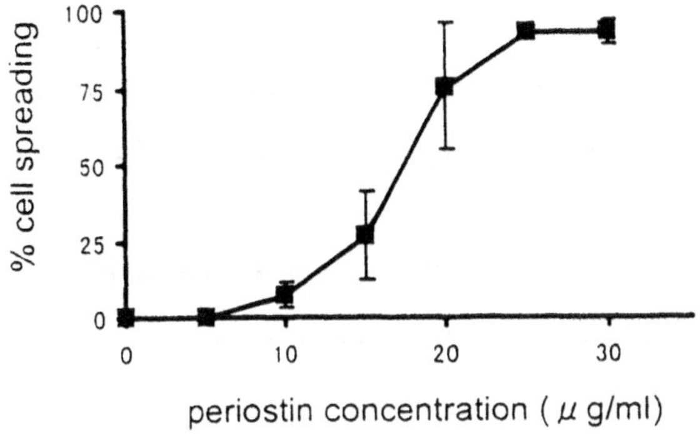


*Figure 4.* Solid phase binding assays showing adhesion of MC3T3-E1 cells to immobilized periostin protein. Trypsinized MC3T3-E1 cells were incubated in wells coated with periostin-Fc recombinant protein at the indicated concentrations. Spread cells and rounded cells were enumerated and the percentage of cell spreading was calculated.

## 3. Conclusion.

The OB-cadherin/cadherin-11 and the periostin/OSF-2 are new molecules appeared as adhesion proteins regulating the osteoblast differentiation and function. Both genes and proteins are unique markers in osteoblast lineage, therefore they will become very important to characterize the osteoblast. To obtain the final conclusion of the function of both genes, the projects of knockout mice are on going.

I would like to thank Drs. Kashima and Machiname in University of Tokyo, and Drs. Amizuka and Ozawa in Niigata University for collaborative works.

References


1. Takeichi M 1991 Cadherin cell adhesion receptors as a morphogenetic regulator. Science 251: 1451-1455.

2. Suzuki S 1996 Structural and functional diversity of cadherin superfamily: Are new members of cadherin superfamily involved in signal transduction pathway ? J Cell Biochem 61: 531-542.

3. Nagafuchi A, Takeichi M 1989 Transmembrane control of cadherin-mediated cell adhesion: a 94 kDa protein functionally associated with a specific region of the cytoplasmic domain of E-cadherin. Cell Regul 1: 37-44.

4. Ozawa M, Baribault H, Kemler R 1989 The cytoplasmic domain of the cell adhesion molecule uvomorulin associates with independent proteins structurally related in different species. EMBO J 8: 1711-1717.

5. Tanihara H, Sano K, Heimark LR, St John T, Suzuki S 1994 Cloning of five human cadherins clarifies characteristic features of cadherin extracellular domain and provides further evidence for two structurally different types of cadherin. Cell Adhes Commun 2: 15-26.

6. Okazaki M, Takeshita S, Kawai S, Kikuno R, Tsujimura A, Kudo A, Amann E 1994 Molecular cloning and characterization of OB-cadherin, a new member of cadherin family expressed in osteoblasts. J Biol Chem 269: 12092-12098.

7. Kawaguchi, J., Takeshita, S., Kashima, T., Imai, T., Machinami, R. and Kudo, A. Expression and function of the splice variant of the human vadherin-11 gene in subordination to intact cadherin-11. J. Bone Mineral. Res. in press

8. Bilezikian JP, Raisz LG, Rodan GA (ed) Principles of Bone Biology. AcademicPress, Calfornia, pp.155-165.

9. Takeshita S, Kikuno R, Tezuka K, Amann E 1993 Osteoblast-specific factor 2:cloning of a putative bone adhesion protein with homology with the insect protein fasciclin I. Biochem J 294:271-278.

10. Zinn K, McAllister L, Goodman CS 1988 Sequence analysis and neuronal expression of fasciclin I in grasshopper and Drosophila. Cell 53:577-587.

11. LeBaron RG, Bezverkov KI, Zimber MP, Pavelec R, Skonier J, Purchio AF 1995 Beta IG-H3, a novel secretory protein inducible by transforming growth factor-beta, is present in normal skin and promotes the adhesion and spreading of dermal fibroblasts in vitro. J Invest Dermatol 104:844-849.

12. Huber O, Sumper M 1994 Algal-CAMs: isoforms of a cell adhesion molecule in embryos of the alga

Volvox with homology to Drosophila fasciclin I. EMBO J. 13:4212-4222.

13. Terasaka K, Yamaguchi R, Matsuo K, Yamazaki A, Nagai S, Yamada T 1989 Complete nucleotide sequence of immunogenic protein MPB70 from Mycobacterium bovis BCG. FEMS Microbiol Lett 49:273-276.

14. Hu S, Sonnenfeld M, Stahl S, Crews ST 1998 Midline Fasciclin: a Drosophila Fasciclin-I-related membrane protein localized to the CNS midline cells and trachea. J Neurobiol 35:77-93.

15. Ducy P, Zhang R, Geoffroy V, Ridall AL, Karsenty G 1997 Osf2/Cbfa1: a transcriptional activator of osteoblast differentiation. Cell 89:747-754.

16. Horiuchi, K., Amizuka, N., Takeshita, S., Takamatsu, H., Katsuura, M., Ozawa, H., Toyama, Y., Bonewald, L.F. and Kudo, A. Identification and characterization of a novel protein, Periostin (OSF-2) with restricted expression to periosteum and periodontal ligament and increased expression by transforming growth factor b. paper is submitted.

# OSTEOCLAST DIFFERENTIATION FACTOR IS A LIGAND FOR OSTEOPROTEGERIN / OSTEOCLASTOGENESIS-INHIBITORY FACTOR

NOBUYUKI SHIMA[1], HISATAKA YASUDA[1], NOBUAKI NAKAGAWA[1], KYOJI YAMAGUCHI[1], EISUKE TSUDA[1], TOMONORI MORINAGA[1], TATSUO SUDA[2], and KANJI HIGASHIO[1]

[1]*Research Institute of Life Science, Snow Brand Milk Products Co. Ltd., Tochigi 329-0512, Japan*

[2]*Department of Biochemistry, School of Dentistry, Showa University, Tokyo 142-8555, Japan*

**ABSTRACT**; Osteoblasts or bone marrow stromal cells support osteoclast development through a mechanism of cell-to-cell interaction with osteoclast progenitors. we purified and molecularly cloned osteoclastogenesis-inhibitory factor (OCIF), which turned out to be identical to osteoprotegerin (OPG). OPG/OCIF is a secreted member of the TNF receptor family that inhibits differentiation and activation of osteoclasts. OPG/OCIF specifically bound to a bone marrow stromal cell line, ST2 treated with $1,25(OH)_2D_3$, but not to untreated cells. To identify a OPG/OCIF ligand, we screened a cDNA expression library of ST2 cells treated with $1,25(OH)_2D_3$ using OPG/OCIF as a probe. The cloned molecule was found to be a member of the membrane-associated TNF ligand family, and it induced osteoclastogenesis *in vitro* in the absence of osteoblasts/stromal cells. Expression of its gene in osteoblasts/stromal cells was up-regulated by osteotropic factors. Antibody against this protein negated bone resorption elicited by osteotropic factors in a fetal mouse long bone culture system. OPG/OCIF abolished the protein-mediated osteoclastogenesis. We conclude that the protein is osteoclast differentiation factor (ODF), a long-sought mediator responsible for an essential signal to osteoclast progenitors for their differentiation into osteoclasts, and that OPG/OCIF acts as a soluble competitor against ODF receptor that is presumably expressed on osteoclasts and osteoclast progenitors.

## 1. A hypothetical membrane-bound factor, ODF

Osteoclast-like cells (OCLs) can be developed *in vitro* from hematopoietic cells by co-culturing with osteoblasts or bone marrow stromal cells in the presence of such stimulators of bone resorption as interleukin (IL)-6, IL-11, parathyroid hormone (PTH), prostaglandin $E_2$ ($PGE_2$), and 1,25-dihydroxyvitamin $D_3$ [$1,25(OH)_2D_3$] (1, 2). These stimulators are classified into three categories in terms of signal transduction pathways: vitamin D receptor-mediated signals [$1,25(OH)_2D_3$]; protein kinase A-mediated signals (PTH, and

$PGE_2$); and gp130-mediated signals (IL-6 and IL-11). Results of co-culture experiments demonstrated that development of osteoclasts is regulated by osteoblasts/stromal cells through a mechanism of cell-to-cell interaction with osteoclast progenitors (1-3). Thus, we proposed that a hypothetical membrane-bound factor(s) expressed on osteoblasts/stromal cells in response to the stimulators of bone resorption induces osteoclastogenesis by signaling to osteoclast progenitors. We named the factor "osteoclast differentiation factor (ODF)" (3).

## 2. Characterization of OPG/OCIF

In 1994, we succeeded in purifying osteoclastogenesis-inhibitory factor (OCIF) from conditioned medium of human fibroblasts. OCIF specifically inhibits in vitro OCL formation elicited through the three distinct signaling pathways stimulated by $1,25(OH)_2D_3$, PTH, and IL-11 (4). The nucleotide sequence analysis of OCIF cDNA (5) revealed that OCIF is a secreted member of the tumor-necrosis factor (TNF) receptor family and is identical to osteoprotegerin (OPG) (6). Analyses of transgenic mice overexpressing OPG/OCIF, animals injected with OPG/OCIF, and OPG/OCIF knockout mice established the physiological role of OPG/OCIF as an osteoclastogenesis-inhibitory factor (5-8).

## 3. OPG/OCIF-binding protein

A mouse bone marrow-derived stromal cell line, ST2, is known to support OCL formation from mouse spleen cells in the presence of $1,25(OH)_2D_3$ and dexamethasone (Dex) (2). We previously showed that OPG/OCIF specifically binds to ST2 cells treated with $1,25(OH)_2D_3$ and Dex, but not to untreated ST2 cells (4, 5). When the binding sites on the treated ST2 cells were occupied by OPG/OCIF, the cells failed to support OCL formation from spleen cells (5). Cross-linking study using $^{125}$I-OPG/OCIF revealed that a 40 k protein on the treated ST2 cells binds to OPG/OCIF (5). These results raised the possibility that the binding protein of 40 k is a ligand for OPG/OCIF and identical to ODF.

## 4. Cloning of mouse ODF cDNA

To identify the OPG/OCIF binding protein, we screened a cDNA expression library of ST2 cells treated with $1,25(OH)_2D_3$ and Dex using $^{125}$I-OPG/OCIF. A cDNA clone encoding 316 amino acids was isolated (9). Hydropathy analysis showed the absence of a signal sequence and the presence of an internal 24-residue hydrophobic domain, which presumably represents a transmembrane domain. This structure is typical of a type II transmembrane protein with an extracellular C-terminal region. A homology search of the GenBank sequence database revealed that the C-terminal 165 residues of the protein had a significant homology to the extracellular domains of the TNF ligand family members. As described below, the protein satisfied the major criteria for ODF in view of its biological activity and its gene expression regulated by bone-resorbing factors. We therefore renamed the protein ODF.

ODF, also designated as OPG ligand (10), has been found to be identical to TRANCE and RANKL, which were cloned as respective regulators of T-cells and dendritic cells (11, 12).

### 5. ODF induces OCL formation from spleen cells

To examine if ODF mediates a cell-to-cell signal responsible for osteoclastogenesis, we carried out an in vitro OCL formation assay by evaluating TRAP activity and calcitonin binding, a combination of which is unique to osteoclasts (9). When COS-7 cells expressing ODF ($COS^{ODF}$) or control COS-7 cells transfected with the empty vector ($COS^{Vec}$) were fixed with paraformaldehyde and then mouse spleen cells were cultured on the fixed cells for 6 days in the presence of M-CSF, TRAP-positive multinucleated cells appeared on the $COS^{ODF}$ cells but not on the $COS^{Vec}$ cells. Concurrent addition of OPG/OCIF to the cultures inhibited the TRAP- and calcitonin receptor-positive cell formation in a dose-dependent manner. These results indicate that ODF mediates the cell-to-cell signaling essential for osteoclastogenesis. M-CSF was indispensable for the ODF-mediated OCL formation. To further examine the biological effect of ODF, we produced a soluble ODF (sODF) consisting of the extracellular domain of ODF and thioredoxin. When mouse spleen cells were cultured in the presence of 10 ng/ml M-CSF and various concentrations of sODF, TRAP-positive multinucleated cells were formed in a dose-dependent manner. OPG/OCIF negated the effect of sODF. Autoradiography using $^{125}$I-calcitonin confirmed the presence of calcitonin receptors on the induced TRAP-positive cells. Furthermore, when these OCLs were cultured on dentine slices for 3 days in the presence of M-CSF and sODF, numerous resorption pits were formed on the slices.

### 6. sODF acts directly on osteoclast progenitors

Mature monocytes and alveolar macrophages, as well as several cell lines of the macrophage lineage, can differentiate into OCLs when co-cultured with stromal cells in the presence of $1,25(OH)_2D_3$. A macrophage cell line, C7, is also capable of differentiating into OCLs in such a co-culture system. sODF induced the formation of TRAP- and calcitonin receptor-positive multinucleated cells from C7 cells in a dose-dependent manner in the presence of M-CSF, indicating that ODF acts directly on osteoclast progenitors. The TRAP-positive multinucleated cells formed from C7 cells produced numerous resorption pits on dentine slices (9). To elucidate the mechanism of human osteoclastogenesis, we further examined the effect of sODF on human peripheral blood mononuclear cells (PBMCs). Treatment of human PBMCs with sODF together with M-CSF induced OCLs, which are TRAP-, vitronectin receptor-, and calcitonin receptor-positive and are capable of resorbing bone (13).

### 7. Expression of ODF gene in stromal cells, osteoblasts, and tissues (9)

Northern blot analysis using a ODF cDNA probe showed that a single mRNA transcript of approximately 2.4-kb was present in ST2 cells treated with $1,25(OH)_2D_3$ and Dex, but not in the untreated cells. Up-regulation of the ODF gene expression was also observed in mouse primary osteoblasts cultured in the presence of $1,25(OH)_2D_3$, IL-11, $PGE_2$, or PTH. Analysis of tissue distribution of the ODF transcript showed that the gene was strongly expressed in trabecular bone, thymus, and lung, and weakly expressed in spleen and bone marrow. High-level expression of ODF gene in trabecular bone suggests a role of ODF for osteoclastogenesis in bone tissue.

## 8. Role of ODF in the microenvironment of bone

To explore the role of ODF in bone tissue, we examined the effect of sODF and OPG/OCIF on bone resorption in a fetal mouse long bone culture system(14). sODF significantly stimulated $^{45}Ca$ release from the labeled bone in a dose-dependent manner and the addition of OPG/OCIF abolished the effect of sODF, suggesting that ODF plays a role for osteoclastogenesis in the microenvironment of bone. To further elucidate whether ODF-mediated signals are involved in bone resorption, we examined the effects of OPG/OCIF and anti-ODF antibody on bone resorption stimulated with various bone-resorbing factors in the organ culture system. $1,25(OH)_2D_3$, $PGE_2$, or PTH significantly stimulated $^{45}Ca$ release and addition of OPG/OCIF or anti-ODF antibody inhibited the effect of these stimulators in a dose-dependent manner. These results led us to conclude that ODF mediates an essential signal for bone resorption induced by various bone-resorbing factors in bone tissue. Taken together with the findings that ODF expression in osteoblastc cells is up-regulated by various bone-resorbing factors, the results suggest that the stimulation of bone resorption by these factors is through up-regulation of ODF on the membrane of osteoblasts/stromal cells.

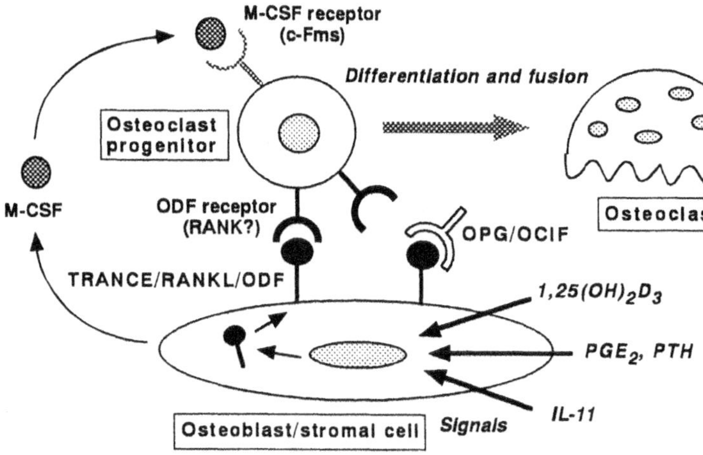


Fig. 1. A model illustrating the mechanism by which osteoblasts/stromal cells regulate osteoclastogenesis.

## 9. Conclusion

The present study indicates that osteoblasts/stromal cells play an essential role in osteoclastogenesis through the expression of ODF (Fig. 1). Three distinct signals stimulated by $1{,}25(OH)_2D_3$, $PGE_2$/PTH, and IL-11 induce ODF expression on osteoblasts/stromal cells. ODF mediates a signal for osteoclastogenesis through ODF receptor expressed on osteoclast progenitors. OPG/OCIF inhibits osteoclastogenesis by interrupting the binding of ODF to ODF receptor. M-CSF produced by osteoblasts or stromal cells is also indispensable for differentiation of osteoclast progenitors.

## 10. References


1. Takahashi, N., Akatsu, T., Uadagawa, N., Sasaki, T., Yamaguchi, A., Moseley, J. M., Martin, T. J., and Suda, T. (1988) Osteoblastic cells are involved in osteoclast formation, *Endocrinology* **123**,2600-2602.
2. Udagawa, N., Takahashi, N., Akatsu, T., Sasaki, T., Yamaguchi, A., Kodama, H., Martin, T. J., and Suda, T. (1989) The bone marrow-derived stromal cell lines MC3T3-G2/PA6 and ST2 support osteoclast-like cell differentiation in cocultures with mouse spleen cells, *Endocrinology* **125**,1805-1813.
3. Suda, T., Takahashi, N., and Martin, T. J. (1992) Modulation of osteoclast differentiation, *Endocrine Rev.* **13**,66-80.
4. Tsuda, E., Goto, M., Mochizuki, S., Yano, K., Kobayashi, F., Morinaga, T., and Higashio, K. (1997) Isolation of a novel cytokine from human fibroblasts that specifically inhibits osteoclastogenesis, *Biochem. Biophys. Res. Commun.* **234**,137-142.
5. Yasuda, H., Shima, N., Nakagawa, N., Mochizuki, S., Yano, K., Fujise, N., Sato, Y., Goto, M., Yamaguchi, K., Kuriyama, M., Kanno, T., Murakami, A., Tsuda, E., Morinaga, T., and Higashio, K. (1998) Identity of osteoclastogenesis inhibitory factor (OCIF) and osteoprotegerin (OPG): a mechanism by which OPG/OCIF inhibits osteoclastogenesis in vitro, *Endocrinology* **139**,1329-1337.
6. Simonet, W. S., Lacey, D. L., Dunstan, C. R., Kelley, M., Chang, M.-S., Luthy, R., Nguyen, H. Q., Wooden, S., Bennett, L., Boone, T., Shimamoto, G., DeRose, M., Eliott, R., Colombero, A., Tan, H.-L., Trail, G., Sullivan, J., Davy, E., Bucay, N., Renshaw-Gegg, L., Hughes, T. M., Hill, D., Pattison, W., Campbell, P., Sander, S., Van, G., Tarpley, J., Derby, P., Lee, R., Amgen EST Program, and Boyle, W. J. (1997) Osteoprotegerin: a novel secreted protein involved in the regulation of bone density, *Cell* **89**,309-319.
7. Mizuno, A., Amizuka, N., Irie, K., Murakami, A., Fujise, N., Kanno, T., Sato, Y., Nakagawa, N., Yasuda, H., Mochizuki, S., Gomibuchi, T., Yano, K., Shima, N., Washida, N., Tsuda, E., Morinaga, T., Higashio, K., and Ozawa, H. (1998) Severe osteoporosis in mice lacking osteoclastogenesis inhibitory factor/osteoprotegerin, *Biochem Biophys Res Commun*. **247**,610-615.
8. Bucay, N., Sarosi, I., Dunstan, C., Morony, S., Tarpley, J., Capparelli, C., Scully, S., Tan, H., Xu, W., Lacey, D., Boyle, W., and Simonet, W. (1998) Osteoprotegerin-deficient mice develop early onset osteoporosis and arterial calcification, *Genes Dev*. **12**,1260-1268.
9. Yasuda, H., Shima, N., Nakagawa, N., Yamaguchi, K., Kinosaki, M., Mochizuki, S., Tomoyasu, A., Yano, K., Goto, M., Murakami, A., Tsuda, E., Morinaga, T., Higashio, K., Udagawa, N., Takahashi, N., and Suda, T. (1998) Osteoclast differentiation factor is a ligand for osteoprotegerin/ osteoclastogenesis-inhibitory factor and identical to TRANCE/RANKL, *Proc. Natl. Acad. Sci. U.S.A.* **95**,3597-3602.
10. Lacey, D., Timms, E., Tan, H., Kelley, M., Dunstan, C., Burgess, T., Elliott, R., Colombero, A., Elliott, G., Scully, S., Hsu, H., Sullivan, J., Hawkins, N., Davy, E., Capparelli, C., Eli, A., Qian, Y., Kaufman, S., Sarosi, I., Shalhoub, V., Senaldi, G., Guo, J., Delaney, J., and Boyle, W. (1998) Osteoprotegerin ligand is a cytokine that regulates osteoclast differentiation and activation, *Cell*. **93**,165-176.
11. Wong, B. R., Rho, J., Arron, J., Robinson, E., Orlinick, J., Chao, M., Kalachikov, S., Cayani, E., Bartlett III, F. S., Frankel, W. N., Lee, S. Y., and Choi, Y. (1997) TRANCE is a novel ligand of the tumor necrosis factor receptor family that activates c-Jun N-terminal kinase in T cells, *J. Biol. Chem.* **272**,25190-25194.
12. Anderson, D. A., Maraskovsky, E., Billingsley, W. L., Dougall, W. C., Tometsko, M. E., Roux, E. R., Teepe, M. C., DuBose, R. F., Cosman, D., and Galibert, L. (1997) A homologue of the TNF receptor ans its ligand enhance T-cell growth and dendritic function, *Nature* **390**,175-179.
13. Matsuzaki, K., Udagawa, N., Takahashi, N., Yamaguchi, K., Yasuda, H., Shima, N., Morinaga, T., Toyama, Y., Yabe, Y., Higashio, K., and Suda, T. (1998) Osteoclast differentiation factor (ODF) induces osteoclast-like cell formation in human peripheral blood mononuclear cell cultures, *Biochem. Biophys. Res. Commun.*. **246**,199-204.
14. Tsukii, K., Shima, N., Mochizuki, S., Yamaguchi, K., Kinosaki, M., Yano, K., Shibata, O., Udagawa, N., Yasuda, H., Suda, T., and Higashio, K. (1998) Osteoclast differentiation factor mediates an essential signal for bone resorption induced by 1 alpha,25-dihydroxyvitamin D3, prostaglandin E2, or parathyroid hormone in the microenvironment of bone, *Biochem. Biophys. Res. Commun.*. **246**,337-341.

# ERYTHROPOIETIN PROTECTS NEURONS FROM ISCHEMIC DAMAGE.

SEIJI MASUDA, MASAYA NAGAO AND RYUZO SASAKI

Division of Applied Life Sciences, Graduate School of Agriculture,
Kyoto University, Kyoto 606-8502, Japan

Erythropoietin (EPO) has been shown to protect primary cultured neurons from *N*-methyl-D-aspartate (NMDA) receptor-mediated glutamate toxicity. Here we report EPO protects neurons against ischemia-induced cell death and the endogenous brain EPO is crucial for neuronal survival. The presence of EPO in neuron cultures did not repress a NMDA receptor-mediated increase in intracellular $Ca^{2+}$, but rescued the neurons from nitric oxide-induced death.

---

EPO regulating erythropoiesis is mainly produced by the kidney in adults and by the liver at fetal stages (1, 2). Stimulation of red blood cell formation was thought to be the sole physiological function of EPO, but a novel function in the central nervous system (CNS) has been proposed (3-7). Primary cultured astrocytes have been shown to produce EPO and low oxygen tension stimulates the production of EPO through an increase in its mRNA (8, 9). EPO mRNA is expressed in the adult rat brain and the expression is hypoxia-inducible (10). Messenger RNAs of EPO and EPOR are also expressed in the primate brain (9). EPO protects primary cultured hippocampal and cerebral cortical neurons from NMDA receptor-mediated glutamate toxicity (3), which is believed to be a major cause of neuron death by ischemia. It remains unknown, however, whether or not the endogenous brain EPO functions *in vivo*.

In this paper, we report that EPO plays an important role in protecting neurons from ischemia-induced cell death. Experimental results on the mechanism underlying the protective effect of EPO on glutamate-induced neuron death are also reported.

## MATERIALS AND METHODS

*Infusion of erythropoietin and soluble erythropoietin receptor.*

Male Mongolian gerbils each weighing 70-80 g were anesthetized. An osmotic minipump was implanted as described (11-15). Recombinant human EPO was dissolved in a vehicle consisting of 0.01 M phosphate-buffered saline, pH 7.5 (PBS) and 0.1% bovine serum albumin. EPO was infused for 7 days into the left lateral ventricle of each normothermic gerbil in which 3-min forebrain ischemia had been induced; control animals received vehicle infusion. Soluble EPOR (sEPOR) was infused for 7 days into the left lateral ventricles of normothermic gerbils in which 2.5-min or 3-min forebrain ischemia had been induced, control ischemic animals received the infusion of vehicle or heat-denatured sEPOR (dsEPOR). The infusion was started at 8 h or at 24 h before the ischemic insult in the EPO-treated groups and in the sEPOR-treated groups, respectively. Occlusion of the common carotid arteries, stimation of learning ability by passive avoidance task and histopathological study of hippocampal CA1 region were performed as described (11-15). The effects of EPO and sEPOR on response latency and CA1 neuronal density were evaluated by the two-tailed Mann-Whitney U-test.

*In vitro experiments.*

Hippocampal and cerebral cortical neurons from brains of 19-day old fetal Wister rats were cultured as described previously (3).

Rate of $^{45}$Ca uptake by cerebral cortical neurons were determined as described (16). Intracellular calcium concentration was determined by the method previously reported (17), using fura-2 loaded cerebral cortical neurons.

Effects of glutamate, Me-TC (a nitric oxide synthase inhibitor), SNP (a nitric oxide-generating agent), and EPO on hippocampalneurons were performed as described (3). The neurons were incubated with 200 μM glutamate or 1 mM sodium nitroprusside (SNP) (Dojin) for 15 min. When effect of S-methyl-L-thiocitrulline (Me-TC) on glutamate toxicity was examined, the neurons were incubated with 25 μM Me-TC 30 min before glutamate challenge. When the neuroprotective effect of EPO was examined, 1 U/ml EPO was added to cultures 24 h before incubation of neurons with test materials.

## RESULTS AND DISCUSSION

*Protective effect of EPO on ischemia-induced neuron death.*

The continuous infusion of EPO at a dose of 2.5, 5 or 25 U/day for 7 days into the lateral ventricle caused a significant prolongation in response latency time (Fig. 1A). Subsequent histological examinations revealed that treatment with EPO at the dose of 2.5, 5 or 25 U/day rescued many ischemic neurons that were destined to degenerate without the treatment (Fig. 1B, G, H). These results indicate that the cerebroventricular infusion of EPO prevents the ischemia-induced learning disability and rescues hippocampal CA1 neurons from lethal ischemic damage.

*Stimulation of neuron death by sEPOR under mild ischemia.*

Ischemia for 2.5 min did not reduce neuronal cell density, while there was a significant reduction in 3-min ischemia. This finding has made it possible to examine whether the infusion of sEPOR capable of binding with EPO provokes neuron damage in 2.5-min ischemia. The continuous infusion of sEPOR at a dose of 5 or 10 μg/day for 7 days into the cerebral ventricles did not show a detrimental effect but the infusion of higher doses (25 or 50 μg/day) caused a significant reduction in response latency time (Fig. 2A) and a significant decrease in hippocampal CA1 neuron density (Fig. 2B), when compared with those of 2.5-min ischemic animals infused with vehicle. Neither the response latency nor neuronal density was altered by the infusion of the inactive dsEPOR (5 or 25 μg/day) in 2.5-min ischemic gerbils. The infusion of sEPOR (25 μg/day) in normal gerbils did not induce neuronal death in the hippocampal CA1 field (data not shown). These findings indicate that brain EPO of possibly astrocyte origin exists as a trophic agent essential for neuronal survival when the brain is loaded with a sublethal ischemic insult.

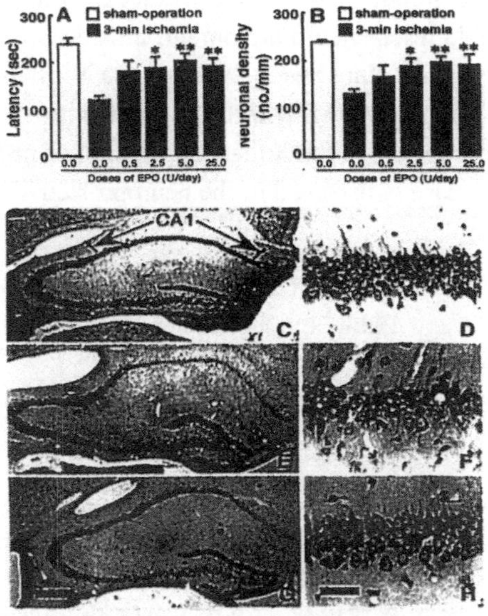


Fig. 1. Effects of intracerebroventricular EPO infusion on the response latency and hippocampal CA1 region of 3-min ischemic gerbils.

A, response latency; B, CA1 neuronal density. Open columns indicate sham-operated (sham-op) animals and closed columns indicate vehicle- or EPO-infused ischemic animals. Each value represents mean ± SE (n = 8-11). $*P < 0.05$ and $**P < 0.01$, significantly different from the corresponding vehicle-infused ischemic group.

C–H, photomicrographs of hippocampal sections stained with cresyl violet; C, E and G, low magnification (bar = 1.0 mm); D, F and H, high magnification (bar = 0.1 mm) corresponding to C, E and G. C and D, a sham-operated animal; E and F; an ischemic animal infused with vehicle; G and H, an ischemic animal infused with 5 U/day of EPO.

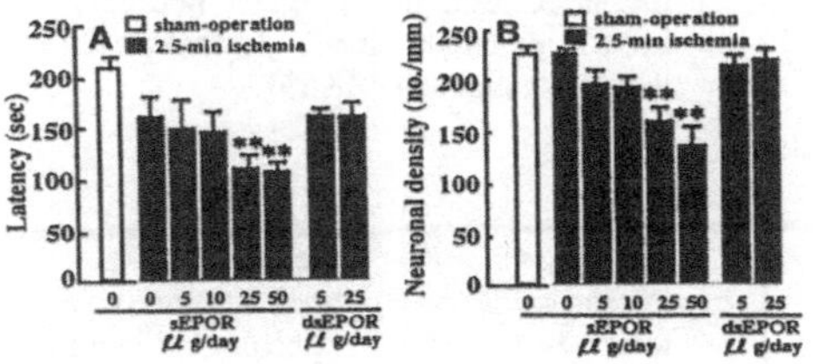


Fig. 2. Effects of intracerebroventricular infusion of sEPOR and heat-denatured sEPOR (dsEPOR) on the response latency and hippocampal CA1 region of 2.5-min ischemic gerbils.

A, response latency; B, CA1 neuronal density. Each value represents mean ± SE (n = 6-8). $**P < 0.01$, significantly different from the corresponding vehicle-infused ischemic group.

*Neuroprotective effect of EPO on cultured neurons.*

For better understanding of the molecular mechanism underlying the neuroprotective action of EPO, we further investigated effect of EPO on cultured neurons. A massive increase of intracellular $Ca^{2+}$ concentration evoked by glutamate-induced NMDA receptor activation plays a critical role in triggering intracellular events that elicit cell destruction. EPO protects primary cultured neurons from NMDA receptor-mediated glutamate toxicity (3). EPO may exhibit its neuroprotective action by repressing glutamate-induced increase in $Ca^{2+}$ concentration. Figure 3 shows effect of EPO on $Ca^{2+}$ uptake by cultured neurons. Glutamate stimulated the rate of $Ca^{2+}$ uptake when compared with the rate in the presence of MK801 (a potent NMDA receptor antagonist), but EPO did not reduce glutamate-mediated stimulation (Fig. 3A). Likewise, glutamate induced an increase in intracellular $Ca^{2+}$ concentration but EPO failed to repress this increase (Fig. 3B).

As previously reported (3), EPO protected neurons from glutamate-induced death (Fig. 4). Me-TC, an inhibitor of nitric oxide synthase, also neutralized the glutamate toxicity almost completely, suggesting that the glutamate toxicity is mediated by nitric oxide. We examined whether or not EPO could rescue neurons from nitric oxide toxicity using SNP, a nitric oxide-generating agent. Incubation of neurons with SNP caused significant neuronal loss, but pretreatment of neurons with EPO resulted in almost complete survival of the neurons. These results suggest that suppression of nitric oxide toxicity but not intracellular calcium increase is involved in the neuroprotective action of EPO.

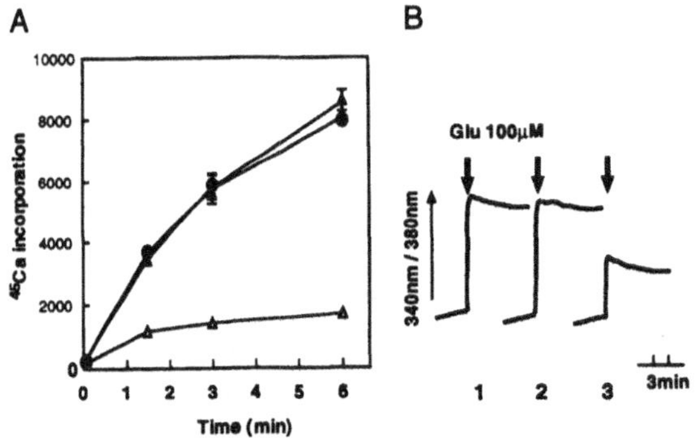


Fig. 3. Effect of EPO on glutamate-induced $Ca^{2+}$ uptake into cultured cerebral cortical neurons.

A, rate of $^{45}Ca^{2+}$ uptake; B, increase in intracellular $Ca^{2+}$ concentration $[Ca^{2+}]_i$. In A, glutamate (100 μM) was added to cultured neurons. MK801 (10 μM), a NMDA receptor antagonist, was added with glutamate. ●, glutamate; ▲, glutamate + EPO; △, glutamate + MK801. Each value is the mean ± SE of triplicate experiments. In B, intracellular $Ca^{2+}$ concentrations were measured using fura-2 loaded neurons. Line 1, glutamate; line 2, glutamate + EPO; line 3, glutamate + MK801.

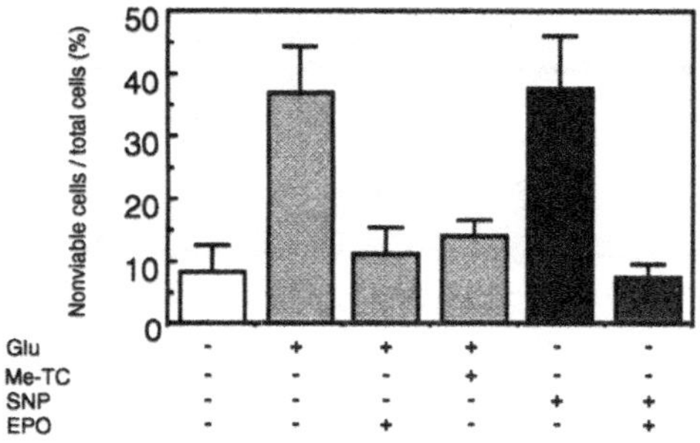


Fig. 4. EPO protects sodium nitroprusside (SNP)-derived nitric oxide toxicity on cultured hippocampal neurons.

Neurons were incubated with test materials as described in "Materials and Methods". Glu, glutamate; Me-TC, a nitric oxide synthase inhibitor. Total and non-viable cell number was counted. Each value is the mean ± SE of triplicate experiments.

Homozygous mice carrying null mutations in the EPO or EPOR genes died around embryonic Day 13 due to deficiency in erythropoiesis in the fetal liver (18), providing no information on the possible functions of EPO in adults. The use of a soluble receptor, as in the present study, would be useful to find a novel function of the ligand, when disruption of the ligand or its receptor gene results in fetal death.

## REFERNCES


1. Krantz, S. B. (1991) *Blood* **77**, 419-434.
2. Jelkmann, W. (1992) *Physiol. Rev.* **72**, 449-489.
3. Morishita, E., Masuda, M., Nagao, M., Yasuda, Y. & Sasaki, R. (1997) *Neuroscience* **76**, 105-116.
4. Masuda, S., Chikuma, M. & Sasaki, R. (1997) *Brain Res.* **746**, 63-70.
5. Masuda, S., Nagao, M., Takahata, K., Konishi, Y., Gallyas, F. Jr., Tabira, T. & Sasaki, R. (1993) *J. Biol. Chem.* **268**, 11208-11216.
6. Morishita, E., Narita, H., Nishida, M., Kawashima, N., Yamagishi, K., Masuda, S., Nagao, M., Hatta, H. & Sasaki, R. (1996) *Blood* **88,** 465-471.
7. Digicaylioglu, M., Bichet, S., Marti, H. H., Wenger, R. H., Rivas, L. A., Bauer, C. & Gassmann, M. (1995) *Proc. Natl. Acad. Sci. USA.* **92**, 3717-3720.
8. Masuda, S., Okano, M., Yamagishi, K., Nagao, M., Ueda, M. & Sasaki, R. (1994) *J. Biol. Chem.* **269**, 19488-19493.
9. Marti, H. H., Wenger, R. H., Rivas, L. A., Straumann, U., Digicaylioglu, M., Henn, V., Yonekawa, Y., Bauer, C. & Gassmann, M. (1996) *Eur. J. Neurosci.* **8**, 666-676.
10. Tan, C. C., Eckardt, K. U., Firth, J. D. & Ratcliffe, P. J. (1992) *Am. J. Physiol.* **263**, F474-F481.
11. Sano, A., Matsuda, S., Wen, T-C., Kotani, Y., Kondoh, K., Ueno, S., Kakimoto, Y., Yoshimura, H. & Sakanaka, M. (1994) *Biochem. Biophys. Res. Comm.* **204**, 994-1000.
12. Wen, T-C., Matsuda, S., Yoshimura, H., Aburaya, J., Kushihata, F. & Sakanaka, M. (1995) *Neuroscience* **65**, 513-521.
13. Wen, T-C., Matsuda, S., Yoshimura, H., Kawabe, T. & Sakanaka, M. (1995) *Neurosci. Lett.* **191**, 55-58.
14. Kotani, Y., Matsuda, S., Wen, T-C., Sakanaka, M., Tanaka, J., Maeda, N., Kondon, K., Ueno, S. & Sano, A. (1996) *J. Neurochem.* **66**, 2197-2200.
15. Matsuda, S., Wen, T-C., Morita, F., Otsuka, H., Igase, K., Yoshimura, H. & Sakanaka, M. (1996) *Neurosci. Lett.* **204**, 109-112.
16. Artalejo, C. R., Garcia, A.G. & Aunis, D. (1987) *J. Biol. Chem.* **262**, 915-926.
17. Katoh, H., Watabe, A., Sugimoto, Y., Ichikawa, A. and Negishi, M. (1995) *Biochim. Biophys. Acta* **1244**, 41-48.
18. Wu, H., Liu, X., Jaenisch, R. & Lodish, H. F. (1995) *Cell* **83**, 59-67.

# MONOVALENT ANTIGEN ACTIVATES ANTIBODY/CYTOKINE RECEPTOR CHIMERA AND CONTROLS HEMATOPOIETIC CELL GROWTH

MASAHIRO KAWAHARA[1], HIROSHI UEDA[1], KOUHEI TSUMOTO[2], KAZUO TODOKORO[3], WALT MAHONEY[4], IZUMI KUMAGAI[2], and TERUYUKI NAGAMUNE[1]
[1]*Department of Chemistry and Biotechnology, Graduate School of Engineering, The University of Tokyo, Tokyo 113-8656, Japan* ; [2]*Department of Biochemistry and Engineering, Graduate School of Engineering, Tohoku University, Sendai, 980-8579, Japan* ; [3]*Tsukuba Life Science Center, The Institute of Physical and Chemical Research (RIKEN), Ibaraki, 305-0074, Japan* ; [4]*Boehringer Mannheim Corporation, Pleasanton, CA 94588-2722, U.S.A.*

**ABSTRACT** : Proliferation and differentiation of hematopoietic cells are controlled by cytokines, which activate their specific receptors expressed on the cell membrane. In this study, we developed a pair of novel chimeric receptors activated by a monovalent antigen. The ligand-recognition domain of erythropoietin receptor(EPOR) was replaced by either $V_H$ or $V_L$ region of anti-hen egg lysozyme(HEL) antibody, HyHEL-10. The two chimeric receptors were expressed in interleukin(IL)-3-dependent Ba/F3 cells. Addition of HEL induced heterodimerization of the two chimeric receptors and proliferation in a dose-dependent manner. HEL stimulation made cells proliferate as efficiently as cytokine stimulation. These results indicate that the action of a cytokine is substituted by a cheap antigen, and that proliferation of cells expressing chimeric receptors is artificially controlled. This approach may be applied to inexpensive protein production and positive selection of gene-transfected cells.

**KEY WORDS :** antibody/cytokine receptor chimera, growth control, hematopoietic cell, hen egg lysozyme

## 1. INTRODUCTION

Cytokines control proliferation or differentiation of animal cells via receptor-mediated signal transduction. Ligand binding causes the specific receptor subunits to dimerize or oligomerize, which triggers the activation of intracellular signal transducers, such as Jak, Stat, Ras, and MAP kinase. Thus, cytokines have been used for growth control and maintenance of animal cells. However, cytokines are so expensive that the use of cytokines has the demerit for industrial applications, such as protein production. To overcome this problem, we investigated whether the ligand-recognition domain of a cytokine receptor could be changed to the binding domain of antibody which recognizes an inexpensive ligand quite different from cytokines.

In this study, we chose hen egg lysozyme(HEL) as the inexpensive ligand and $V_H$ or $V_L$ region of anti-HEL antibody, HyHEL-10, as its binding domain. The reason why we used this model system is that HyHEL-10 has a noteworthy characteristic that interaction between $V_H$ and $V_L$ is weak($Ka<10^6$ [$M^{-1}$]) in the absence of HEL whereas interaction of the complex composed of $V_H$, HEL and $V_L$ is strong($Ka=10^9$[$M^{-1}$])[1]. To utilize this characteristic for growth control, we should select signal transducing receptors which are inactive in the absence of ligand and activated by dimerization in the presence of ligand. As a receptor satisfying these conditions, we selected murine erythropoietin receptor(EPOR) which promotes growth and differentiation of erythroid cells in response to its cognate ligand, erythropoietin. The ligand-recognition domain of EPOR was replaced by either $V_H$ or $V_L$ region of HyHEL-10, and the two chimeric receptors were expressed in IL-3-dependent Ba/F3 cells. Whether the addition of HEL could rescue the transfectants in the absence of IL-3 as a result of the $V_H$-HEL-$V_L$ complex formation of the receptor ectodomain and the dimerization in the endodomain, or not, was investigated.

## 2. MATERIALS AND METHODS

*Plasmid construction*

HyHEL-10 $V_H$ or $V_L$ gene was originated from pKTN2[2]. By substituting the N-terminal half of the EPOR ectodomain gene of pME-ER[3] with either $V_H$ or $V_L$ gene, the chimeric receptor expression vectors pME-$V_H$ER and pME-$V_L$ER were made. Final constructs are composed of SRα promoter, immunoglobulin H chain signal peptide gene, $V_H$ or $V_L$ gene, intron, EPOR encoding sequence from Val 118 to C-terminus, SV40 early poly A and rest of vector sequence containing G418 and ampicillin resistance genes.

*Transfection into Ba/F3 cells*

A murine IL-3-dependent pro-B cell line, Ba/F3, was maintained in RPMI1640 medium supplemented with 10% fetal bovine serum, 30μg/ml of kanamycin, and 2ng/ml of murine IL-3(Genzyme). $2x10^6$ cells were transfected with 5μg each of pME-$V_H$ER and pME-$V_L$ER by electroporation. The transfectants were selected in medium containing 350μg/ml of G418, and then in medium containing 10μg/ml of HEL. The selected cells were cloned in medium containing 1μg/ml of HEL by limiting dilution in the presence of excess Ba/F3 cells, and named Ba/HEL. Ba/HEL cells were maintained in medium containing 500ng/ml of HEL. Ba/F3 cells were also transfected with wild type EPOR gene(pME-ER) as a control. The transfectants were named Ba/EPOR.

*Formation of ligand-receptor complex*

Ba/HEL cells were washed three times with phosphate-buffered saline(PBS) containing 1mg/ml of BSA(BSA-PBS), and incubated with BSA-PBS containing biotinylated HEL at

4°C for 90min. Cells were washed three times with BSA-PBS and lysed. After centrifugation, the supernatant was mixed with streptavidin immobilized on agarose beads(Sigma) at 4°C for 60min to capture the biotinylated ligand-receptor complexes. Beads were washed, and then boiled in Laemmli sample buffer. Captured proteins were resolved by SDS-PAGE, transferred to nitrocellulose membrane, probed with anti-mouse EPOR C-terminus antibody(Santa Cruz) and HRP-conjugated anti-rabbit IgG(Biosource), and detected by ECL system(Amersham).

*Signal transmission to Stat5b*

Ba/HEL cells were starved for 12h in medium without HEL and IL-3. Cells were stimulated with HEL at 37°C for indicated minutes, and the reaction was stopped and dephosphorylation was inhibited by addition of ice-cold PBS containing final concentration of 1mM $Na_3VO_4$. After cell lysis the lysate was immunoprecipitated with anti-mouse Stat5b antibody(Santa Cruz) and blotted with HRP-conjugated anti-phosphotyrosine antibody(Transduction Laboratories). Reprobing was performed with anti-Stat5b antibody and HRP-conjugated anti-rabbit IgG.

*Cell Proliferation Assay*

Cells were washed twice in medium without IL-3, HEL and EPO, and seeded into 24 well plates at concentration of $3x10^3$ cells/ml with indicated concentrations of HEL(for Ba/HEL) or EPO(for Ba/EPOR). Cell concentration was measured by trypan blue exclusion assay.

## 3. RESULTS

*Formation of ligand-receptor complex*

Western blot analysis showed that transfectants express both $V_H$-EPOR and $V_L$-EPOR chimeric receptor chains(data not shown). To make sure that the two chimeric receptors form the dimer in response to HEL, the experiment was performed as described in materials and methods. As shown in Fig.1, there were two bands. The slower-migrating band corresponds to $V_H$-EPOR, and faster one corresponds to $V_L$-EPOR. The density of two bands increased with an increase in biotinylated HEL concentration. Furthermore, the density of two bands in each lane was almost the same. These results indicate that HEL induced heterodimer formation of $V_H$-EPOR and $V_L$-EPOR in a dose-dependent manner.

*Signal transmission to Stat5b*

It was investigated whether the chimeric receptors could be functionally activated and transmit the signals to cellular kinases. Transcription factor Stat5b is one of signal transducers phosphorylated in signal transduction pathway from EPOR. Stat5b was tyrosine-phosphorylated with HEL stimulation, but not without HEL stimulation(Fig.2). Five minutes' stimulation was unlikely to be enough for Stat5b activation, and in the case of ten minutes' stimulation, Stat5b was phosphorylated in a HEL concentration-dependent manner. This result indicates that the chimeric receptors recognized HEL and transmitted similar signal with EPOR.

*Cell proliferation assay*

The proliferation response of Ba/HEL cells to HEL stimulation was examined, and was compared with EPO responsive proliferation of Ba/EPOR cells. Ba/HEL cells proliferated in a HEL concentration-dependent manner, although they proliferated without HEL(Fig.3A). The viable cell concentration obtained by optimal stimulation with HEL was 5 times as high as that without HEL addition. Ba/EPOR cells proliferated in an EPO concentration-dependent manner(Fig.3B). But the maximum proliferation rate of Ba/HEL was higher than that of Ba/EPOR, indicating the efficient activation of chimeric receptors by HEL addition. Thus, antigen stimulation made the chimeric receptor-bearing cells proliferate as efficiently as cytokine stimulation.

## 4. DISCUSSION

It had been shown that growth inhibition of hybridoma cells by IL-6 addition enhanced protein productivity[4] and that efficient production of proteins was achieved by inhibition of apoptosis when cells were under arresting cell cycle[5]. Therefore, controlling cell proliferation as well as inhibition of apoptosis is very important for protein production, and cytokines have been used for this purpose. We created a pair of novel chimeric receptors which transduces the growth signal in hematopoietic cells in response to the inexpensive ligand, HEL and can control cell proliferation. Our research revealed that action of cytokine can be replaced by an inexpensive antigen, so this approach can be applied to reduce production cost of proteins.

Seeing from other viewpoints, this approach can be used for positive selection of gene-transfected cells. The receptor which transduces the proliferation signal in response to HEL, that is, "HEL receptor", doesn't exist in normal cells, so only the HEL receptor gene-bearing cells should be specifically selected in the presence of HEL. When the HEL receptor gene and the gene of interest are transfected at the same time, one might expect the positive selection of transfectants. As for gene therapy, the gene-transfected cells should be selected in vitro or even in vivo to enhance the therapeutic effect. For hematopoietic diseases, hematopoietic stem cell(HSC) is the most desirable target cell. In this case, antigen promotes gene-transfected HSC amplification, which might enable to maintain and control the therapeutic effect for a long time.

The major problem so far is that Ba/HEL cells showed some growth without HEL addition. It is known that there are weak interactions between $V_H$-$V_L$ or $V_L$-$V_L$[6], and if the concentration of the fragments is high, such dimers may be formed in the absence of HEL. The fact that transfectants which was transfected only with $V_L$-EPOR gene could proliferate in the absence of IL-3 and HEL(not shown) implies that the proliferation without HEL is due to the formation of $V_L$-EPOR dimer. Recent report shows that single amino acid substitution at the interface can affect stability of $V_L$ dimer[6]. If the proliferation should be completely controlled, it might be preferable to use the $V_L$ mutant which does not form homodimer and retains strong affinity to form $V_H$-HEL-$V_L$ complex.

Since the combination of antigen-antibody is virtually infinite, the applicability of current approach is tremendous. However, so far we have only limited knowledge on the mechanism of antigen-induced formation of $V_H$-antigen-$V_L$ complex and the presence of antibody except HyHEL-10 which has suitable characteristics for this approach. To create antibody/cytokine receptor chimera activated by various antigens and to enlarge further applications along this approach, we are looking for other functional antigen-antibody systems and trying to make clear the mechanism behind this phenomenon.

**REFERENCES**


1. Ueda, H. et al. (1996) *Nature Biotechnol.* **14**, 1714-1718.
2. Tsumoto, K. et al. (1994) *J. Biol. Chem.* **269**, 28777-28782.
3. Chiba, T. et al. (1993) *Nature* **362**, 646-648.
4. Makishima, F. et al. (1992) *Cytotechnology* **10**, 15-23.
5. Terada, S. et al. (1997) *Cytotechnology* **25**, 17-23.
6. Raffen, R. et al. (1998) *Protein Eng.* **11**, 303-309.


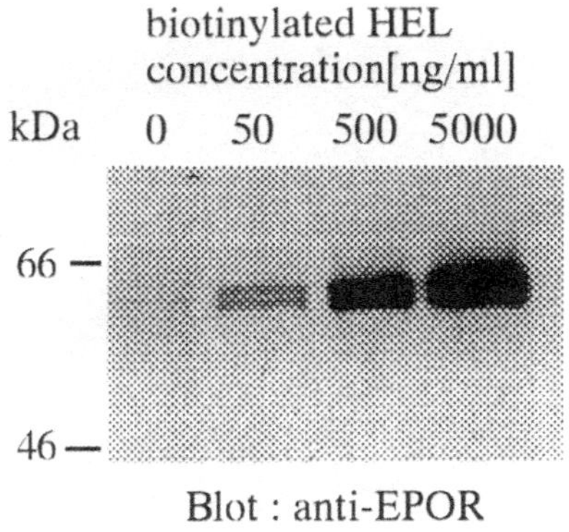


Figure 1. Formation of ligand-receptor complex

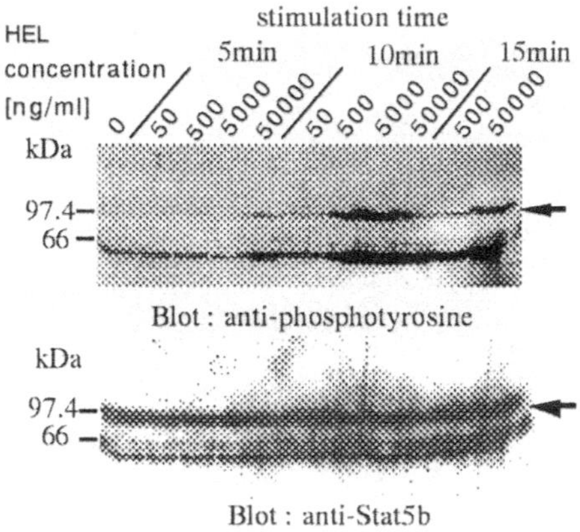


Figure 2. Signal transmission to Stat5b

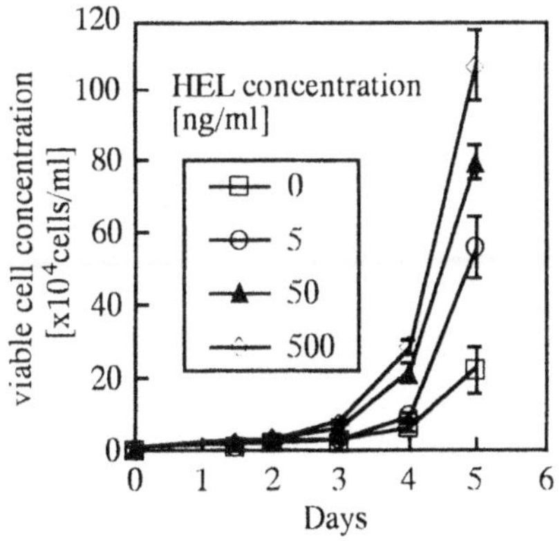


Figure 3A. HEL-dependent proliferation of Ba/HEL cells.

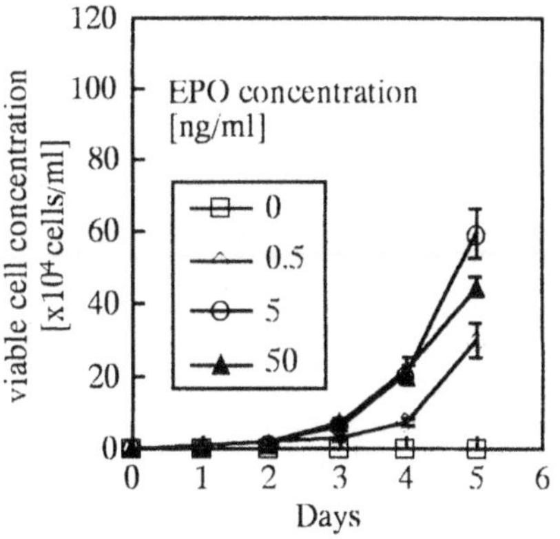


Figure 3B. EPO-dependent proliferation of Ba/EPOR cells.

# MOLECULAR CLONING OF THE GENES THAT DETERMINE THE INVASIVE ACTIVITIES OF RAT ASCITES HEPATOMA CELLS

Y. MIURA, S. TOTSUKA, Q. LEE AND K. YAGASAKI
*Department of Applied Biological Science,*
*Tokyo Noko University, Fuchu, Tokyo 183-8509, Japan*

Key Word Hepatoma, Invasion, Stathmin, Tap, TCP-1

## 1. Introduction

It is well known that metastasis is a unique characteristic of tumor cells. Tumor cells metastasize via complicated and sequential steps, among which invasion is thought to be most important. We have been using rat ascites hepatoma cell line of AH109A as a model to study the mechanisms of invasion and have found that there are two populations in AH109A cells with different invasive activities (Miura *et al.*, 1998). One population is adhesive AH109A cells (AH109A-Ad) which adhere to the culture dishes in the presence of rat serum, and the other population is floating AH109A cells (AH109A-Fl) which remain in a floating state in the presence of rat serum. AH109A-Ad showed the higher invasive activity than AH109A-Fl when assessed by the *in vitro* invasion assay we reported (Miura *et al.*, 1997). The molecular cloning of the genes that express at the different levels in these two populations of AH109A will give us much information about the molecular mechanisms of tumor invasion. Using the differential display method (Liang and Pardee, 1992), we have cloned two genes that express at a different level in these two cells. One gene was rat stathmin which was expressed at the higher level in AH109A-Ad than in AH109A-Fl; stathmin is a cytosolic protein involved in microtubule assembly. We found by transfection experiments of the antisense DNA of stathmin into AH109A-Ad that stathmin is also involved in cell extension induced by rat serum. The other gene, clone #T3-3, is a novel gene that has strikingly high homology with YHR74 gene, a hypothetical gene in yeast genome.

In this study, we report the more detailed functions of these two genes. We also report the molecular cloning of 12 more genes by the differential display method.

## 2. The effect of antisense DNA of stathmin on the invasion of AH109A

As stated, stathmin is involved in the microtubule assembly, and a change in the state of the microtubule assembly may induce a morphological change in cells. In fact, we have already ascertained that the transfection of antisense DNA of stathmin significantly inhibited the extension of AH109A cells induced by rat serum. This morphological change may be related to cell motility or invasion. We next examined the effect of antisense DNA of stathmin on the invasion of AH109A cells.

The sequence of sense, antisense and nonsense DNA of stathmin was described by Di Paolo *et al.* (1996). Fifteen $\mu$g of these DNAs were respectively transfected in 4 X $10^5$ cells of AH109A by transfection reagent (Tfx-20$^{TM}$, Promega, WI, USA). After 24 hr, AH109A cells were collected and overlaid on a monolayer culture of mesentery-derived mesothelial cells (M-cell) at the density of 2.4 X $10^5$ cells / dish. The number of invaded cells and colonies underneath of M-cell monolayer was counted after 48 hr under a phase contrast microscope.

The transfection of antisense DNA of stathmin reduced the mRNA content of stathmin to the level of 50% of control cells transfected by liposome only, while the transfection of sense and nonsense DNA of stathmin did not affect the mRNA content of stathmin. The invasive activity of antisense DNA-transfected AH109A, however, was unaltered compared with those of other cells. This suggests that the cell extension step is not profoundly involved in the invasion step of AH109A cells, and that the higher expression of stathmin gene in AH109A-Ad than in AH109A-Fl is involved in the morphological change induced by rat serum.

### 3. Molecular cloning of the genes that express at a different level between AH109A-Ad and AH109A-Fl

After isolation of stathmin and clone #T3-3, we continued molecular cloning of the genes that express at a different level between AH109A-Ad and AH109A-Fl using the differential display method. At present, we have cloned 12 more genes and 2 of them were already-reported genes. One is rat TCP-1 gene which expresses at a higher level in AH109A-Fl than in AH109A-Ad. TCP-1 belongs to the CCT chaperonin family and is reported to have the ability to induce the folding and assembly of actin and tubulin. It is thought to be involved in the folding of the cytoskeletal proteins (Kubota *et al.*, 1994, Chen *et al.*, 1994). The other gene is Tap, Tip associated protein. Tip is a tyrosine kinase interacting protein expressed in transformed T cell. Tap also expresses at a higher level in AH109A-Fl than in AH109A-Ad. Tap is known to express ubiquitously in human tissues and when overexpressed in human lymphocytes, it induces the aggregation of these cells by interacting with Tip protein (Yoon *et al.*, 1997). As reported (Miura *et al.*, 1998), AH109A-Fl cells often form cell aggregates during their growth; the higher expression of Tap in AH109A-Fl may be related to this characteristic. We have constructed the cDNA library of AH109A and are now trying to clone the full-length cDNAs of these genes to clarify their precise functions.

Ten other genes that we have cloned have no or only weak homologies with already-reported genes (TABLE 1). Some of them are only reported as ESTs. Three of them (#FA1-2, #FA1-3 and #FA1-5) were shown to be identical. They express at a higher level in AH109A-Ad than in AH109A-Fl. As shown in TABLE 1, this gene (#FA1-2, #FA1-3 and #FA1-5) expresses highly in intestine and heart, and weakly in kidney, lung, thymus and brain. It is interesting that this gene obtained from hepatoma cells does not express in their host organ, liver. We are now trying to clone the full-length cDNA of these ten genes to study their structures and functions.

TABLE 1. Summary of cloned genes

| CLone | Expression level | Homology | Tissue distribution |
|---|---|---|---|
| #AA 1-3 | Ad > Fl | Putative RNA binding protein | ND |
| #FA 1-2<br>#FA 1-3<br>#FA 1-5 | Fl > Ad | EST | Intestine, Heart (High)<br>Kidney, Lung<br>Thymus, Brain (weak) |
| #AA 4-2 | Ad >Fl | Salivary protein (Ves 2) gene | ND |
| #AA 4-3 | Ad > Fl | Precollagen P, Extensin | ND |
| #FA 8-1 | Fl > Ad | EST | ND |
| #AG 8-1 | Ad >Fl | ID transcript repetitive sequence | ND |
| #FC 6-3 | Fl > Ad | None | ND |
| #FC 8-1 | Fl > Ad | Putative RNA binding protein | ND |

## 4. Summary

We cloned 14 genes by comparing AH109A-Ad and AH109A-Fl. Although 11 of them are novel genes the structures and functions of which are not known, three of them have already been reported: Tap, stathmin and TCP-1. Tap is thought to be involved in cell aggregation or cell-cell interaction. Stathmin and TCP-1 may be involved in morphological change and cell motility. Both cell-cell interaction and motility are known to play important roles in the invasion and metastasis of tumor cells, so, these genes may regulate tumor cell invasion and metastasis in concert.

We should determine the structures and more precise functions of these genes, in that way gaining greater understanding of the molecular mechanisms of invasion and metastasis.

## 5. References

Chen, X., Sullivan, D. S. and Huffaker, T. C. (1994) Two yeast genes with similarity to TCP-1 are required for microtubule and actin function. *Proc. Natl. Acad. Sci. USA*, **91**, 9111-9115.

Di Paolo, G., Pellier, V., Catsicas, M., Antonsson, B., Catsicas, S. and Grenningloh, G. (1996) The phosphoprotein stathmin is essential for nerve growth factor-stimulated differentiation. *J. Cell Biol.*, **133**, 1383-1390.

Kubota, H., Hynes, G., Carne A., Ashworth, A. and Willison K. (1994) Identification of six Tcp-1-related genes encoding divergent subunits of the TCP-1 containing chaperonin. *Current Biology*, **4** 89-99.

Liang, P. and Pardee, A. B. (1992) Differential display of eukaryotic messenger RNA by means of the polymerase chain reaction. *Science*, **257**, 967-971.

Miura, Y., Shiomi, H., Sakai, F. and Yagasaki, K. (1997) Assay systems for screening food components that have anti-proliferative and anti-invasive activity to rat ascites hepatoma cells: *In vitro* and *ex vivo* effects of green tea extract. *Cytotechnology*, **23**, 127-132.

Miura, Y., Miyauchi, M., Kubota, K., Li, Q., Tachikawa, H., Fujimoto, D. and Yagasaki, K. (1998) Characterization of the subpopulations of rat ascites hepatoma AH109A cells with different invasive and metastatic activities. *"Proceedings of the Tenth Annual Meeting of the Japanese Association for Animal Cell Technology"* Ed. by Iijima, S. Kluwer Academic Publishers, Dordrecht, The Netherlands, in press.

Yoon, D-K., Lee, H., Seol. W., DeMaria, M., Rosenzweig, M. and Jung, J. U. (1997) Tap : a novel cellular protein that interacts with Tip of herpesvirus Saimiri and induces lymphocyte aggregation. *Immunity*, **6**, 571-582.

# ANTITUMOR PROTEIN (AP) FROM A MUSHROOM INDUCED APOPTOSIS TO TRANSFORMED HUMAN KERATINOCYTE BY CONTROLLING THE STATUS OF pRB, c-MYC, CYCLIN E-CDK2, and $p21^{WAF1}$ IN THE GI/S TRANSITION

**Yukio KAWAMURA , Mariko MANABE, and Kazumi KITTA**
**National Food Research Institute**
**2-1-2 Kannon-dai, Tsukuba, Ibaraki 305-8642, Japan**

**ABSTRACT:** Antitumor protein (AP) from a mushroom, induced the morphological changes typical to apoptosis such as nuclear condensation, aneuploidity , and DNA fragmentation at concentrations as low as 5-20 ng/ml to cancer cells. Molecular alterations related to cell cycle, especially G1/S transition were investigated with a human keratinocyte transformed with oncoproteins, E6 and E7 of human pappiloma virus(HPV)-16. AP didn't alter significantly an oncosuppressor p53 level, but induced hyperphosphorylation of pRb. Time-dependent change of G1 cyclins, cdk2 and cdk4 after addition of AP showed that expression level of cdk inhibitors, INK4 family, and $p27^{KIP1}$ did not altered, while that of $p21^{WAF1}$ was downregulated.

## Introduction

Evidence has been accumulating that change or loss in function of the tumor-suppressing gene products, oncosuppressors such as p53 and retinoblastoma protein(pRb), are the most frequently observed phenomenon in many human cancers. p53 plays a central roll in protecting cells from malignant transformation by retarding the progression of cell cycle from G0/G1 to S phase or inducing apoptosis depending on degree of the lesion(1-4). The pRb functions as a tumour suppressor in a concerted fashion with p53 proteins through regulation of transcription(5-9).

Selective cytotoxicity against tumor or transformed cells in comparison with the normal counterpart cells gives an effective strategy for screening of antitumorigenic or tumoricidal substances.

In the prersent study, to screen tumoricidal substances in biological materials, we chose the cells of a normal mouse fiblobrast cell line NIH/3T3 and its SV40-transformant SV-T2 by using preferential cytotoxicity against the transformed cell as an antitumor index, and an antitumorigenic protein was found in a mushroom. Some molecular characterization and the action profile of the antitumor protein to SV40-transformed NIH/3T3 cell(SV-T2) and a human HPV transformed keratinocyte(10) were discussed.

**Materials and Methods**

Cell Culture: Cells(A31 and SV-T2 ,and a HPV16-transformed human keratinocyte ) were maintained in DMEM or the complete MCDB152 medium supplemented with appropriate serums, growth factors , and antibiotics at 37 C in a 5% CO2-humidified incubator.

Cell Viability : Cellular viability was determined by MTT assay(11). Exponentially growing cells were harvested, counted, and inoculated at $2x10^4$/well into 96-well flat-bottomed tissue culture plates. After 24h, the antitumor protein was applied to triplicate culture wells, and the cultures were incubated for another 24h at 37 C. The cells were reincubated for an additional 4h in DMEM containig 3-[4,5-dimethyl-2-thiazolyl]-2,5-diphen -2H-tetrazolium bromide (500mg/ml) . After solubilization of the formazan crystals produced, the absorbance of each well was determined using a microplate reader to calculate the percentage of cell viability.

Flowcytometric Analysis: Cellular DNA was stained with propidium iodide by the methods of Lars L. Vindeloveta(12) and was analyzed by fowcytometry with the FACSort/Cell FIT DNA system(BectonDickinson Co. USA).

Cytochemical Change: Changes in morphology of nuclei were evaluated by staining with acridine orange. Cells were fixed in 3% fomaldehyde and washed three times in phosphate buffered saline (PBS). After treated with 0.5 % tritonX 100 in PBS, those were stained with 1mg/ml of acridine orange in PBS and observed by fluorescence microscopy(Orympus).

Determination of DNA Fragmentation: Cells were harvested, washed with PBS and suspended in a lysis buffer( 50mM Tris-HCl,pH7.9, 5mM EDTA, 5%

SDS ,and 0.1 mg/ml of proteinase K. The intact genome DNA was removed by centrifugation for 20min at 14,000 x g. The DNA in the supernatant was extracted with phenol and chloroform-isoamylalcohol(24:1) and precipitated with 70% ethanol. The precipitates were treated with RNaseA ( 1mg/ml ) at 37C for 12h, analyzed by electrophoresis in 5% agarose gel ,and visualized by ethidium bromide staining.

Northern Blotting: Cells were seeded at 2 x $10^4$/cm2 with or without the antitumor protein (30ng/ml) and harvested at the indicated times. Total RNA was extracted by adding guanidinum thiocyanate-phenol-chloroform method. RNA(10 $\mu$ g/ml ) was electrophoresed in a 1.2% agarose and transfer to HybondN membranes (Amersham) followed by crosslinking with UV irradiation. The blott was prehybridized at 42 C for 2h and hybridised at 42 C for 12 to 16h. After washing the blot, $[^{32}]$p DNA probes were prepared by randompriming. The *c-myc* expression was analyzed with human cDNA fragment. Glyceraldehyde 3-Phosphate dehydrogenase (G3PDH ) expression was analyzed with 1.1kb human cDNA probe. Quantification of autoradiograms was carried out with Bas 2000.

Immunoprecipitation and Immunoblotting: Cells were lysed in an extraction buffer containing surfactants and protease inhibitors. The total protein concentration was detemined by a dye-binding assay kit. Equal aliquots( 2mg) were separated in SDS-PAGE(13) and electrotransferred to a PVDF membrane. The blotted membrane was incubated with antibodies for each protein and then with the second antibodies . The antibody protein complexes were detected by ECL detection system.

**Results**

Apoptosis induction

We first examined influence of AP on the viability of PHK16-0b cells. Cytotoxicity assay by MTT demonstrated that AP had intense cytotoxicity against PHK16 cells and that its LD50 was 15ng/ml(Fig 1).

In the cells treated with AP, the followings was observed : disruption of the cellular membrane, blebbing from cytoplasm, cell shrinkage by light microscopy, and chromatin condensation by fluorescence microscopy (acridine orange staining). These changes were distinctive features of apoptosis(14).

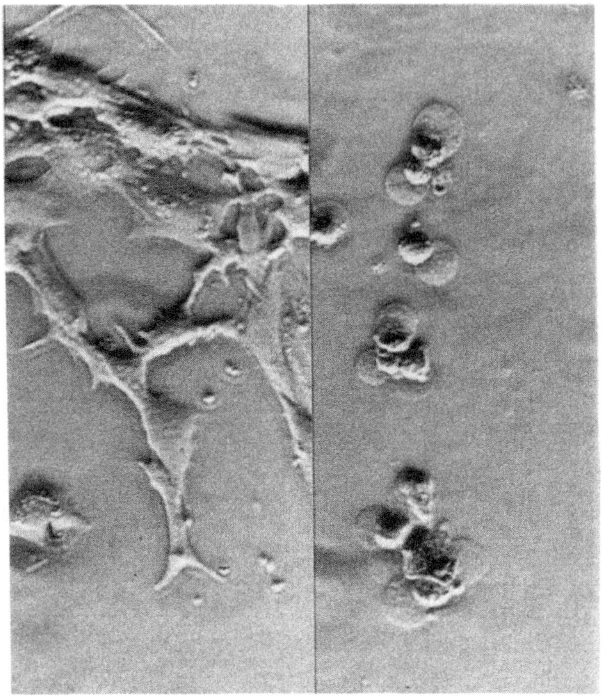

Fig.1 Action of AP on mouse NIH/3T3 cell (left) and on theSV40 transformed cell (SV-T2)(right). Ten ng of Ap was added to culture medium. After 12hr, cell death was obvious in SV-T2, but not in NIH/3T3.

Degradation of the nuclear DNA is another hallmark of apoptosis(15). To confirm the type of the cell death by AP, it was determined whether internucleosomal cleavage could be detected. Cells treated with mitomycin C were used as a positive control of apoptotic cell death because mitomycin C inhibits DNA synthesis and induces apoptosis in epithelium cells(16).
Dgradation of nuclear DNA was observed in PHK16-0b cells treated with AP and with mitomycin C for 24h, though typical DNA 1adders were not detected in these cases. These results suggested that AP induced apoptotic death inPHK16-0b cells.

*c-myc* expression in PHK16-0b

Cellular oncogenes appear to play key roles in the biochemical pathways controlling cell proliferation and death(17). Especially, activation of *c-myc* expression is implicated in proliferation of epithelial cells(18-20). The involvement of c-myc was examined in cell death induced by AP. The decrease in the level of expression of c-myc mRNA was observed at 12h after addition of AP,and it was dose-dependent.

We also observed that AP induced the Gl arrest of SV-T2 cells temporarily prior to cell death (data not shown). The down-regulation of *c-myc* by AP in PHK16-0b might include the G1 arrest before cell death

Phosphorylation status of pRb.

A number of studies have indicated that the transcription of *c-myc* can be modulated by pRb through the transcriptional activity of E2F(21-22). The level of phosphorylation of pRb showed a dose-dependent increase in PHK16-0b cells treated with AP. The pRb gradually sifted to hyperphosphorylated position until 12h after addition of AP, indicating the AP-induced temporary phosphorylation of pRb in PHK16-0b. Although the phosphorylation of pRb could be the event leading to increase of *c-myc* expression in normal cell cycle advance of keratinocytes, the level of *c-myc* was depressed with time by AP.

Table 1 The changes in cell cycle related molecules by the addition on AP

| cell | A31 | | A31 | | SV-T2 | |
|---|---|---|---|---|---|---|
| time | 3h | 6h | 3h | 6h | 3h | 6h |
| Ip | | | | | | |
| | w/ cyclin D1 | | w/ cdk4 | | w/ cyclin D1 | |
| cyclin D1 | ↓ | ↓ | ND | | ↓ | ↓ |
| cdk4 | → | → | → | → | → | → |
| p21 | → | → | → | → | ND | |
| p27 | → | ↓ | ↓ | ↓ | ↓ | ↓ |
| p15&p16 | ND | | ND | | ND | |
| cyclin E | → | → | → | → | → | → |
| Simple western | | | | | | |
| pRb | → | → | → | → | → | → |
| p53 | ND | | → | ↓ | ↓ | ↓ |
| c-myc | → | → | | | ↓ | ↑ |
| NFkB | → | → | | | → | → |
| IkB | → | → | | | → | → |

The status of cdks

We monitored the levels of p53,pRb,G1 cyclins, cdk4, cdk2 and ckIs in response to AP(Table 1). The level of cyclinD was slightly decreased after addition of AP. The level of cyclinE was decreased at 6h after addition of AP,which seemed correlated with the decreased level of hyperphosphorylated pRb in the cells. No significant change was observed in the amount of cdk4 and cdk2. However, the change of the ratio of the hyperphosphorylated form in cyclinE-associated cdk2s was consistent with the phosphorylation state of pRb. These results indicated that the levels of cyclinD, cyclinE, cdk4 and cdk2 were independent of the hyperphosphorylation of pRb, but that of cyclinE-associated cdk2 did not.

**Summary**

We have presented evidence that a HPV16-tansformed human cell treated with AP undergo apoptosis. To clarify the mechanism of induction of apoptosis by AP, we measured the changes of the levels of the cell cycle-related factors in the cell after treatment of AP. It was found that AP induced phosphorylation of pRb temporarily because of activation of cyclin E- associated cdk2 and depressed the *c-myc* expression. AP is likely to show the paradoxical behaviors in the same time as promoting and arresting cell cycle in the cell. In the previous study, we also observed that a mouse cancer cell was arrested in Gl-phase transiently by AP. As soon as the arrested cell escaped from GI phase to S phase, it underwent apoptosis. Taken together,these results suggested that this unbalanced control of cell cycle induced by AP might lead to apoptosis in tumorigenic cells, although the signal transduction to apoptosis by AP has to still be explored in more detail.

**References**

1. Levine, AJ. , Momand, J., and Finlay, CA.(1991) *Nature* **351**, 453-456
2. Levine, AJ. (1997) *Cell* **88**, 323-331
3. Ludlow, JW. (1993) *FASEB J.* **7**, 866-871
4. Ludlow, JW, and Skuse, GR. (1995) *Virus Res.* **35**, 113-121
5. Cordon-Cardo (1995) *Am. J. Pathol.* **147**, 545-560
6. Ko, LJ, and Prives, C. (1996) *Genes Dev.* **10**, 1054-1072

7. Gottlieb, TM., and Oren, M. (1996) *Bioch. Biophys. Acta* **1287**, 77-102

8. Lukas, J., Mu"ller, H., Bartkova, J., Spitkovsky, D., Kjerulff, AA., Jansen-Du"rr, P., Strauss, M., and Bartek, J. (1994) . *J. Cell Biol.* **125**, 625-638

9. Lukas, J., Bartkova, J., Rohde, M., Strauss, M., and Bartek, J. (1995) *Mol. Cell. Biol.* **15**, 2600-2611

10. Yasumoto, S., et al.(1992) *Tissue Cule. Res. Commun.* **11**, 13-24 .

11. Mosmann, T. (1983) *J.Immunol.Meth.*, **65**, 55-63

12. Vindelov, L.L., Christensen, I.J. & Nissen, N.I.(1983) *Cytometry* , 323-327 .

13. Laemmli, U.K.(1970) *Nature* **227**, 680-685

14. Wyllie, A.H.(1986) *Int. Rev. Cyt. Suppl.* **17**, 755-785 .

15. Wyllie, A.H.(1980) *Naturer* **284**, 555-556 .

16. Fritsche, M., Haessler, C. & Brandner, G.(1993) *Oncogene* **8**, 307-318 .

17. Grassilli, E., et al.(1992) *Biochem. Biophys. Res. Commun.* **188**, 1261-1266

18. Seth, A., Gupta, S. & Davis, R.J. (1993) *Mol. cell. Biol.* **13**, 4125-4136

19. Jansen-Durr, P., et al. (1993) *Proc. Natl. Acad. Sci. U.S.A.* **90**, 3685-3689

20. Heikkila, R., et al. (1987) *Nature* **328**, 445-449.

21.Pagano, M., Theodoras, AM., Tam, SW., and Draetta, GF. (1994) . *Genes Dev.* **8**, 1627-1639

22. Henriksson, M., and Lu"scher, B. (1996) *Adv. Cancer Res.* **68**,109-182

# TRANSFORMATION OF NIH/3T3 AND C3H/10T1/2 CELLS BY CITRININ AND OCHRATOXIN A

*Cell Transformation by Mycotoxins*

ALKA MEHTA and NAOFUMI KITABATAKE
*Research Institute for Food Science, Kyoto University,*
*Uji, Kyoto 6110011, Japan*

**Abstract**

The citrinin and ochratoxin A are the toxic secondary metabolites of common food contaminating fungi. The genotoxicity of citrinin and ochratoxin A was studied using transformation assay. 3-Methylcholanthrene (MCA) was used as positive control. Cytotoxicity assay of citrinin, ochratoxin A and MCA was performed using MTT assay. Citrinin (20 $\mu$g/ml ) and ochratoxin A (10 $\mu$g/ml ) induced the formation of foci and subsequently colonies in soft agar after a single 24 hour exposure. They also showed some synergistic effect when applied subsequently in the same culture. These results show that both citrinin and ochratoxin A have the ability to transform cell, and are thus potential carcinogens.

## Introduction

The in-vitro cell transformation of fibroblast cells has provided an important model for studying the carcinogenesis of environmental carcinogens at cellular level and to understand the molecular mechanism of it (Chen and Heidelberger, 1969a;b; Reznikoff et al.1973a;b). Citrinin and Ochratoxin A; toxic secondary metabolites of many Penicillium and Aspergillus species which are natural food contaminants (Krogh et al. 1973; Scudamore et al. 1997). Toxicity of these mycotoxins in animals (Krogh et al. 1974) and in cultured cells (Kitabatake et al. 1993) is well studied. There are controversial reports about the carcinogenicity of these mycotoxins in animal experiments (Arai and Hibino,1983; Kanisawa,1984) and in-vitro experiments using Ames and Rac assay (Ueno and Kubota, 1976; Maureen et al. 1978;

Ueno et al. 1978; Wehner et al. 1978).

In the present study, the genotoxicity of citrinin and ochratoxin A were studied using a transformation test on the NIH/3T3 and C3H/10T1/2 cell lines with a focus assay. These cell lines are highly sensitive to post confluence inhibition of cell division. The results were confirmed by colony formation in soft agar. Experiment were performed using protocol given by Reznikoff et al. (1973A,b). Citrinin and ochratoxin A are cytotoxic even in very low concentrations (Kitabatake et al. 1993). Hence the treatment dose chosen which give less than 50% cytotoxicity in 24 hour exposure.

## Material and Methods

### CHEMICALS

Cell culture medium; Eagels basal medium (BME), Dalbecco's modified Eagles medium (DME), 3-methylcholanthrene (MCA), Ochratoxin A, were purchased from Sigma chemicals Co. (St Louis, M O USA). Trypsin, fetal bovine serum (FBS), calf serum (CS) were purchased from Gibco/BRL Life technologies, INC. Gaithersburg, M.D. USA. Citrinin was prepared in the laboratory from the toxigenic strain of *Penicillium citrinum*, isolated during the survey of food contaminating fungi (Trivedi 19880. Purity was checked by HPLC and NMR.

### CELL LINES AND CULTURE CONDITIONS

C3H/10T1/2 clone 8 and NIH/3T3 clone 5611 were obtained from Japanese Cancer Research Bank (JCRB), Japan. NIH/3T3 cells were passaged in DME medium supplemented with 10%CS in a 60 mm plastic dish (Nunc). C3H /10T1/2 cells were passaged in BME medium supplemented with 10% heat inactivated FBS (55°C for 30 min) in a 60 mm plastic dish (Falcon, Oxnard,Calif.). Cultures were maintained in humidified incubator with an atmosphere of 5% $CO_2$ in air at 37°C.

### TRANSFORMATION ASSAY

To assay the transformation ability of citrinin and ochratoxin A, the protocol of Reznikoff et al. (1973A,b)was followed with some slight modifications. MCA was used as a positive control. Five thousand and 1000 cells of NIH/3T3 and C3H/10T1/2 respectively were seeded in a 60 mm dish containing 5 ml medium and incubated at 37°C in 5% $CO_2$ atmosphere. Cells were treated with test compounds after 48 hr of incubation and exposed for 24 hr after which medium replaced with fresh medium that did not contain compound. Plates divided in two groups: in one group the second 24 hr treatment with test compounds was done on the 7th day and on 4th day of incubation to NIH/3T3 and C3H/10T1/2 respectively. Culture medium changed every third day throughout the incubation period. Foci were observed after 2 and 4 weeks of incubation.

*Scoring of foci:*

Foci were scored as the opaque area in contrast to flat background, and categorized as Type I, Type II and Type III according to the criteria described by Reznikoff et al. (1973A,b). Dishes were dehydrated and stained with Giemsa, the foci having multilayered cells stained dark.

*Replating*

To confirm the focus formation the cells from the foci were replated. The parent dish washed with PBS and foci was trypsinized with 0.25% trypsin. Cells were then transferred to a well of 24 well plate in 1 ml complete medium. After 24 hr incubation plates were stained with Giemsa.

*Soft agar colony assay:*

Cells from sister plate of Giemsa stained were used for soft agar colony assay. Cells were collected by trypsinization, and washed with phosphate buffer saline (PBS). The colony assay was performed as described by Kuratomi et al. (1978). Three mililitre of complete medium containing 0.53% low melting agarose (FMC Bio-products, USA) was placed in the 60 mm dish. The top agar layer of 0.53% agar containing 1000 cells/ml was placed on solidified bottom agar layer. Colonies were scored after 2 weeks of culture.

**Results**

Cytotoxicity of citrinin, ochratoxin A and MCA was assayed by MTT method as describe previously (Trivedi et al. 1990). The amount of test compound which gave less then 50% cytotoxicity in 24 hr treatment was used for transformation experiments i.e. 10 $\mu$g/ml of ochratoxin A and 20 $\mu$g/ml of citrinin. The cytotoxicity of MCA is found quite low in comparison to test mycotoxins; citrinin and ochratoxin A. Hence, 10 $\mu$g/ml was used same as used by Reznikoff et al (1973a,b).

Citrinin, ochratoxin A and MCA showed the foci in NIH/3T3 and C3H/10T1/2 cells. Citrinin gave more foci of Type I than either ochratoxin A or MCA. In all cases, Type II and III foci were also observed with citrinin treatment. These foci are the patches of cells which lost the post confluence inhibition. Table 1 show that citrinin and ochratoxin A gave foci in NIH/3T3 in single 24 hr treatment with test compounds. When cells from these foci replated , the positive control MCA showed maximum number of foci. While ochratoxin A gave less foci in replating. This show that the acquired changes could not be transmitted in the next generation from all the transformed cells.
itrinin gave less foci in 60 mm dish but after replating more foci were noted. In two step treatment also the foci were noted in all cases and number of foci increased.

On average one focus per plate (Type I) was observed in the negative control. Cells were transferred to soft agar and the colonies that formed were observed. All the treated plates show many colonies of more than 20 cells. Some very big colonies of more than 40 cells were also observed in ochratoxin A and MCA treated plates. In the negative control none of the plates showed colonies.

C3H/10T1/2 cells also gave foci 6 weeks after seeding. All of the dishes that were treated with citrinin, ochratoxin and MCA showed two or more foci/dish with the one step treatment. In two-step treatment, no increase was noted in

Table 1: Effects of Citrinin and Ochratoxin A on the Formation of foci and colonies from NIH/3T3

| Treatment Step1 Step2 | No. Of foci[1)] in 60 mm dish | No. Of foci[2)]; in 25 mm dish | No. Of colonies[3)] in 0.53% agar |
|---|---|---|---|
| none | 1 | - | - |
| MCA[4)] | 3 | 20 | 15 |
| CIT[5)] | 2 | 15 | 14 |
| OTA[6)] | 12 | 9 | n.d.[7)] |
| CIT → OTA | 1 | 13 | n.d. |
| OTA → CIT | 6 | 16 | n.d. |
| CIT → MCA | 5 | - | n.d. |
| MCA → CIT | 14 | 5 | n.d. |
| MCA → OTA | 13 | 16 | n.d. |
| OTA → MCA | 3 | 1 | n.d. |

1) Number of foci/60 mm dish (n=6)
2) Number of foci after replating
3) Number of colonies/plate in soft agar, size 20-40 cells/colony (n=6)
4) 3-Methylcholanthrene
5) Citrinin
6) Ochratoxin A
7) n.d.; not determined

foci formation when the treatment with citrinin followed by ochratoxin A or vice-versa. Cells from the foci of the treated dishes produced many colonies in soft agar. In C3H/10T1/2 cells, the two step treatment apparently had no effect on the formation of Type II and III foci. In contrast , NIH/3T3 cells gave higher number of Type II and III foci two step treatment than in one step treatment.

## Discussion

In the present study, both citrinin and ochratoxin A showed the ability to transform NIH/3T3 and C3H 10T1/2 cells in a single 24 hr exposure. Citrinin and ochratoxin are often found simultaneously in the food material, hence it is important to see the synergistic effect of these mycotoxins. The two step treatment with citrinin followed by ochratoxin A and vice-versa enhanced the formation of type II and III foci, as compared to one step treatment in NIH/3T3 cells, while no remarkable difference in foci and colony formation was observed in C3H/10T1/2 cells. In general, NIH/3T3 cells transform more readily than C3H/10T1/2 cells indicating that NIH/3T3 is more sensitive to toxin tested than C3H/10T1/2. The previous reports on animal experiments (Shinohara, 1976; Imaida, et al. 1982; Kanisawa, 1984) have shown that citrinin has no initiating activity, but instead works as promoter. Hence, we expected more of the foci in the two step treatment, especially in case of ochratoxin A followed by citrinin. But such results have not been observed, particularly in case of C3H/10T1/2 cells. On the contrary citrinin alone in the one step treatment showed the transformation ability.

Table 2: Effects of Citrinin and Ochratoxin A

| | Citrinin | Ochratoxin A |
|---|---|---|
| In-vivo carcinogenicity | | |
| (Mouse) | Positive[a] | Unknown[a]<br>Positive[b] |
| (Rat) | Negative | Positive |
| Initiating | Positive[d)] | Positive[d)] |
| Promoting | - | Positive[d)] |
| DNA attacking ability (rac assay) | Positive[a)] | Negative[a)] |
| Mutagenicity (Ames test) | Negative[e,f)] | Negative[f)] |
| Transformation (NIH/3T3, C3H/10T1/2) | Positive[g)] | Positive[g)] |

a, Ueno and Kubota (1976); b, Kanisawa and Suzuki (1978);
c, Shinohara et al. (1978); d, Imaida et al. (1982);
e, Wehner et al. (1978); f, Maureen et al. (1978);
g, Our observation

In in-vitro experiment for DNA attacking ability with the recombination deficient mutant of *Bacillus subtilis* citrinin was found rac positive while ochratoxin A was negative (Ueno and Kubota, 19760. Citrinin and ochratoxin A both were not mutagenic in Ames rest using *Salmonella typhimurium* (Maureen et al. 1978). It is generally accepted fact that carcinogens are mutagen. However, the carcinogenesis is a multi step phenomenon and it is difficult to conclude by testing at any one stage. The compilation of data (Table2) shows that the citrinin and ochratoxin A showed tumor formation in rat and mice however no mutagenic activity in in-vitro experiment. The fibroblast cell transformation assay system is more close to whole animal system in terms of presence of S9 enzyme activation system. Although citrinin and ochratoxin A showed toxicity even without S9 enzyme activation (Kitabatake et al. 1993). Our positive results from the transformation of C3H/10T1/2 and NIH/3T3 support the previous reports of tumor formation in animal experiments by citrinin and ochratoxin A.

Further studies are needed to know the cellular and molecular mechanism of transformation by these mycotoxins. These studies shows that citrinin and ochratoxin A are the potential carcinogens and the hazards from these mycotoxins should be re-evaluated

## References


Arai, M.; Hibino,T.**1983**, *Cancer Lett.* 17, 281-287.

Chen, T. T.; Heidelberger, C. **1969a**, *Int. J. Cancer.* 4, 166-178.

Chen, T.T., Heidelberger, C. **1969B**, *J. Natl. Cancer Inst.* 42, 915-925.

Imaida, K.; Hirose. M.; Ogiso,T.; Kurata, Y.; Ito,N. **1982**, *Cancer Lett.* 16, 137-143.

Kanisawa, M. In: Toxigenic fungi, their toxins and health hazards, pp. 245-254. Edts; H. Kurata and Y. Ueno. Elsvie Amsterdam. **1984** ;

Kitabatake, N.; Doi, E.; Trivedi, A. B. **1993** ,*Comp. Biochem. Physiol.*, 105, 429-433.

Krogh, P.; Hald, B.; Pederson, E. **1973**, *Acta Pathol. Microbiol. Scand. Sect.B.* 81, 689-695.

Krogh,P.; Axelsen,N.H.; Elling,F.; Gryd-Hansen,F.; Hald,B.; Hyldgaard-Jensen, J.; Larsen,A. E.; Madsen,A.; Mortensen,H.P.; Moller,T.; Petersen,O.K.; Ravnskov,V.; Rostgaard,M.; Aalund,O.**1974**, *Acta Pathol.Microbiol. Scand.SectA*, 246, 1-21.

Kuratomi, Y.; Ono, M.; Yasutake, C.; Mawatari, M.; Kuwano,(1987). *Cell Physiol.*, **130**, 51-57.

Martin, W.; Lorkowski, G.; Creppy, E. E.; Dirheimer, G.; Roschenthaler, R.(1986) *Appl. Environ. Microbiol.* **52** 1273-1279

Maureen, H. K.; Benson, P. M.; Heath, H. Hayes, A. W.**1978** *Mutation Res.*, 53, 11-20.

Reznikoff, C. A.; Brankow, D. W.; Heidelberger, C.**1973A** *Cancer Res.*,33, 3231-32319.

Reznikoff, C. A.; Bertram, J. S.; Heidelberger, C.**1973b** *Cancer Res.*,33, 3239-3249.

Shinohara,Y.; Arai, M.; Hirao, K.; Sugihara, S.; Nakanishi, K.; Tsunoda, H.; Ito, N. **1976**, *Gann*, 67, 147-155.

Scudamore, K. A.; Hetmanski, M. T.; Chan, H. K.; Collins, S. **1997**, *Food additives and contaminants*, 14(2), 157

Trivedi, A. B. **1988**, Study of toxic food contaminants and their spoilage metabolites. Ph. D. Thesis, University of Sausor, India.

Ueno, Y.; Kubota, K.**1976** *Cancer Res.*,36, 445-451.

Ueno, Y.; Kubota, K.; Ito, T.; Nakamura, Y.**1978** *Cancer Res.* , 38, 536-542.

Wehner; F. C.; Thiel, P. G.; Rensburg, S. J. V.; Demasius, I. P. C.**1978** *Mutation Res.*, 58, 193-203.

# SENESCENCE INDUCTION IN CANCER CELLS

*Y. KATAKURA, T. MIURA, N. UEHARA, E. NAKATA, AND S. SHIRAHATA*
*Division of Bioresource and Bioenvironmental Sciences, Graduate School of Genetic Resources Technology, Kyushu University*
*6-10-1 Hakozaki, Higashi-ku, Fukuoka 812-8581, Japan*

## 1. Introduction

We show here the strategies to induce cellular senescence in cancer cells by using human lung adenocarcinoma cell line A549 cells as model cell line, by which we are aiming for the establishment of the novel method for cancer therapy. Until now, we have succeeded in inducing cellular senescence in human lung adenocarcinoma cells, A549 cells in several independent experiment. Firstly, we established A5DC7 cells which showed MHC class II inducibility against IFN-γ stimulation. This cell showed normal senescent cell phenotypes and repressed telomerase activity, thus was vested with finite replicative life-span concomitant with the telomere shortening. These results showed that A549 cells contains the cells which has an ability to senesce by the exogenous signals. Secondary, we also observed morphological change by the treatment of hydrogen peroxide and established AST-9 cells. Although AST-9 cells have strong telomerase activity, AST-9 cells shortened the telomere depending upon the number of passagings. We predict that this telomere shortening occurred by the functional impairment of telomerase, which was caused by the reduced amount of telomere binding protein on the telomere. The reduced amount of telomere binding protein on the telomere of AST-9 cells was confirmed by electrophoretic mobility shift assay. These two experiments demonstrate that there exist several and independent pathways to induce cellular senescence in cancer cells. In this study, we show here another successful example of senescence induction in cancer cells by TGF-β.

## 2. Materials and methods

### 2.1 β-GALACTOSIDASE ACTIVITY

β-galactosidase (β-Gal) staining was performed according to the method described by

Dimri *et al* (1). Briefly, cells were fixed in 3% formaldehyde and then incubated with fresh senescence-associated β-Gal stain solution. Staining was carried out at 37°C for 16 hr.

### 2.2 TELOMERE LENGTH

Telomere length distributions were analyzed at a series of time points during serial passagings. Length of the terminal restriction fragments (TRF) was determined by Southern blot analysis with a telomeric sequence probe and used to measure the length of the telomere as described previously (2).

### 2.3 TELOMERASE ACTIVITY

We used PCR-based TRAP (Telomeric Repeat Amplification Protocol) assay for detecting telomerase activity with some modifications (2). The TRAP assay was quantitative when lysate equivalent to a range from 50 to 1000 cells per reaction were used. TRAP assay products were visualized by SYBR Green I (TAKARA, Shiga, Japan) staining.

### 2.4 RT-PCR

hTERT and TEP1 mRNAs were detected by RT-PCR method. hTERT mRNA was amplified using oligonucleotide primers LT5 (CGGAAGAGTGTCTGGAGCAA) and LT6 (GGATGAAGCGGAGTCTGGA) for 26 cycles (94°C for 45 s, 60°C for 45 s, 72°C for 90 s), and TEP1 mRNA was amplified using oligonucleotide primers TEP1.1 (TCAAGCCAAACCTGAATCTGAG) and TEP1.2 (CCCGAGTGAATCTTTCTAC-GC) for 26 cycles at the same condition as above.

## 3. Results and discussion

### 3.1 TGF-β INDUCES TELOMERE SHORTENING-INDEPENDENT PREMATURE SENESCENCE IN A549 CELLS

When A549 cells was culture with 10 ng/ml of TGF-β, A549 cells immediately change its morphology into flat and enlarged morphology characteristic to senescent cells within one week (Fig. 1). Senescence-associated β-galactosidase activity was enhanced upon treatment of TGF-β for one week. Telomere length remained unchanged even after the treatment. From these immediate and telomere shortening independent phenotypic changes, we can conclude that TGF-β can induce premature senescence in A549 cells. However, TGF-β treated A549 cells did not arrested the growth, which is one of the key feature of premature senescence, but continued to grow. Thus we can not judge whether

TGF-b can vest A549 cells with finite life-span.

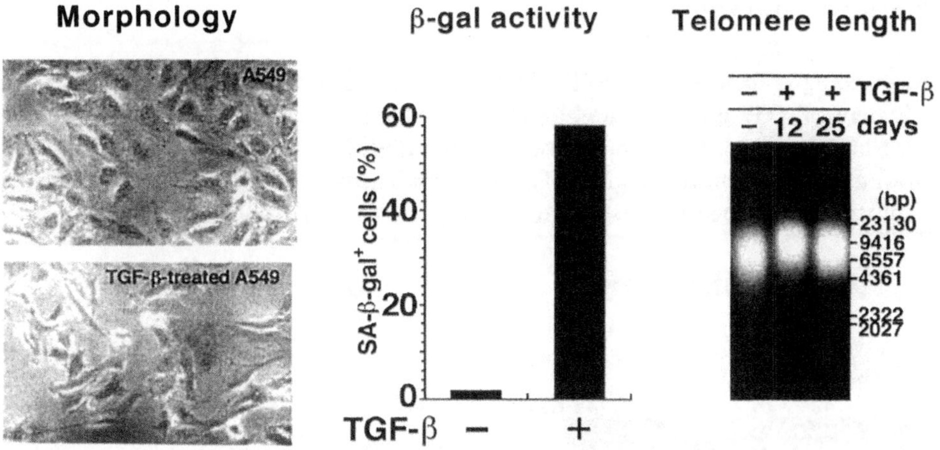


**Fig. 1 TGF-β induces premature senescence**

## 3.2 TGF-β REPRESSES hTERT TRANSCRIPTION, RESULTING IN SUPPRESSION OF TELOMERASE ACTIVITY IN A549 CELLS

Telomerase compensates for the infinite replicative potential of immortal cells. Thus we investigated the effect of TGF-β on telomerase activity and transcription of telomerase component. The results demonstrate that TGF-β represses hTERT transcription. hTERT is the catalytic subunit of telomerase, and its expression is known to completely correlated with telomerase activity. Corresponding to this result, TGF-β treated A549 cells gradually lose its telomerase activity after its addition (Fig. 2).

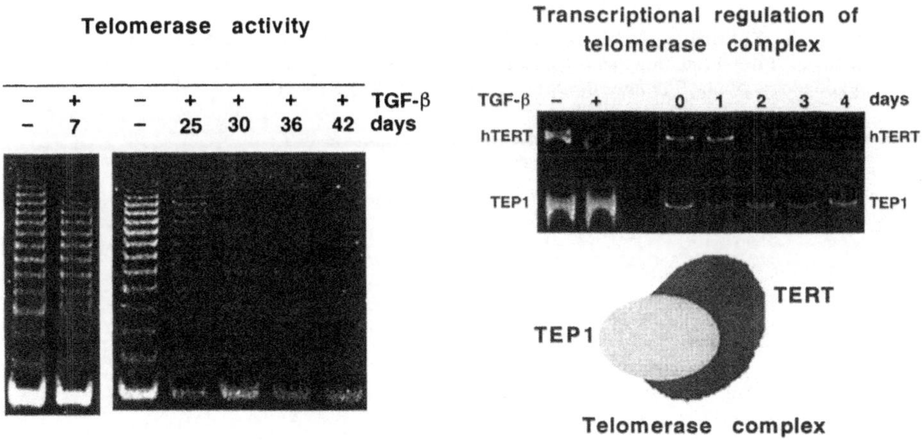


**Fig. 2 TGF-β represses telomerase activity and hTERT transcription**

## 3.3 TGF-ß-TREATED A549 CELLS SHORTENED THE TELOMERE CONCOMITANT WITH THE SUPPRESSION OF TELOMERASE RESULTING IN GROWTH INHIBITION

Corresponding to the previous results showing that the telomerase repression byTGF-ß, telomere of TGF-ß-treated A549 cells gradually shortened depending upon the number of passagings. Furthermore, A549 cells cultured with TGF-ß for a long term showed growth inhibition (Fig. 3). These results indicate that TGF-ß induces the replicative senescence as well as the premature senescence in A549 cells.

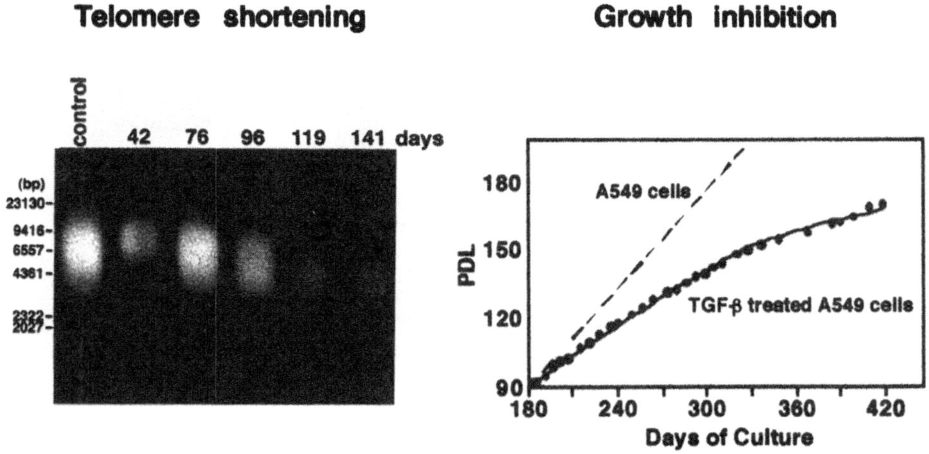


**Fig. 3 TGF-ß-treated A549 cells entered into replicative senescent state**

## 4. Reference

1. Dimri, G. P., Lee, X., Basile, G., Acosta, M., Scott, G., Roskelley, C., Medrano, E. E., Linskens, M., Rubelj, I., Pereira-Smith, O., Peacocke, M., and Campisi, J. (1995) A biomarker that identifies senescent human cells in culture and in aging skin *in vivo*. *Proc. Natl. Sci. Acad. USA*, **85**, 5112-5116.
2. Katakura, Y., Yamamoto, K., Miyake, O., Yasuda, T., Uehara, N., Nakata, E., Kawamoto, S. and Shirahata, S. (1997) Bidirectional regulation of telomerase activity in a subline derived from human lung adenocarcinoma, *Biochem. Biophys. Res. Commun*, **237**, 313-317.

# GROWTH CHARACTERISTICS OF NIH3T3 CELLS EXPRESSING bFGF AND/OR IGF-I AND/OR IGF-II

**DAJIANG LI[1], SIMON HETTLE, JOHN MCLEAN AND CAROLINE MACDONALD.**
***Department of Biological Sciences, University of Paisley, PA1 2BE, Scotland.***
***[1]Present address, Wistar Institute, Philadelphia.***

The traditional requirement for serum for the growth of cell lines in culture can often be provided by serum-free media containing a variety of growth factors. An alternative approach, which is investigated here, is to engineer the cells to express growth factors in an endogenous fashion. We report here on the growth of NIH3T3 cells transfected with basic fibroblast growth factor (bFGF) and/or insulin-like growth factors I and II (IGF-I and IGF-II). Morphological changes and the growth rate were compared with growth in normal serum-containing medium. The experimental data suggested that the expression of either bFGF alone, or the co-expression of bFGF and either IGF-I or II could improve the survival of NIH3T3 cells in low serum or serum-free media. The use of such lines could decrease the use of serum in cell culture and thus both reduce the costs involved in this technique and simplify the down-stream purification procedure in protein harvest. Hence, such lines may be of value in both experimental and industrial applications.

## Introduction

The media used to grow mammalian cell lines in culture traditionally relied on the use of serum supplements to promote cell growth. However, the inclusion of serum leads to problems of reproducibility of growth patterns, regularity of supply, purification difficulties and potential for the transmission of infectious agents and pathogens. In order to overcome these limitations, associated with the use of serum, many defined media have been developed which contain the growth promoting agents in the form of purified or recombinant growth factors. An alternative approach, described here, is to engineer the cell line by introducing the genes for growth promoting factors directly into the cell so that growth can be stimulated directly in an autocrine fashion.

Mitogens have been classified into three groups by Phillips and Cristofalo (1988):

- class 1 or competence factors which are capable of stimulating quiescent cells to leave Go e.g. platelet derived growth factor (PDGF), epidermal growth factor (EGF), fibroblast growth factor (FGF) and thrombin;

- class 2 or progression factors which allow cells to progress into S phase e.g. insulin-like growth factor-I (IGF-I ); multiplication stimulating activity (MSA) and insulin;
- class 3 factors which are required for maximal mitogenic stimulation e.g. dexamethesone and hydrocortisone.

In this study we chose to introduce basic fibroblast growth factor (bFGF), insulin-like growth factor I (IGF-I) and insulin-like growth factor II (IGF-II) into cells both singly and in combination. These factors were chosen as representatives of different classes of mitogens and because there was evidence for a role in cell proliferation for both bFGF (Hirobe 1992; Kitchens *et al*, 1994) and the IGFs (Macaulay 1992; Cullen *et al*, 1992).

**Materials and Methods**

NIH3T3 cells were cultured in Dulbecco's Modified Eagles Medium (DMEM) with 10% FBS (foetal bovine serum), 7.5% FBS, 5% FBS, 2.5% FBS, or without FBS, and containing penicillin (50 IU/ml) and streptomycin (50 µg/ml). Soft agar media, made from DMEM with 20% FBS, with penicillin (100 IU/ml) and streptomycin (100 µg/ml), consisted of bottom agar (2 x DMEM medium containing 0.66% agar) and top agar (2 x DMEM medium containing 0.33% agar).

For transfection 5 x $10^5$ NIH3T3 cells were seeded in 10 ml of medium in 10 cm dishes and incubated overnight at 37°C. HEPES-buffered saline (0.5 ml of 2x) was added dropwise to 0.25 M $CaCl_2$ (0.5ml) and plasmid DNA (1mg/ml in TE) in a sterile 15 ml Falcon tube with constant swirling and the suspension left at room temperature for 20 mins before addition to the cells. After incubation for 4-6 hours at 37°C the medium was aspirated and the cells rinsed with 10 ml of PBS. Glycerol (2 ml of 15%) was added for 1 minute, the cells rinsed twice with 10 ml of PBS, twice with 10 ml of growth medium and then fed with fresh growth medium. After incubation at 37°C for 24 hours the cells were trypsinized and split at appropriate ratio (>1:10). After a further 24 hours growth clones were selected which were expressing the transfected sequences. Cells were incubated for 9 days (with media changes every 3 days) in media containing geneticin at 500 µg/ml or hygromycin at 200 µg/ml as appropriate. Dishes were screened by inverted phase microscopy and clones were picked into 24 well culture plates for recovery.

Cell growth rates were measured in 25cm$^2$ flasks seeded at a density of 4 x $10^5$ cells/flask in triplicate. The medium was replaced after 3 days and the cells harvested and counted on day 6.

The vectors constructed for the expression of bFGF, IGF-I and IGF-II have been described previously (Li, 1997) and were based on either the pLNCX non-splicing retrovirus vector or the pCMV hyg vector. In each case the exogenous growth factor gene was expressed from the CMV immediate early promoter and the gene for the selectable marker (neo$^r$ for the FGF vector and hyg$^r$ for the IGF vectors) from the MoMLV LTR promoter. Expression of bFGF, IGF-I and IGF-II was confirmed by immunoblotting. Transfected and non-transfected NIH3T3 cells (1 x $10^8$) were lysed in 1 ml of buffer containing 50 mM of TrisHCl buffer (pH 7.5), 400 mM NaCl, 1 mM $MgCl_2$, 1% Nonidet P-40, and 1μM phenylmethylsulfonal fluoride. Nuclei and cell debris were removed by centrifugation at 16000 x g for 10 min at 4°C and the supernatants were analysed for bFGF, IGF-I, and IGF-II by sodium dodecyl sulphate/polyacrylamide gel electrophoresis (SDS/PAGE) and Western blotting on nitrocellulose. The Western blots were probed with affinity-purified mouse anti-bFGF (Sigma), IGF-I, and IGF-II antibodies (Serotec Inc.). The second antibody for luminescence development was horseradish peroxidase-conjugate rabbit Ig against mouse Ig and was used as described by the manufacturer (Amersham Lifescience, ECL system).

## Results

Cell lines expressing bFGF alone (3B) IGF-I alone (3I) and IGF-II alone (3II) and bFGF plus IGF-I (3BI) and IGF-II (3BII) were isolated and compared with the NIH3T3 control line. In each case the transfected cell lines showed changes in morphology compared with the parental line often growing more densely and in a less organised fashion. All of the transfected cell lines acquired the ability to form clones in soft agar (Table 1) with the lines in which two growth factors were being expressed showing the highest number of colonies. As far as growth in low serum medium was concerned it is apparent that bFGF is important for good growth and that cells co-expressing bFGF and IGF-I grow best in low serum. In the absence of serum, however, all of the lines show significantly reduced growth over a 6 day period (Table 2).

Table 1: Number of colonies formed after 10 days growth from 1000 cells seeded into a 6 cm diameter petri disk.

| *CELL LINE* | *NUMBER OF COLONIES* |
|---|---|
| 3T3 | 0 |
| 3B | 238 |
| 3I | 224 |
| 3II | 98 |
| 3BI | 742 |
| 3BII | 672 |

No reports have been published describing the expression of bFGF and IGF-I or bFGF and IGF-II in the same cell line and the effects of such expression on the growth of the cells. IGF-I and -II have been reported to have similar biological functions even in different cell types, although IGF-I exerts a stronger effect on cells than IGF-II (Froger-Gaillard *et al*, 1989; Taylor *et al*, 1988; Kemp *et al*, 1988). IGF-I plays the main role in growth and metabolic promotion in adult mammalian cells, and IGF-II in foetal tissues (Vetter *et al*, 1986). However, in this study, differences were observed in the growth of cells transfected with bFGF/IGF-I or bFGF/IGF-II, and these could be considered to be opposite rather than "strong" or "weak" effects. Thus, we conclude that the coexpression of bFGF/IGF-I or bFGF/IGF-II have different effects on the growth of transfected cell lines.

Table 2: Number of cells ($x10^{-5}$) recovered after 6 days growth in DMEM plus serum at the concentration indicated.

| | 10% | 7.5% | 5.0% | 2.5% | 0% |
|---|---|---|---|---|---|
| **NIH3T3** | 20 | 13 | 12 | 7 | 1 |
| **3B** | 37 | 36 | 33 | 21 | 6 |
| **3BI** | 36 | 23 | 16 | 13 | 5 |
| **3BII** | 17 | 21 | 10 | 10 | 2 |
| **3I** | 23 | 26 | 24 | 18 | 5 |
| **3II** | 17 | 14 | 13 | 7 | 2 |

## Conclusion

This preliminary study indicates that worthwhile improvements in the growth of cell lines in low serum can be obtained by introducing genes for growth factors in to the cells themselves. This approach, used in conjunction with medium optimisation, offers considerable promise for the reduction of variability and cost and the simplification of product recovery in the cell culture based systems used in the biotechnology industry.

## References

Cullen K J, Lippman M E, Chow D, Hill S, and Rosen N. (1992) Insulin-like growth factor II overexpression in MCF-7 cells induces phenotypic changes associated with malignant progression. *Molecular Endocrinology* **6,** 91-100.

Froger-Gaillard B, Hossenlopp P, Adolphe M, and Binoux M. (1989) Production of insulin-like growth factors and their binding proteins by rabbit articular chondrocytes: relationships with cell multiplication. *Endocrinology* **124,** 2365-2372.

Hirobe T (1992) Basic fibroblast growth factor stimulates the sustained proliferation of mouse epidermal melanoblasts in a serum-free medium in the presence of dibutyryl cyclic AMP and keratinocytes. *Development* **114,** 435-445.

Kemp S F, Kearns G L, Smith W G and Elders M J. (1988) Effects of IGF-I on the synthesis and processing of glycosaminoglycan in cultured chick chondrocytes. *Acta Endocrinology* **119,** 245-250.

Kitchens D L, Snyder E Y and Gottlieb D I. (1994) FGF and EGF are mitogens for immortalized neural progenitors. *Journal of Neurobiology* **25,** 797-807.

Li D J. (1997) University of Paisley: Studies on the expression of bFGF and IGF-I and -II in 3T3 cells, PhD thesis.

Macaulay VM. (1992) Insulin-like growth factors and cancer. *British Journal of Cancer* **65,** 311-320.

Phillips P D and Cristofalo V J. (1988). Classification system based on the functional equivalency of mitogens that regulate WI-38 cell proliferation. *Experimental Cell Research* **175,** 396-403.

Taylor A M, Dandona P, Morrell D J and Preece M A. (1988) Insulin-like growth factor-I, protein kinase-C, calcium and cyclic AMP: partners in the regulation of chondrocyte mitogenesis and metabolism. *FEBS Letters* **236,** 33-38.

Vetter U, Zapf J, Heit W, Helbing G, Heinze E, Froesch E R and Teller W M. (1986) Human foetal and adult chondrocytes. Effects of insulin-like growth factors I and II, insulin, and growth hormone on clonal growth. *Journal of Clinical Investigation* **77,** 1903-1908.

# TARGETED DISRUPTION OF A MITOCHONDRIAL TRANSCRIPTION FACTOR A IN THE CHICKEN DT40 CELL LINES

Yuichi Matsushima[1], Kiyoshi Matsumura[2], Shohji Ishii[2], Yasuo Kitagawa[1,2]
[1]Nagoya University BioScience Center, [2]Graduate School of Agricultural Sciences, Nagoya University, Nagoya 464-8601, Japan

**Abstract**

Mitochondrial transcription factor A (mtTFA) is a key regulator of mammalian mitochondrial DNA (mtDNA) transcription and replication. mtTFA is a nucleus-encoded high mobility group-box protein, which binds upstream of the light- and heavy-strand promoters of mtDNA and promotes transcription of mtDNA. We cloned and sequenced chicken mtTFA. The overall amino acid sequence of chicken mtTFA has 44% homology to human mtTFA. A chicken gene for mitochondrial transcription factor A was disrupted by gene targeting in the chicken B-cell line DT40 which has high level of homologous recombination and rapid growth rate. Heterozygous knockout DT40 cells exhibited about 50% reduction of mtDNA copy number. Similarly, the mitochondrial transcripts were reduced in the heterozygous knockout DT40 cells.

## 1. Introduction

Mitochondria are one of one essential organelles in eucaryotic cells where cellular ATP is generated through the process of oxidative phosphorylation. Proteins components of the respiratory chain are gene products of both mitochondrial and nuclear genes. The mitochondrial genome itself encodes 13 respiratory chain subunits, but the vast majority of the subunits are encoded in nuclear genes. Similarly, the processes of replication and transcription of mitochondrial DNA are regulated by nuclear genes except for two rRNA genes and 22 tRNA genes are encoded in mitochondrial genome. Mitochondrial transcription factor A (mtTFA) is a nuclear encoded high mobility group box protein, which binds upstream of the light and heavy strand promoters and promotes transcription of mitochondrial DNA [1, 2]. mtTFA also may plays a role in regulation of mitochondrial DNA replication [1, 2]. When leading strand of mitochondrial DNA replication starts, it requires RNA primer transcripted from the light strand promoter. Therefore, mtTFA is necessary for this primer synthesis. In this study, we tried to disrupt the chicken gene for mtTFA by gene targetting in DT40 cell [3].

## 2. Materials and Methods

### 2.1 Cell culture and transfection

DT40 cells were maintained in DMEM medium, supplemented with 10% fetal bovine serum at 37°C and 5% $CO_2$. Cells were routinely split to $1 \times 10^5$ per ml every three days. For each transfection $10^7$ cells were suspended in 0.5% PBS containing 30 μg linearized plasmid and electroporated with Gene Pulser apparatus (BioRad) at 550 V and 25 μF. Following electroporation, cells were transferred to 20 ml fresh medium lacking drugs and incubated for 24 hr. Cells were then resuspended in 90 ml medium containing the appropriate drugs and divided into five 96-well plates. After 7-10 days, drug resistant colonies were transferred to 50 ml T flask. Concentrations of drugs used in this study were: 30 μg/ml mycophenolic acid and 2 mg/ml G418.

### 2.2 DNA and RNA extraction.

Genomic DNA was purified with DNAZOL (GIBCO-BRL). Total RNA was extracted from DT40 cells with acid guanidinium/thiocyanate/phenol/chloroform method.

### 2.3 Screening and sequencing of cDNA and Genomic clones.

DT40 genomic DNA library in λ FIXII vector, gifted from Dr. Tateo Nakayama, and chicken 5days embryo cDNA library in λ ZAPII vector, gifted from Dr. Atsushi Kuroiwa, were used in this study. The human mtTFA cDNA was used as a probe. Approximately $1 \times 10^6$ phages from each library were screened in a buffer containing 50 mM Tris-HCl (pH 7.5), 5x standard saline citrate (SSC), 5x Denhardt's solution, 1% sodium dodecylsulfate (SDS) and 100 mg/ml salmon sperm DNA. After hybridization at 53°C for 16 h, filters were washed three times with 2x SSC containing 0.1% SDS at room temperature and once with 0.5x SSC containing 0.1% SDS at 53°C for 30 min and exposed to x-ray film (Fuji). The genomic DNA clones were analyzed by restriction endonuclease mapping and Southern blot analysis. Pertinent restriction fragments were subcloned into pBluescript and sequenced by the dideoxynucleotide method using a DNA sequencing system model 373S of Applied Biosystems.

2.4 Northern and Southern blot analysis.

Genomic DNA (5 to 15 μg) were separated by 0.8% agarose gel electrophoresis after enzyme digestion and transferred to Hybond $N^+$ (Amersham). Total RNA (10 μg) were separated by 1.0% formaldehyde agarose gel electrophoresis and transferred to Hybond $N^+$ (Amersham).Hybridization with the probes was carried out in a buffer containing 50 mM Tris-HCl (pH 7.5), 5x standard saline citrate (SSC), 5x Denhardt's solution, 1% sodium dodecylsulfate (SDS) and 100 mg/ml salmon sperm DNA. After hybridization at 65°C for 16 h, filters were washed three times with 2x SSC containing 0.1% SDS at room temperature and once with 0.2x SSC containing 0.1% SDS at 65°C for 30 min and analyzed with a Fuji Film BAS 2000 Image Analyzer. All probes were obtained by RT-PCR or PCR.

## 3. Results and discussion

We first cloned chicken mtTFA (*cmtTFA*) cDNA and genomic DNA. This cDNA have a open reading frame of a 264 amino acid sequence. The overall amino acid sequence of chicken mtTFA has only 44% and 46% homology to human and mouse, respectively (data not shown) [1]. The *cmtTFA* gene contains five introns (Fig. 1A). *cmtTFA* is a single copy gene as determined by Southern blot analysis (data not shown).

To begin targeted disruption of *cmtTFA* , mycophenolic acid and neomycin-resistance genes, each driven by the chicken b actin promoter, were inserted into the first exon of *cmtTFA* genomic DNA, giving the constructs Mpa-mtTFA and Neo-mtTFA, respectively (Fig. 1A). DT40 cells were transfected with Mpa-mtTFA and 30 mycophenolic acid resistant clones were isolated and screened for homologous recombination by Southern blotting. Hybridization with probe 1 (Fig. 1A) showed that *Eco*RI and *Sal*I double digested DNA from 12 of the 30 clones gave rise to a new 5-kb band, the size expected following homologous recombination (Fig. 1B). This result indicates that targeted disruption of one allele of the *mtTFA* gene occurred at a frequency of 40%.

To investigate mtDNA copy number in these clones, each three +/*mtTFA*$^-$ and +/+ clones were further analyzed by Southern blotting. The copy numbers of mitochondrial DNA were reduced by approximately 50% in +/*mtTFA*$^-$ clones (Fig. 2A). In +/*mtTFA*$^-$ clones, transcripts of mtTFA and ATPase subunitd 6 and 8, which are encoded by mitochondrial genome [4], were reduced by approximately 50% (Fig. 2B).

There was no significant change in transcripts of mitochondrial DNA polymelase [5] which encoded in nuclear genome (Fig. 2B).

If mtTFA is not essential for cell viability, a second round of gene targeting should disrupt the second mtTFA allele. To test this, two of 12 heterozygous clones were transfected with Neo-mtTFA and selected in medium containing both mycophenolic acid and G418. DNAs from over 50 resistant clones were analyzed by Southern blot. The results showed that none of them underwent homologous recombination. To confirm that Neo-mtTFA was capable of homologous recombination, we transfected wild-type DT40 cells with this construct. Out of 30 G418 resistant clones, 14 had targeted disruption of one *mtTFA* allele. To provide further evidence that the second *mtTFA* allele could not be disrupted, heterozygous

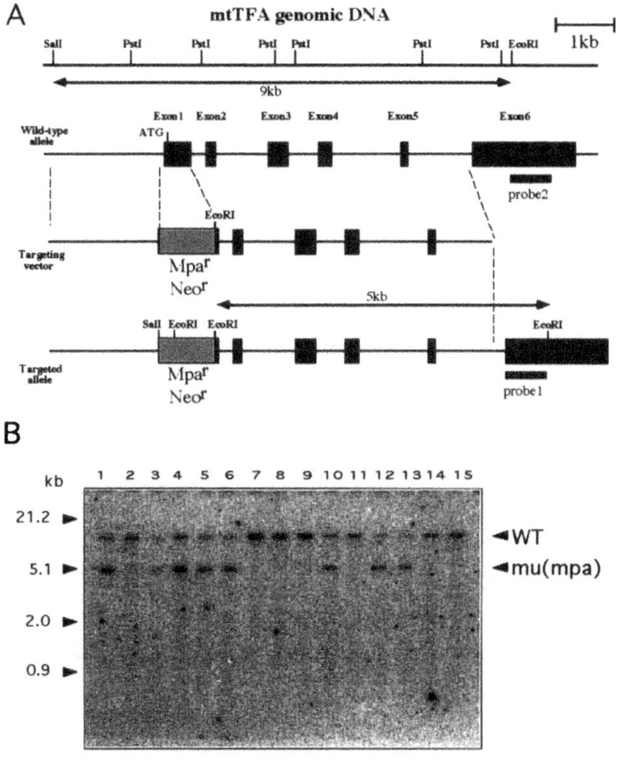


Figure. 1 High frequency disruption of a single *mtTFA* allele.
(A) Restriction map of the chicken *mtTFA* locus and constructs used for targeted disruption. Shaded box represent exons of the chicken *mtTFA* gene. (B) Wild type DT40 cells were transfected with the construct Mpa-mtTFA. Genomic DNAs from 15 mycophenolic acid-resistant clones were digested with *Eco*RI and *Sal*I and hybridized with probe 1. Eight clones gave rise to a 5-kb band consistent with homologous recombination. Positions of frabments from wild-type (WT) and Mpa$^r$ disrupted alleles are indicated.

DT40 cells with one allele disrupted by Neo-mtTFA were transfected with Mpa-mtTFA. None of 40 resistant clones underwent homologous recombination. Taken together, these data suggest that mtTFA is an essential gene in DT40 cells. One of the possible explanation for this lethality is that mitochondrial DNA and its transcription is essential for DT40 cell viability [6].

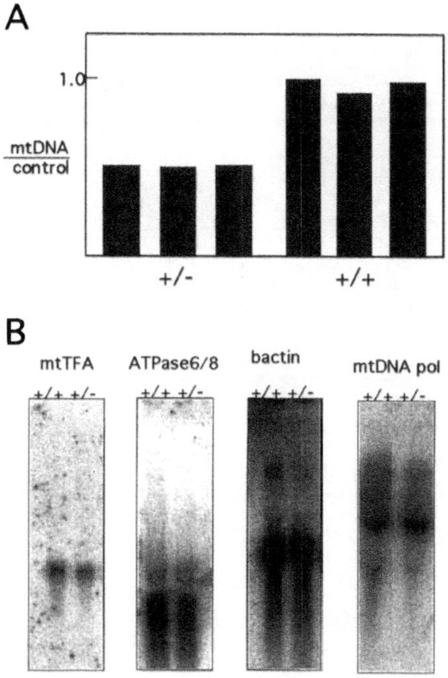


Figure. 2 Characterization of levels of mtDNA and mitochondrial transcripts.
(A) Southern blot analysis of mtDNA levels in wild type (+/+) and heterozygous (+/-) DT40 clones. The membrane were first hybridized with a mtDNA probe and then re-hybridized with probe 2. Hybridization signals were quantitated with a Fuji Film BAS 2000 Image Analyzer. (B) Northern blot analysis of transcript levels in wild type (+/+) and heterozygous (+/-) DT40 clones.

## 4. References

1. Parisi, M.A. and Clayton, D.A. (1991) Similarity of human mitochondrial transcription factor 1 to high mobility group proteins, *Science* **252**, 965-969.

2. Dairaghi, D.J., Shadel, G.S., and Clayton, D.A. (1995) Addition of a 29 residue carboxyl-terminal tail converts a simple HMG box-containing protein into a transcriptional activator. *J Mol Biol* **249**,11-28.

3. Buerstedde, J.M. and Takeda, S. (1991) Increased ratio of targeted to random integration after transfection of chicken B cell lines. *Cell* **67**, 179-88.

4. Desjardins, P. and Morais, R. (1990) Sequence and gene organization of the chicken mitochondrial genome. A novel gene order in higher vertebrates. *J Mol Biol* **212**, 599-634.

5. Ropp,P.A. and Copeland,W.C. (1996) Cloning and characterization of the human mitochondrial DNA polymerase, DNA polymerase gamma. *Genomics* **36**, 449-458.

6. Zinkewich-Peotti, K., Parent, M., and Morais, R. (1990) Mitochondrial DNA modulation of the anchorage-independent phenotype of transformed avian cells. Cancer Res **50**, 6675-82.

# EXPRESSION AND CHARACTERIZATION OF MOUSE EPIDERMAL TRANSGLUTAMINASE IN BACULOVIRUS-INFECTED INSECT CELLS

Kiyotaka Hitomi, Koji Ikura* and Masatoshi Maki

Laboratory of Molecular and Cellular Regulation, Graduate School of Bioagricultural Sciences, Nagoya University, Nagoya 464-8601, Japan.
* Department of Applied Biology, Faculty of Textile Science
Kyoto Institute of Technology, Kyoto 606-8585, Japan.

**ABSTRACT.** Transglutaminase (TGase) is a calcium ion-dependent enzyme that catalyze the formation of isopeptide cross-links in proteins between the $\gamma$-carboxamide groups of glutamine residues and $\varepsilon$-amino groups of lysine residues. Epidermal-type TGase (TGase 3) is involved in the formation of the cornified cell envelop, which serves a vital barrier function for the skin, by cross-linking of a variety of structural proteins in the epidermis. This enzyme is activated from zymogen form(77kDa) by limited proteolysis into 50kDa and 27kDa molecules with unknown proteases during epidermal terminal differentiation.

It has been difficult to isolate sufficient quantities of native enzymes from tissues or to obtain recombinant proteins in a bacterial expression system for biochemical studies on its properties of TGase 3. In this study, we circumvented these problems by expressing recombinant full-length mouse TGase 3 in baculovirus-infected insect cells. The expressed TGase 3 was purified to homogeneity by successive chromatography and HPLC. Treatment of the purified recombinant protein with dispase, which was bacterial protease known to activate TGase 3 zymogen, produced the activated TGase 3. The migration of TGase 3 zymogen in SDS-polyacrylamide gel electrophoresis was anomolous when proTGase 3 was incubated with calcium ion. This possibly results from intrinsic enzymatic activity. GTP inhibited the enzymatic activity of recombinant TGase 3, as in the case of tissue-type TGase (TGase2).

## 1. Introduction

Transglutaminases are $Ca^{2+}$-dependent enzymes that catalyze the post-translational modification of proteins by transamidating the glutamine residues. TGases constitute a large protein family and are distributed in various tissues and cells [1, 2].

In the epidermis, four TGases had been reported to exist in different amounts. Two of these are the TGase 1 (TGase K) and TGase 3 (TGase E), which are involved in the formation of cornified cell envelop(CE) by cross-linking various structural protein in the terminally differentiated keratinocyte [3]. The third enzyme, TGase 2 (TGase C) stabilize the dermoepidermal junction in basal cell layer [4]. Recently, the fourth member was discovered in keratinocyte by reverse transcription-PCR [5].

TGase 3 is expressed in the last stage of terminal differentiation of the epidermis [6]and the inner root sheath and the medullary layers in hair follicle [7], but not expressed in cultured keratinocytes [8]. TGase 3 accounts for >75% of total TGase activity in epidermis, although there is far less TGase 3 proteins than TGase 1 and TGase 2 [9]. In hair follicle, TGase 3, rather than TGase 1, preferentially cross-link major structural protein such as trichohyalin and keratin intermediate filaments [10]. Native proTGase 3 from guinea pig skin was purified and characterized by Kim *et.al* .[9]. ProTGase 3, synthesized as 77kDa zymogen, are proteolysed into a 50kDa containing catalytic domain and 27kDa molecules upon activation. The proteolysis mechanisms and the enzymes responsible for the activation are not well understood. The cDNA of human and mouse were cloned by PCR using degenerated primers based on the peptide sequences of guinea pig enzyme [8]. Both proteins contain 692 amino acids of molecular mass about 77kDa from the deduced amino acid sequence and are activated by proteolysis.

Since too little TGase 3 is expressed in the skin to study its biochemical properties and activation mechanisms, it was necessary to construct a recombinant protein. Expressed protein of proTGase 3 in *E.coli* was recovered as inclusion body. Therefore, we decided to express recombinant proTGase 3 in a baculovirus system. In this system, recombinant baculoviruses are used as vectors to express heterologous genes under the forceful polyhedrin promoter in insect cells, providing correctly folded, modified proteins in large amounts.

In this study, the unexpected intracross-linking of zymogen form of TGase 3 was observed by the incubation with calcium and the inhibitory effects of GTP on its enzymatic activity were also discovered.

## 2. Results and Discussion

### 2.1 PURIFICATION OF RECOMBINANT TRANSGLUTAMINASE 3 EXPRESSED IN INSECT CELLS.

Insect cells were infected with recombinant baculovirus encoding full-length mouse proTGase 3 cDNA under the control of the polyhedrin promoter. The maximum expression level was observed from 36 to 48 h post-infection at a multiplicity of infection of 2-5.

Although the 77 kDa protein of recombinant proTGase 3 was highly expressed in infected Sf9 cells, less than one third of the produced proTGase 3 was recovered in the soluble fraction. For the purification, the suspension-cultured cells were collected and lysed with hypotonic buffer containing protease inhibitors. ProTgase 3 were purified from lysate by DEAE-Sephacel chromatography and heparin-Sepharose affinity chromatography based on the content of 77 kDa band. To obtain highly purified proTGase 3, contaminating proteins were excluded by anion ion-exchange chromatography and size fractionation chromatography (SMART system). After these purification steps, approximately 20 μg of proTGase 3 was obtained from 100 ml of suspension culture of Sf9 cells ($2x10^8$).

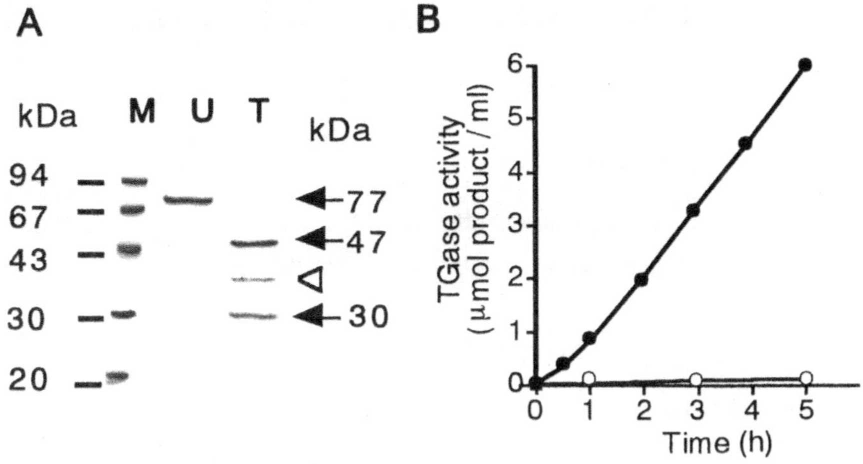


**Fig.1 Activation of recombinant purified proTGase3.**

(A) SDS-PAGE analysis of approximately 2 μg of purified proTGase before(lane 1) and after(lane 2) digestion with 1 unit of dispase for 10 min at 37 ℃. The processed bands (47 kDa, 30 kDa) are indicated by arrows. Lane U and T indicate untreated and dispase-treated zymogen, respectively. The open arrowhead indicates the dispase added to the reaction mixture. (B) The enzymatic activity of proTGase3 before(open circle) and after(closed circle) proteolysis by dispase were assayed.

## 2.2 ACTIVATION OF ZYMOGEN FORM OF TRANSGLUTAMINASE 3

To confirm that the purified recombinant proTGase 3 had enzymatic activity, proteolytic activation was performed by protease digestion according to the previous report [9]. Dispase, which was neutral bacterial protease, was examined for the activation. As shown in Fig.2A, 77 kDa of proTGase 3 was proteolysed into 47 kDa and 30 kDa molecules by the dispase treatment. The smaller fragment migrated at slightly larger position than those of guinea pig TGase 3 reported, probably due to the differences in primary structures of TGase 3 between guinea pig and mouse. Both the digested and undigested proTGase 3 preparations were measured for the TGase activity using CBZ-Gln-Gly and hydroxylamine as substrates (Fig.1B) [11]. The proteolysed TGase 3 had an apparent transamidating activity, whereas no enzymatic activity was observed in the untreated preparation.

## 2.3 MOBILITY-SHIFT OF ZYMOGEN OF TRANSGLUTAMINASE 3 BY TREATMENT WITH CALCIUM ION

During the course of purification, we observed that the treatment with high concentration of $Ca^{2+}$ caused the mobility-shift of proTGase 3 on SDS-PAGE (Fig.2).

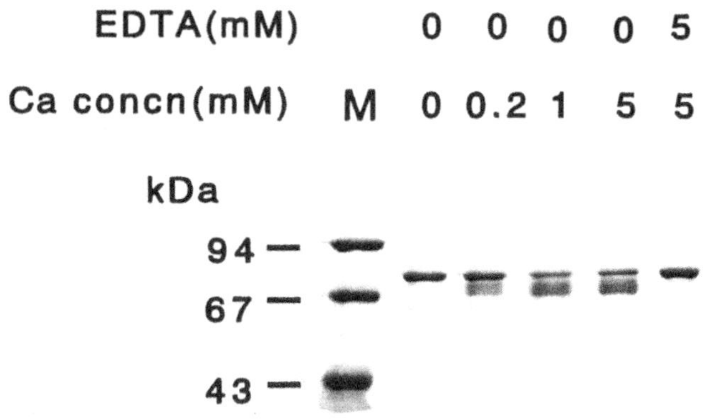


**Fig.2 $Ca^{2+}$-dependent mobility-shift of proTGase3.**
Incubation at a final concentration of 0,0.2,1,5 mM of CaCl2. To verify the $Ca^{2+}$ specific reaction, EDTA were added at a final concentration of 5 mM to chelate the $Ca^{2+}$.

The 77 kDa molecule migrated slightly faster when incubated in the solution containing $Ca^{2+}$ at 30 °C for 5 min. This migration was also dependent on the concentrations of $Ca^{2+}$, but aberrant migration was not observed when adding 5 mM EDTA in the solution(Fig.2 lane 5). The TGase activity was responsible for the aberrant migration because the migration were liable to occur at higher pH and by blocking active site cysteine residue(data not shown). Active-site mutant of proTGase 3 was prepared in baculovirus system and compared with the wild type. The mobility of mutant proTGase 3 was not affected by the presence of high concentration of $Ca^{2+}$ (data not shown). These results suggested that the intrinsic TGase activity was responsible for the aberrant mobility caused by $Ca^{2+}$. The mobility-shift was assumed to be intramolecular cross-linking, leading to the structural changes in zymogen form of TGase 3.

## 2.4 INHIBITION OF TRANSGLUTAMINASE 3 ACTIVITY BY GTP

In recent years, tissue-type TGase (TGase 2) has been reported to be GTP-binding protein and the enzymatic activity is specifically blocked by the presence of GTP. These results tempted us to examine the effect of GTP on the enzymatic activity of TGase 3. As shown in Fig.3 , both GTP and GDP were found to exhibit concentration-dependent inhibition of the TGase 3 activity at 25-200 μM. Concentrations of GTP and

GDP required for 50% inhibition were 10 μM and 50 μM, respectively. Furthermore, weak inhibitions by GMP and ATP were commonly observed as in the case of TGase2 (data not shown). This inhibition was not the result of chelating effect of GTP against Ca-dependent enzymatic activity because ATP showed weak inhibition at the same concentration. GTP-binding motifs were not seen in the primary sequence of proTGase 3 as in the case of TGase2. More recently, presence of GTP was found to prevent the $Ca^{2+}$-dependent proteolysis of TGase2 by *in situ* analysis [12]. Effects of GTP on the degradation of proTGase 3 by various proteases are under investigation.

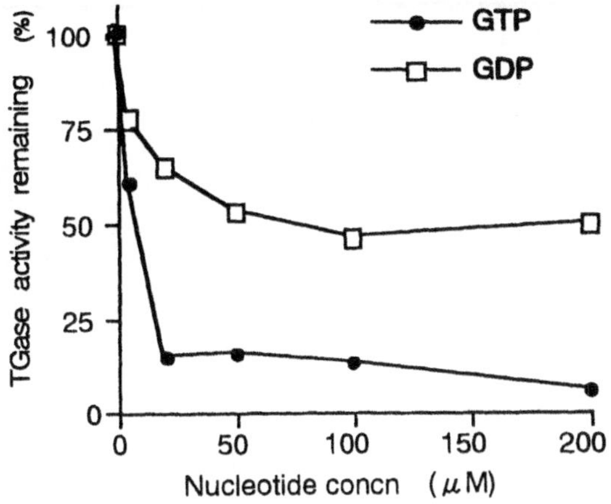


**Fig.3 Effects of GTP and GDP on mouse TGase3 activity.** Increasing concentrations of GTP(●) and GDP(□) were pre incubated for 5 min with activated mouse recombinant TGase 3 (2 μg) before TGase reaction for 30 min. The remaining enzymatic activities were calculated based on the amount of the products without nucleotide.

## REFERENCES

1. Greenberg, C. S., Birckbichler, P. J., and Rice, R. H. (1991) Transglutaminases: multifunctional cross-linking enzymes that stabilize, *FASEB Journal*, **5**, 3071-3077.
2. Aeschlimann, D., and Paulsson,M. (1994) Transglutaminase:Protein cross-linking enzymes in tissues and body fluids, *Thrombosis and Haemostasis,* **71**,402-415.
3. Rice, R. H., Rong, X., and Chakravarty, R.(1990) Proteolytic release of keratinocyte transglutaminase, *Biochem.J.*, **265**, 351-357.
4. Lichti, U., Ben, T., and Yuspa, S. H.(1985) Retinoic acid-induced transglutaminase in mouse epidermal cells is distinct, *J.Biol.Chem.,* **260**, 1422-1426.
5. Aeschlimann, D., Koeller, M. K., Allen-Hoffmann, B. L.,and Mosher, D. F. (1998) Isolation of a cDNA encoding a novel member of the transglutaminase gene family from human keratinocytes., *J.Biol.Chem.,* **273**, 3452-3460.
6. Kim, S. Y., Chung, S. I., and Steinert, P. M. (1995) Highly active soluble processed forms of the transglutamiase 1 enzyme in epidermal keratinocytes, *J.Biol.Chem.,* **270**, 18026-18035.
7. Lee, S. C., Kim, I. G., Marekov, L. N., O'Keefe, E. J., and Parry, D. A. (1993) The structure of human trichohyalin. Potential multiple roles as a, *J.Biol.Chem..,* **268**, 12164-12176.
8. Kim, I. G., Gorman, J. J., Park, S. C., Chung, S. I., and Steinert, P. M., (1993) The deduced sequence of the novel protransglutaminase E (TGase 3) of human, *J.Biol.Chem.,* **268**, 12682-12690.
9. Kim, H. C., Lewis, M. S., Gorman, J. J., Park, S. C., Girard, J. E., and Folk, J.E. (1990) Protransglutaminase E from guinea pig skin. Isolation and partial characterization, *J.Biol.Chem.*, **265**, 21971-21978.
10. Candi, E., Melino, G.G., Mei, E., Tarcsa, S. I., Chung, L. N., Marekov, and P. M. Steinert. (1995) Biochemical, Structual, and transglutaminase substrate properties of human loricrin the major epidermal cornified cell envelope protein, *J.Biol.Chem.,* **270**, 26382-26389.
11. Folk, J. E. (1970) Transglutaminase(Gunea pig liver), Methods in Enzymology **174**, 889-895.
12. Zhang, J., Lesort, M., Guttmann, R. P., and Johnson, G. V. (1998) Modulation of the in situ activity of tissue transglutaminase by calcium and GTP, *J.Biol.Chem.*, **273**, 2288-2295.

# EFFECT OF TEA POLYPHENOLS ON DEGRANULATION IN HUMAN MATURE BASOPHILS DIFFERENTIATED WITH IL-4

Hirofumi TACHIBANA, Yousuke SUNADA, Takashi HARA, and Koji YAMADA
*Graduate School of Bioresources and Bioenvironmental Science, Kyushu University, Hakozaki 6-10-1, Fukuoka, 812-8581, Japan*

**Abstract.** We examined the effects of tea polyphenols on the activation process of human mature basophils by stimulation with $Ca^{2+}$ ionophore A23187 or IgE receptor crosslinking. IL-4 was used to induce differentiation in human leukemia cell line KU812 into the morphological and functional human mature basophils used in this study. Among the tea polyphenols examined, (-)-epicatechin gallate (ECG) and (-)-epigallocatechin gallate (EGCG) inhibited degranulation caused by IgE receptor crosslinking as well as by A23187. Thus, our findings suggest that the effects of ECG and EGCG may act on signaling pathway which regulates degranulation in human basophils after the increase of intracellular $Ca^{2+}$ concentration.

## 1. Introduction

As effector cells in the IgE-dependent immune response system, mast cells and basophils appear to play an important role in allergic reactions (1). The release of histamine and other inflammatory mediators by these cells is the primary event in the reaction (2). Basophils express the high affinity IgE receptor (FcεRI) on the cell surface and possess cytoplasmic granules containing various potent inflammatory mediators such as histamine, proteases and chemotactic factors (3). When a specific allergen associates with the FcεRI-bound IgE, elevation of the intracellular $Ca^{2+}$ concentration occurs, and then basophils can release preformed mediators and metabolites of arachidonic acids (4). Inhibition of any steps of these sequential reactions leads to the attenuation of allergic symptoms. The mechanism that regulates the FcεRI-mediated degranulation remains largely unknown in human basophils (5).

Tea polyphenols have been reported to inhibit degranulation in rat mast cells and rat basophilic leukemia RBL-2H3 cells (6, 7). When measuring the sensitivity of degranulation induced stimuli, some differences between human and rat mast cells have been reported (8, 9). However, the effect of tea polyphenols on human basophils has not yet been studied. In the present study, we examined the effect of various tea polyphenols on histamine release from human mature basophilic cells derived from the human leukemia cell line KU812 (10). KU812 cells were originally established from a patient with chronic myelogenous leukemia and is available for studies on differentiation and signal transduction (10). However, the cells are generally thought to be an immature basophilic cell line able to express only a very low level of FcεRI (11). Recently, we demonstrated that IL-4 can induce differentiation of KU812 cells into morphologically and functionally mature human basophilic cells (12). These IL-4-treated KU812 cells were then used as human mature basophils in this study. Here, we found that tea polyphenols may act on a point of the signaling pathway for degranulation after the increase of intracellular $Ca^{2+}$ concentration occurs.

## 2. Materials and Methods

### 2.1. CHEMICALS

The following tea polyphenols; (-)-epigallocatecin-3-gallate (EGCG), (-)-epigallocatechin (EGC), (-)-epicatechin-3-gallate (ECG), (-)-epicatechin (EC), (+)-catechin (C), were purchased from Kurita Water Industries Ltd (Tokyo, Japan). The chemical structures of these compounds are shown in Fig. 1. The calcium ionophore A23187 was obtained from Sigma (St. Louis, MO). Histamine dihydrochloride and o-phthalaldehyde (OPT) were purchased from Wako (Osaka, Japan). Tyrode buffer was consisted of 137 mM NaCl, 2.7 mM KCl, 1.8 mM $CaCl_2$, 1.1 mM $MgCl_2$, 11.9 mM $NaHCO_3$, 0.4 mM $NaH_2PO_4$, 5.6 mM glucose, pH 7.2.

### 2.2. CELLS AND CELL CULTURE

KU812 cells were obtained from the Japanese Cancer Research Resources Bank (Tokyo, Japan). The cells were maintained in RPMI-1640 (Nissui, Tokyo, Japan) supplemented with 10% fetal bovine serum (Intergen, Purchase, NY), 100 U/ml penicillin G, 100 mg/ml streptomycin and 10 mM HEPES buffer. To prepare mature basophils, the KU812 cells were stimulated with 1 ng/ml of IL-4 for 14 days (12).

### 2.3. HISTAMINE RELEASE

IL-4-treated KU812 cells (1.0 x $10^6$ cells) were washed and resuspended in 2 ml of Tyrode's buffer. 0.5 ml of 25 μM A23185 (5 μM final concentration) or 0.5 ml of 10 μg/ml goat anti human IgE antibody (2 μg/ml final concentration) was added to the cell suspension. Before the addition of the IgE antibody, 0.5 ml of 25 μgl/ml human myeloma IgE was added to the cell suspension and incubated a 37°C for 4 hours to sensitize the cells. After addition of either A23187 or the anti-human IgE antibody, the mixture was incubated at 37°C for 20 min and the reaction was terminated by cooling at 4°C for 15 min. The cell suspension was then centrifuged at 300 x g for 10 min, and the amount of histamine in the supernatant was measured by fluorometric assay as described below.

### 2.4. MEASUREMENT OF HISTAMINE

The histamine content was measured by the fluorometric assay. After mixing 2 ml of the supernatant with 0.75 g of NaCl and 0.5 ml of 1 N NaOH, 5 ml of a 3:2 (V/V) mixture of n-butanol and chloroform was added and mixed for 5 min. After centrifugation at 270 x g for 5 min, 4 ml of the organic solvent layer was recovered , mixed with 0.15 ml of 1 N NaOH and 0.1 ml of 0.2% OPT, and let stand for 5 min at room temperature. The reaction was terminated by adding 0.14 ml of 0.5 N $H_2SO_4$, and then the fluorescence intensity was measured using a spectrofluorophotometer (RF500, Shimazu, Kyoto, Japan) with the excitation at 360 nm and the emission at

450 nm. The percentage inhibition of histamine release was calculated as follows: inhibition (%) = 100-(test - negative control)/(positive control - negative control) x 100. Negative control was the histamine content without stimulation, positive control was after stimulation, and test was after stimulation with sample.

## 3. Results

### 3.1. EFFECT OF TEA POLYPHENOLS ON HISTAMINE RELEASE FROM HUMAN MATURE BASOPHILS

Green tea contains polyphenols, which include flavanols, flavandiols, flavonoids, and phenolic acids. Most of the green tea polyphenols are flavonols, commonly known as catechins (Fig. 1). We investigated the effect of catechins on histamine release from human mature basophils derived from IL-4-stimulated KU812 cells (12). As shown in Table 1, ECG (69% inhibition) and EGCG (55% inhibition) significantly inhibited histamine release which was induced by 100 μM $Ca^{2+}$ ionophore A23187. However, C, EC, and EGC were not shown to be effective. Next we examined the dose-response by testing the effect of 1 to 100 μM of EGCG on histamine release. As shown in Table 2, EGCG inhibited, in a dose dependent manner, histamine release from KU812 cells stimulated with anti human IgE antibody. When cells were stimulated by IgE receptor crosslinking, the inhibitory effect of EGCG was 29 % at 10 μM and 41% at 100 μM. Like wise, when the cells were stimulated with A23187, the inhibitory effect of EGCG was 69% at 10 μM and 74% at 100 μM.

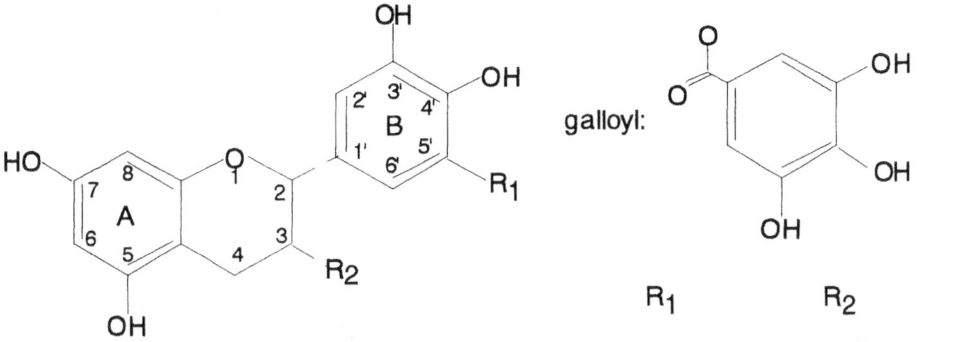


| | $R_1$ | $R_2$ |
|---|---|---|
| (+)-Catechin (C) | H | H |
| (-)-Epicatechin (EC) | H | OH |
| (-)-Epigallocatechin (EGC) | OH | OH |
| (-)-Epicatechin gallate (ECG) | H | galloyl |
| (-)-Epigallocatechin gallate (EGCG) | OH | galloyl |

## 4. Discussion

In this study, we demonstrated that ECG and EGCG inhibit histamine release from a human mature basophilic cell line. In rat RBL-2H3 cells or rat peritoneal exudate cells, the inhibitory effect of ECG on the histamine release was shown to be much lower than that of EGCG (6, 7). Among the tea polyphenols, 3-O-galloyl containing compounds such as ECG and EGCG inhibited histamine release from human basophils, while compounds lacking the galloyl structure such as C, EC, and EGC were not effective. In rat basophilic leukemia RBL-2H3 cells and rat peritoneal exudate cells, the triphenol structure has been shown to play a role in the inhibition of histamine release (6, 7). These results suggest that sensitivity to the tea polyphenols on histamine release is different between rat and human basophils.

**Table 1.**
Effect of various tea polyphenols on A23187-mediated histamine release from basophilic-differentiated KU812 cells. KU812 cells were stimulated with A23187 for 20 min in the presence of various tea polyphenols (100 μM). Histamine released into the supernatants as determined by fluorometric assay. Results are mean ± SE (n=3).

| Tea polyphenols | Histamine release (ng/$10^6$ cells) | Histamine release inhibition (%) |
|---|---|---|
| None | 59.9 ± 0.3 | - |
| C | 61.4 ± 0.8 | 0 |
| EC | 60.1 ± 3.4 | 0 |
| EGC | 57.4 ± 2.8 | 5.0 |
| ECG | 26.1 ± 1.3 | 68.9 |
| EGCG | 33.2 ± 0.8 | 54.5 |

**Table 2.**
Effect of EGCG on A23187- or IgE-mediated histamine release from basophilic-differentiated KU812 cells. Human IgE-sensitized KU812 cells or non-sensitized cells were stimulated with anti human IgE or A23187 for 20 min, respectively, in the presence of various concentration of EGCG (1-100 μM). Histamine released into the supernatants as determined by fluorometric assay. Results are mean ± SE (n=3).

| | IgE receptor crosslinking | | A23187 | |
|---|---|---|---|---|
| EGCG (μM) | Histamine release (ng/$10^6$ cells) | Histamine release inhibition (%) | Histamine release (ng/$10^6$ cells) | Histamine release inhibition (%) |
| 0 | 81.4 ± 2.0 | - | 110.7 ± 0.7 | - |
| 1 | 64.9 ± 1.9 | 20.3 | 81.0 ± 1.0 | 36.4 |
| 10 | 57.8 ± 2.1 | 29.1 | 54.1 ± 2.0 | 69.2 |
| 100 | 48.2 ± 10.4 | 40.8 | 49.9 ± 2.2 | 74.4 |

FcεRI cross-linking induces tyrosine phosphorylation of phospholipase C (PLC). The reaction products of PLC, which induces inositol 1, 4, 5-triphosphate, and diacylglycerol, stimulates the $Ca^{2+}$ release from intracellular storage sites and activates protein kinase C (PKC), respectively. Histamine release from the intracellular secretory granules is induced by the increase of intracellular $Ca^{2+}$ concentration and the activation of PKC (13). If the inhibitory effect of EGCG on histamine release from KU812 cells is due to the inhibition of the PKC activation pathway, histamine release induced by A23187, which by passes FcεRI cross-linking, should not be inhibited by EGCG. EGCG was shown not to inhibit the increase of intracellular free $Ca^{2+}$ in KU812 cells after stimulation with the anti IgE antibody. EGCG also did not inhibit the increase of intracellular $Ca^{2+}$ in cells stimulated with ionomycin (data not shown). Therefore, EGCG and ECG may affect the signaling pathway at a point after the signaling from the increase of intracellular $Ca^{2+}$ concentration.

K252b, an ectoprotein kinase inhibitor, has been shown to inhibit the histamine release from RBL-2H3 cells and human basophils through the binding to the cell surface molecule rather than by permeation into the cells (5). If EGCG binds to a cell surface molecule, it would be interesting to identify the molecule that act as an intermediate the inhibitory signal on deguranulation in human basophils.

Tea polyphenols are major component of green tea representing 30-40% of the dry matter. Among them, EGCG is the most abundant component (40%-60%) followed by EGC (10-20%), EC (4-6%), and C (2-4%). In humans, orally administrated catechins can be detected in blood and are excreted in urine and feces (14). Clarification of the mechanism for the inhibition of histamine release through ECG and EGCG may provide a new approach to suppress IgE-mediated allergic reactions.

## 5. Acknowlegments

This work was supported in part by grants from Program for Promotion of Basic Research Activities for Innovative Biosciences.

## 6. References

1. Daeron, M. (1997) *Annu. Rev. Immunol.* **15**, 203-234.
2. Beaven, M. A., and Metzger, H. (1993) *Immunol. Today* **14**, 222-226.
3. Schroeder, J. T., Kagey-Sobotka, A., and Lichtenstein, L. M. (1995) *Allergy* **50**, 463-472.
4. Beaven, M. A., Rogers, J., Moore, J. P., Hesketh, T. R., Smith, G. A., and Metcalfe, J. C. (1984) J. Biol. Chem. **259**, 7129-7136.
5. Teshima, R., Saito, Y., Ikebuchi, H., Rajiva De Silva, N., Morita, Y., Nakanishi, M., Sawada, J., and Kitani, S. (1997) *J. Immunol.* **159**, 964-969.
6. Matsuo, N., Yamada, K., Yamashita, K., Shoji, K., Mori, M., and Sugano, M. (1996) *In Vitro Cell. Dev. Biol.* **32**, 340-344.
7. Matsuo, N., Yamada, K., Shoji, K., Mori, M., and Sugano, M. (1996) *Allergy* **52**, 58-64.
8. Igarashi, Y., Kurosawa, M., Ishikawa, O., Miyachi, Y., Saito, H., Ebisawa, M., Iikura, Y., Yanagida, M., Uzumaki, H., and Nakahata, T. (1996) *Clin. Exp. Allergy* **26**, 597-602.
9. Butchers, P. R., Fullartton, J. R., Skidmore, I. F., Thomson, L. E., Vardey, C. J., and Wheeldon, A. (1979) *Br. J. Pharmacol.* **67**, 23-32.
10. Kishi, K. (1985) *Leuk. Res.* **9**, 381-390.
11. Gauchat, J-F., Henchoz, S., Mazzei, G., Aubry, J. P., Brunner, T., Blasey, H., Life, P., Talabot, D., Flores-Romo, L., Thompson, J., Kishi, K., Butterfield, J., Dahinden, C., and Bonnefoy, J-Y. (1993) *Nature* **365**, 340-343.
12. Hara, T., Yamada, K., and Tachibana, H. (1998) *Biochem. Biophys. Res. Commun.* **247**, 542-548.
13. Lindau, M., and Gomperts, B. D. (1991) *Biochem. Biophys. Acta* **1071**, 429-471.
14. Das, N. P. (1971) *Biochem. Pharmacol.* **20**, 3435-3445.

# COUNTERACTION OF THE ACTIVE FORM OF VITAMINS A AND D ON UP-REGULATION OF ADIPOCYTE DIFFERENTIATION WITH PPARγ LIGAND, A THIAZOLIDINEDIONE, IN 3T3-L1 CELLS

YOSHIFUMI HIDA, TERUO KAWADA, SHUN KAYAHASHI, TOMOMI ISHIHARA# and TOHRU FUSHIKI
*Laboratory of Nutrition Chemistry, Division of Applied Life Sciences, Graduate School of Agriculture, Kyoto University, Kyoto 606-8502, Japan*
*#Lead Optimization Research Laboratory, Tanabe Seiyaku Co.Ltd., Saitama 355, Japan*

*Key Words:* adipocyte differentiation, peroxisome proliferator activated receptorγ, retinoic acid, 1,25-dihydroxyvitamin $D_3$, thiazolidinedione

The active form of vitamin A, retinoic acid (RA), and that of vitamin D, 1,25-dihydroxyvitamin $D_3$ (1,25 $(OH)_2$ $D_3$) inhibited adipocyte differentiation of 3T3-L1 preadipocytes in the presence of thiazolidinedione, a specific ligand for peroxisome proliferator-activated receptorγ (PPARγ). These fat-soluble vitamins repressed the up-regulated protein expression of PPARγ2 during the first 40 hours of initiation of 3T3-L1 preadipocyte differentiation. Compared with RA, 1,25 $(OH)_2$ $D_3$ inhibited PPARγ2 expression more effectively and caused concomitantly a greater inhibition of adipocyte differentiation. These results suggest that the inhibitory action of adipocyte differentiation by RA or 1,25 $(OH)_2$ $D_3$ is exhibited through direct repression of the expression of PPARγ2 protein, even in the presence of its ligand. They also raise the intriguing possibility that attenuation or amplification of the pharmacological effects of thiazolidinedione that are dependent on PPARγ in adipose cells is caused by alteration of the levels of these fat-soluble vitamins.

## 1. Introduction

The adipocyte differentiation program is regulated by the sequential expression of transcriptional activators, namely CCAAT / enhancer binding protein (C/EBP) and PPAR families (1). These transcription factors coordinate the expression of genes involved in creating and maintaining the adipocyte phenotype (2). PPAR families are members of the nuclear hormone receptor superfamily. These receptors heterodimerize with retinoid X receptor (RXR) and become transcriptionally active when bound to a ligand. PPARγ2 is expressed at the highest level in adipose tissue and adipocyte cells, and at low levels, or not at all, in other tissues and cell lines (3).

The synthetic antidiabetic thiazolidinediones have been shown to enhance the differentiation of 3T3-L1 preadipocytes (3). The recent finding that thiazolidinediones and a lipid metabolite, 15-deoxy-$\Delta^{12,14}$ prostaglandin $J_2$, are ligands of PPARγ (3, 4) reveals a novel signaling pathway that directly links these compounds to processes involved in glucose homeostasis and lipid metabolism including adipocyte differentiaton.

RA and 1,25 $(OH)_2$ $D_3$, have been found to inhibit the differentiation of adipocytes in cultured cells (5, 6, 7). Adipose tissue is also known to be a target organ for RA and 1,25

(OH)2 D3. The inhibitory effects of fat-soluble vitamins on adipogenesis observed *in vitro* can also be seen *in vivo*. Thus, it has been reported that low levels of dietary fat-soluble vitamins, especially vitamin A and carotenoid, actively stimulate the development of adipose tissue, namely, bovine marbling (8). In addition, the actions of RA and 1,25 (OH)2 D3 are reported to be mediated by specific nuclear receptors, that is, retinoic acid receptor (RARα, RARβ, RARγ subtypes) and vitamin D receptor (VDR), which are members of the steroid and thyroid/retinoid receptor superfamily. These receptors function in the form of heterodimers with RXRs as well as PPARs. We have further shown that both of RARγ and VDR mRNA levels are up-regulated by RA and 1,25 (OH)2 D3, respectively, in 3T3-L1 cells (9, 10). Therefore, we speculate that RA and 1,25 (OH)2 D3 prevent the differentiation of preadipocytes to adipocytes through a mechanism by which retinoic acid receptor (RAR) or vitamin D receptor (VDR) competes with PPARγ for the common heterodimeric partner RXRα (11). In this study, we examined the effects of fat-soluble vitamins, vitamins A and D, on the differentiation of preadipocytes to adipocytes in terms of their influence on the protein expression of PPARγ, a key regulator of adipogenesis.

## 2. Materials and methods

### 3.1. MATERIALS

1,25 (OH)2 D3 was purchased from Wako Pure Chemical Industries, Ltd. (Osaka, Japan) and all-trans RA from Sigma (St. Louis, Mo, USA). Thiazolidinedione T-174 ((±)-5-[[2-(2-naphthalenylmethyl)-5-benzoxazolyl]methyl]-2,4-thiazolidinedione), a specific ligand for PPARγ (12), was obtained from Tanabe Seiyaku Co. Ltd. (Osaka, Japan). All other chemicals were of guaranteed reagent grade or tissue culture grade.

### 3.2. CELL CULTURE

3T3-L1 cells were routinely grown as described before (7). Under these conditions, confluence was reached within 3 days. For adipocyte differentiation, cells were switched to differentiation medium (DM) containing 10% fetal bovine serum (FBS), 0.25 mM dexamethasone (DEX), 0.5 mM 1-methyl-3-isobutylxanthine (MIX), and 10 μg of insulin per ml 2 days after cell confluence, referred to as day 0. Thiazolidinedione (T-174) was added in dimethyl sulfoxide at less than 0.1% v/v in DM. RA and 1,25 (OH)2 D3 were dissolved in a solution containing ethanol, whose concentration in each well was less than 0.1% v/v, and added at the same time. From 40 h after initiation of differentiation by DM, differentiating cells were maintained in post-DM containing 10% FBS and 5 μg of insulin per ml, with a change of post-DM every 2-3 days. Glycerophosphate dehydrogenase (GPDH) activity, a marker of adipocyte differentiation, was measured as described previously (7).

### 3.3. WESTERN BLOTTING ANALYSIS

3T3-L1 cells were lysed in Dulbecco's phosphate buffered saline (PBS) containing 0.02% EDTA. Equal volumes of total cell lysate from each well were subjected to SDS-PAGE, then electroblotted onto polyvinylidene difluoride membranes. The blots were then incubated with primary antibody, polyclonal anti-PPARγ2 antibody (Affinity Bioreagents. Inc.) or polyclonal anti-RXRα antibody (Santa Cruz Biotechnology, Inc.) in TBS containing 0.1% Tween 20. The former antibody recognizes both PPARγ1 and PPARγ2. After washing, the membranes were incubated with pig anti-rabbit IgG conjugated with horseradish peroxidase (DAKOPATTS a/s, Denmark), and signals were developed with the enhanced chemiluminescence (ECL) system (DuPont, USA).

## 3. Results

Addition of T-174, a specific ligand for PPARγ, along with DM during 40-h period of initiation of differentiation of 3T3-L1 preadipocytes enhanced GPDH activity in a dose-dependent manner, as shown in Figure 1a, and resulted in much greater promotion

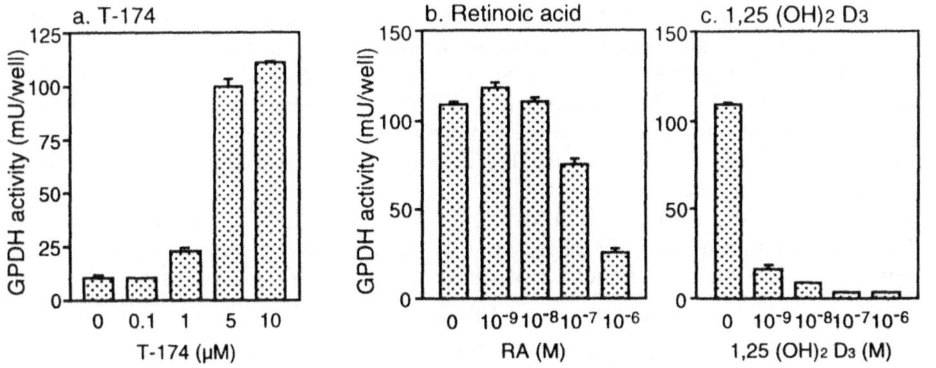


*Figure 1.* (a) Thiazolidinedione T-174, activator of PPAR γ, enhances GPDH activity of 3T3-L1 cells. (b, c) Retinoic acid and 1,25 (OH) 2 D3 inhibit GPDH activity induced by T-174 in 3T3-L1 cells. 3T3-L1 cells were treated with 10 μM T-174 along with DM, supplemented with each concentration of retinoic acid or 1,25 (OH) 2 D3 on day 0. After 40 h, the medium was replaced with post-DM. GPDH activity was measured 7 days after the start of treatment. The values are means ± SEM for three independent cultures.

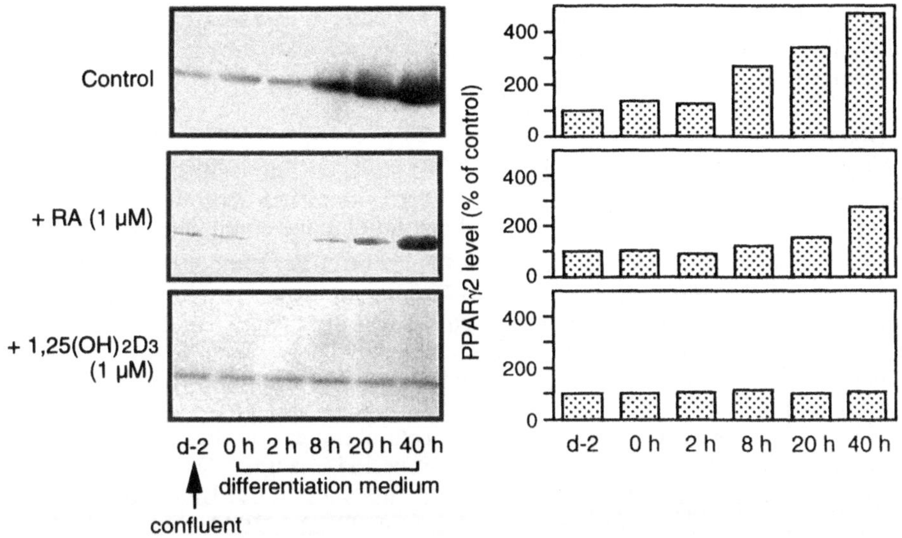


*Figure 2.* Retinoic acid and 1,25 (OH) 2 D3 inhibit the induction of PPAR γ2 protein during a 40 h period of initiation of 3T3-L1 preadipocytes differentiation. 3T3-L1 cells were treated for the indicated times with 10 μM T-174 in DM, supplemented with retinoic acid or 1,25 (OH) 2 D3 at 1 μM. Whole cell lysates were prepared and subjected to SDS-PAGE, followed by Western blotting using anti-PPAR γ2 antibody. Data from Western blotting experiments were quantitated by densitomer (the bar chart shown *right* ). Protein levels of PPAR γ2 are expressed as percentages of the values detected on day -2 in the respective treatments. These values are for one representative culture of three independent cultures with nearly identical results.

of differentiation than that of only DM. Interestingly, the addition significantly decreased the concentration of glucose in the medium (less than 10% of control at 7 days after the addition). Figure 1b, c indicates that RA and 1,25 $(OH)_2$ $D_3$ both inhibit the T-174-stimulated differentiation of 3T3-L1 preadipocytes to adipocytes in a dose-dependent manner. However, while RA strongly inhibited the differentiation at a concentration of 1 μM, 1,25 $(OH)_2$ $D_3$ exhibited remarkable inhibition even at a concentration of 1 nM.

The relationship between the inhibitory effect on adipocyte differentiation of these fat-soluble vitamins and the expression of PPARγ2 was analyzed by immunoblotting during the first 40 h of induction of the terminal differentiation of preadipocytes into adipocytes. The protein expression of PPARγ2 began to increase 8 h after the initiation of adipocyte differentiation induced by T-174 in DM and continued to increase up to 40 h (Figure 2, Control). These results strengthen the correlation between the induction of PPARγ and the acquisition of adipocyte phenotype after terminal differentiation in 3T3-L1 cells, in agreement with previous reports (2). However, treatment with 1 μM RA or 1,25 $(OH)_2$ $D_3$ during the first 40 h of adipocyte differentiation inhibited the rise in the expression of PPARγ 2 protein, suggesting that RA and 1,25 $(OH)_2$ $D_3$ influence adipogenesis at an early stage of differentiation. A slight increase of PPARγ2 protein expression occurred at 40 h after the initiation of differentiation in the presence of RA, but 1,25 $(OH)_2$ $D_3$ exhibited a stronger inhibitory effect, with PPARγ protein expression remaining essentially at the basal level throughout the 40 h. These results suggest that the inhibition of adipocyte differentiation in 3T3-L1 cells by RA or 1,25 $(OH)_2$ $D_3$ may be mediated by suppression of the induction of PPARγ as a master regulator in the control of adipogenesis.

## 4. Discussion

PPARγ is a dominant trigger of adipocyte differentiation, acting through transactivation of adipose-specific genes in a PPARγ-activator-dependent manner. Recent data have demonstrated that antidiabetic thiazolidinedione compounds are specific and have high-affinity for PPARγ (4, 12). Furthermore, these compounds can function as potent inducers of adipogenesis (2, 12, 13). In the present study, thiazolidinedione potently stimulated the differentiation and glucose utilization of 3T3-L1 cells as compared with the usual treatment by mixtures of hormonal stimulants. Accordingly, to analyze the mechanism of inhibition of adipocyte differentiation, we employed this PPARγ-specific ligand, T-174, as a tool.

Our approach was to examine PPARγ protein expression during the first 40 h of initiation of adipocyte differentiation, a crucial period in adipocyte differentiation when mitotic arest occurs and other transcription factors such as C/EBPα and CHOP-10 are induced. Adachi et al. reported the repression of PPARγ mRNA by retinoid, but they examined the stage after terminal differentiation in 3T3-L1 cells (14). We also studied PPARγ mRNA expression by Northern blotting analysis, but found no detectable expression during the first 40 h of initiation of adipocyte differentiation (data not shown). This result might suggest that PPARγ mRNA is relatively unstable in 3T3-L1 cells or that its turnover cycle is rapid. Consequently, the use of protein expression level to assess the induction of PPARγ during the course of adipose conversion of 3T3-L1 preadipocytes is significant not only for following what is actually taking place, but also for its advantage in detection sensitivity.

PPARγ2 protein expression was enhanced by T-174 in DM in a time-dependent manner in 3T3-L1 cells. When the DM was replaced by post-DM at 40 h after the initiation of adipocyte differentiation, the enhancement of PPARγ2 protein expression gradually declined in subsequent culture (data not shown).

As we have found that RA or 1,25 $(OH)_2$ $D_3$ increase RARγ or VDR mRNA, as members of the receptor subfamily including PPARs, in 3T3-L1 cells (9, 10), we investigated whether PPARγ2 expression is regulated by the same mechanism of autoregulated amplification in a ligand-dependent manner or not. The induction by DM (a basal medium supplemented by DEX, MIX and insulin) alone in 3T3-L1 preadipocytes enhanced the PPARγ2 protein expression to the same degree as did that by DM containing T-174 (data not shown). When only T-174 was added to the basal medium without DEX, MIX and insulin, the expression level of PPARγ2 protein remained steady, that is, the enhancement of PPARγ2 protein expression was not recognized. After all, we were unable to reach a conclusion concerning the mechanism of autoregulated amplification of PPARγ2. Furthermore, we examined the effect of individual components of DM, DEX, MIX and insulin, on PPARγ2 protein expression during adipogenesis, and found that treatment with MIX alone enhanced the PPARγ2 expression level (data not shown).

In conclusion, our results indicate that RA and 1,25 $(OH)_2$ $D_3$ inhibit the protein expression of PPARγ 2 during the early stage of 3T3-L1 preadipocyte differentiation in the presence of a PPARγ -specific ligand. Furthermore, 1,25 $(OH)_2$ $D_3$ was much more effective in inhibiting adipocyte differentiation than was RA. Since PPARγ is known to be a master regulator of the adipocyte differentiation program, these results suggest that the inhibitory action of fat accumulation by these fat-soluble vitamins, RA and 1,25 $(OH)_2$ $D_3$, is exhibited through direct repression of the expression of PPARγ 2 protein itself.

## 5. References

[1] O.A. MACDOUGALD and M.D. LANE, Annu. Rev. Biochem. **64** 345-373 (1995)

[2] P. TONTONOZ, E. HU, R.A. GRAVES, A.I. BUDAVARI and B.M. SPIEGELMAN, Genes Dev. **8** 1224-1234 (1994)

[3] B.M. FORMAN, P. TONTONOZ, J. CHEN, R. BURN, B.M. SPIEGELMAN and R. M. EVANS, Cell **83** 803-812 (1995)

[4] J.M. LEHMANN, L.B. MOORE, T.A. SMITH-OLIVER, W.O. WILKISON, T.M. WILLSON and S.A. KLIEWER, J. Biol. Chem. **270** 12953-12956 (1995)

[5] R.L. STONE and D.A. BERNLOHR, Differentiation **45** 119-127 (1990)

[6] M. SATO and A. HIRAGUN, J. Cell. Physiol. **135** 545-550 (1988)

[7] T. KAWADA, N. AOKI, Y. KAMEI, K. MAESHIGE, S. NISHIU AND E. SUGIMOTO, Comp. Biochem. Physiol. **96A** 323-326 (1990)

[8] S. TORII, T. MATSUI and H. YANO, Animal Sci. **63** 73-78 (1996)

[9] Y. KAMEI, T. KAWADA, R. KAZUKI and E. SUGIMOTO, Biochem. J. **293** 807-812 (1993)

[10] Y. KAMEI, T. KAWADA, R. KAZUKI , T. ONO, S. KATO and E. SUGIMOTO, Biochem. Biophys. Res. Commun. **193** 948-955 (1993)

[11] T. KAWADA, Y. KAMEI and E. SUGIMOTO, Int. J. Obesity **20** S52-57 (1996)

[12] J. MIZUKAMI and T. TANIGUCHI, Biochem. Biophys. Res. Commun. **240** 61-64 (1997)

[13] A. CHAWLA, E.J. SCHWARZ, D.D. DIMACULANGAN and M.A. LAZAR, Endocrinology **135** 798-800 (1994)

[14] H. ADACHI, M.I. DAWSON and A.M. JETTEN, Mol. Cell. Differentiation **4** 365-381 (1996)

# CCK regulation of monitor peptide gene expression in pancreatic acinar AR42J cells.

TOSHI KINOUCHI, SATOSHI TSUZUKI, CHIEKO MINAMI, YOSHIHIRO HAYASHI, ETSURO SUGIMOTO and TOHRU FUSHIKI.
*Division of Applied Life Sciences, Graduate School of Agriculture, Kyoto University, Kyoto 606-8502, Japan*

The mechanism(s) by which the cholecystokinin (CCK) stimulation of AR42J rat pancreatoma cells results in an increased mRNA expression of a CCK-releasing peptide (monitor peptide, MP) were explored. Using a established assay system by means of reverse transcription-polymerase chain reaction, CCK was shown to increase the level of MP mRNA by about 9-fold. When protein synthesis was blocked by the addition of cycloheximide, the MP mRNA level remained unchanged in the presence of CCK. Inhibition of the transcription with actinomycin D showed a half-life for the MP mRNA of approximately 17 h, and this rate remained unchanged by CCK treatment, suggesting that CCK may regulate the MP mRNA level by influencing gene transcription. A23187, bombesin, substance P and carbachol increased the MP mRNA level. $CoCl_2$ abolished both the CCK and A23187 actions on the MP mRNA expression. Neither dibutyryl cAMP and forskolin nor secretin and VIP had any effect on the MP mRNA expression. Both TPA and PDB failed to increase the MP mRNA. It was therefore proposed that the CCK stimulates the MP mRNA expression of AR42J cells in a calcium-dependent and protein kinase C-independent manner.

## INTRODUCTION

Prolonged high protein intake leads to increases in the concentrations and mRNA levels of proteases in the pancreas (2, 4, 7). Cholecystokinin (CCK) is believed to be involved in the adaptive increase in pancreatic proteases as a mediator in response to prolonged intake of a high protein diet (6, 8). The release of CCK is mediated, at least in part, by a CCK-releasing peptide (monitor peptide, MP), which was originally found in the bile-pancreatic juice of rats, and is synthesized in the pancreas and secreted into the small intestine (5). CCK release was increased in rats after adaptation to a high protein diet (3). This increased capacity to release CCK can be partially accounted for by the increasing level of MP secreted into the lumen. Therefore, characterizing the regulatory mechanisms of the MP expression would be of great help for understanding the overall pancreatic adaptation to a high protein diet.

The results of our previous study showed that both the concentration in zymogen granules and the mRNA level of the MP as well as the pancreatic proteases in the rat pancreas increased in response to prolonged high protein intake (7). We also reported that the MP mRNA level was increased after administration of CCK in rat pancreas and in the pancreatic cell line AR42J (8). Taking these two results together, we have proposed a cycle in which the plasma CCK increases the MP in the pancreas, which in turn induces a high plasma CCK level. A question arises as to how CCK exerts its effect on the MP mRNA expression. This study was designed to characterize the molecular mechanisms underlying the CCK-regulated MP mRNA expression in the pancreas. In order to further understand the CCK action on the MP mRNA, the effect of various secretagogues and insulin on the MP mRNA expression was also examined.

## MATERIALS AND METHODS

*Materials.* The rat pancreatic acinar cell line AR42J was purchased from Seiyaku (Osaka, Japan); CCK-Octapeptide (sulfated form), bombesin, VIP, and secretin from the Peptide Institute Inc. (Osaka, Japan); carbamylcholin chloride (carbachol), substance P, insulin, A23187, forskolin, and dibutyryl cAMP from Sigma (St. Louis, MO, USA); $CoCl_2$ and chlorpromazine from Nacalai Tesque (Kyoto, Japan); Heat-TUFF DNA polymerase from Clontech (Palo Alto, CA, USA); M-MLV Reverse Transcriptase and dNTP from GIBCO BRL (Gaithersburg, MD, USA); RNasin from Toyobo (Osaka, Japan); [$\gamma$-$^{32}$P] ATP (4000 Ci/mmol) from ICN Biomedicals Inc. (Costa Mesa, CA, USA); all other reagents from Wako Pure Chemicals (Osaka, Japan).

*Cell Culture and Secretagogue Treatments* The conditions for the AR42J cell cultures were as described previously (8). Cells were seeded in 6-well dishes at a density of 7.5 x $10^5$ cells per well and grown for 48 h. Following this period, the culture medium was replaced by fresh medium containing various secretagogues or insulin with or without 10 nM CCK, and then used for RNA preparation after 12 h. Neither serious damage of cells nor significant differences in the amount of extracted RNAs were observed during the 12 h incubation of any secretagogues or insulin.

*Drug Treatment.* AR42J cells were seeded in 6 well dishes at a density of 7.5 x $10^5$ cells per well and grown for 48 h. Half of each cell culture was preincubated with 10 nM CCK for 24 h and then incubated with actinomycin D (at a final concentration of 10 μg/ml) in the presence of CCK. The other half of each culture was incubated with the above concentration of actinomycin D without addition of CCK. The cells were lysed at various times for subsequent measurements of the MP mRNA levels. For cycloheximide treatment experiments, AR42J cells grown for 48 h as described above were incubated with 10 μM cycloheximide in the presence or absence of 10 nM CCK and then used for RNA preparation after 24 h. Two days later, the culture medium was replaced by fresh medium containing A23187 and/or TPA, PDB, forskolin or $Bt_2$-cAMP, and the cells were then incubated for an additional 12 h, followed by subsequent RNA preparation. The schedule of incubation for W-7, chlorpromazine and $CoCl_2$ with or without 10 nM CCK was the same as described above.

*RNA Isolation and Reverse Transcription-Polymerase Chain Reaction (RT-PCR).* Two μg of AR42J total cellular RNA isolated by the acid guanidinium thiocyanate-phenol-chloroform (AGPC) method of Chomczynski *et al.* (1) were used for cDNA synthesis using 4 p mol of downstream primer specific to the MP with the following sequence: GCTAAACATTATGGTAGCCCTAAA and 20 U of reverse transcriptase in 4 μl. PCR was performed with 4 μl of the cDNA as a template with 20 pmol of upstream primer with the following sequence: CCTAATTGCCCTAAGCAAATTATG, 16 pmol of down-stream primer, and 0.5 U of Heat-TUFF DNA polymerase in a Program Temp Control System PC-700 (Astec, Fukuoka, Japan). Samples were heated to 94°C for 3 min. Subsequent cycles were conducted in three temperature steps: 1 min at 94°C, 2 min at 65°C and 3 min at 72°C. After 22 cycles, the samples were incubated at 72°C for an additional 7 min. The PCR products were quantified by DNA slot-blot hybridization with the $^{32}$P-labeled MP-specific oligonucleotide as a probe (7).

## RESULTS and DISCUSSION

*RT-PCR determination of the MP mRNA in AR42J cells.* As shown in Fig. 1, a linear relationship was observed between the amount of applied RNA and corresponding PCR product up to 22 cycles. Under the present experimental conditions, CCK at a final concentration of 10 nM increased the MP mRNA level of AR42J cells; by 12 h, an 8- to 9-fold increase compared to unstimulated cells was observed, and the level of the MP mRNA remained unchanged up to 48 h (Data not shown).

*Effects of cycloheximide and actinomycin D treatments on the CCK-stimulated increase in the MP mRNA expression.* Cycloheximide alone had no effect on the MP

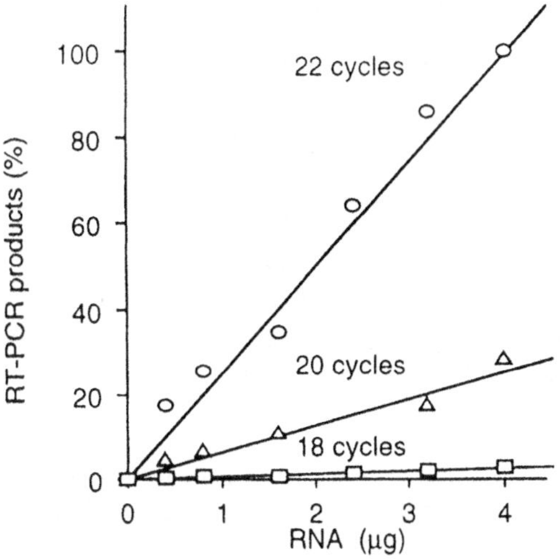


Fig. 1. Reverse transcription-polymerase chain reaction (RT-PCR) determination of monitor peptide mRNA in AR42J cells. A linear relationship between amount of applied RNA andcorresponding RT-PCR products after various PCR cycles are shown.

mRNA (Fig. 2). Furthermore, when administered, it did not prevent the induction of the MP mRNA expression by CCK (923 ± 119% of control; n=4) (Fig. 2). The MP mRNA in the CCK-unstimulated cells decayed rapidly with an estimated half-life of 16.9 h (Fig. 3). This decay remained unchanged 24 h after CCK pretreatment, suggesting that the change in transcription rate may be the predominant factor responsible for the CCK-stimulated increase in the MP mRNA expression.

*CCK-stimulated increase in the MP mRNA Expression via a Calcium-Dependent Pathway.* One µM A23187, a calcium ionophore, caused a significant increase (6.9-fold) in the MP mRNA (Table 1), indicating that the modulation of intracellular $Ca^{2+}$ is involved in the up-regulation of the MP mRNA. The addition of 2 mM $CoCl_2$ alone caused a decrease by 67% in the MP mRNA level. An addition of 2 mM $CoCl_2$ together with 1 µM A23187 or 10 nM CCK to AR42J cells abolished the effect of the latter compounds on the level of the MP mRNA (Fig. 4). These findings therefore confirm that CCK increases the MP mRNA by a calcium-dependent mechanism. Neither the addition of 0.5 mM dibutyryl cAMP nor of 25 mM forskolin, showed any remarkable effect on the MP mRNA levels (Table 1). These results indicated that the cAMP pathway might not be involved in the regulation of the MP mRNA expression.

Both 1 µM TPA and 1 µM PDB, a potent activator of protein kinase C, did not increase the MP mRNA expression and failed to potentiate the A23187-induced MP mRNA accumulation (Fig. 5)

Fig. 6 shows that the treatment of AR42J cells with 100 µM carbachol, 10 nM bombesin, and 100 nM substance P resulted in a 3.2-fold, 10.2-fold, and 3.5-fold increase in the MP mRNA, respectively. Furthermore, substance P and carbachol showed a synergistic effect when applied to the cells together with 10 nM CCK. In contrast, the combined effect of CCK and bombesin was less than additive (Data not shown). Neither 10 nM secretin nor 100 nM VIP had any effect on the steady state of the MP mRNA level. Furthermore, no significant potentiation in the MP mRNA level was shown when AR42J cells were incubated with secretin or VIP in the presence of CCK. The addition of insulin had no effect on the MP mRNA expression under our experimental conditions. The CCK action on the MP mRNA was not influenced by a concomitant incubation with insulin (Data not shown).

Table 1 . *Effect of various protein kinase activators on expression of MP mRNA*

| Additions | MP mRNA Relative Level |
|---|---|
| None | 1.00 |
| 10 nM CCK | 8.60 ± 1.34 * |
| 1 µM A-23187 | 6.88 ± 1.51 * |
| 25 µM Forslolin | 1.24 ± 0.02 |
| 500 µM Dibutyryl-cAMP | 1.14 ± 0.35 |

Results are means ± SD from 4 separate experiments in which eachsample was assayed in triplicate. *P < 0.0001 compared with un-treated cells.

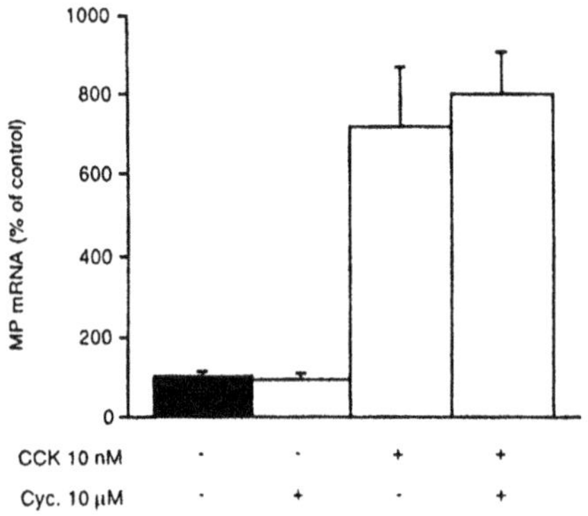


Fig. 2. Effect of cycloheximide on CCK-stimulated increase in MP mRNA.

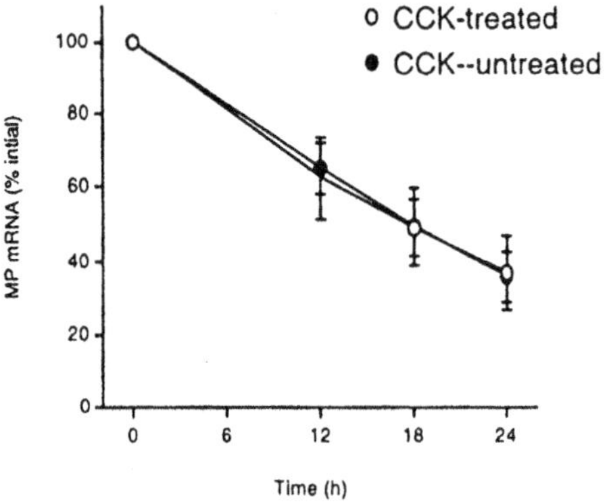


Fig. 3. Effects of CCK on MP mRNA half-life.

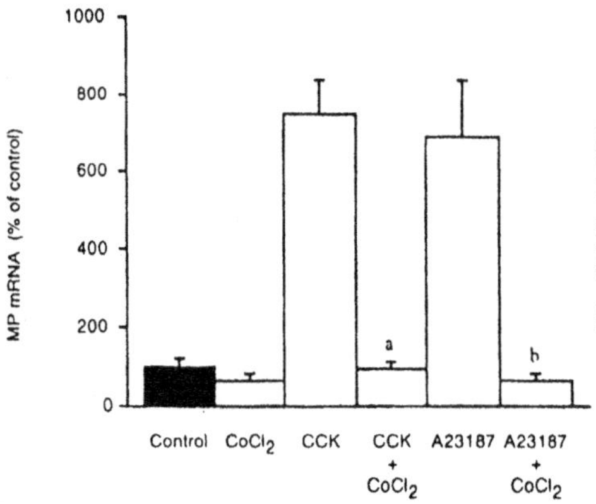


Fig. 4. Effect of $CoCl_2$ on CCK- and/or A-23187-stimulated increases in MP mRNA.

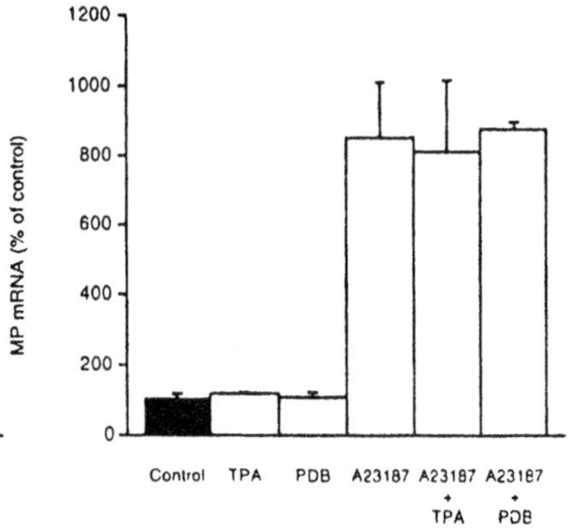


Fig. 5. Effect of 12-O-tetradecanoylphorbol-13-acetate (TPA) and phorbol 12,13-dibutyrate (PDB) on expression of MP mRNA.

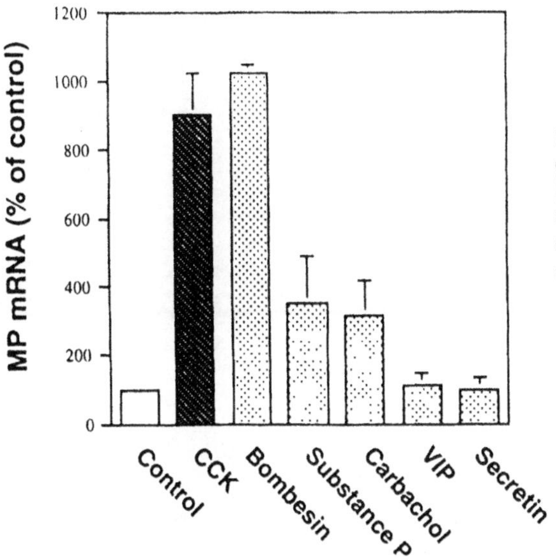


Fig. 6. Effect of various secretagogues on expression of MP mRNA. Results are means ± SD from 4 separate experiments in which each sample was assayed in triplicate.

In summary, we demonstrated that the CCK-stimulated increase in the MP mRNA is calcium-dependent and protein kinase C-independent, and that it may occur at the transcriptional level. These findings are informative for understanding the mechanism of many CCK actions, e.g., in the dietary adaptation, protein synthesis and gene expression of exocrine pancreas.

## REFERENCES


1 Chomczynski, P. and N. Sacchi. Single-step method of RNA isolation by acid guanidinium thiocyanate-phenol-chloroform extraction. *Anal. Biochem.* 162: 156-159, 1987.

2. Girgi, D., W. Renaud, J. P. Bernard, and J. C. Dagorn. Regulation of proteolytic enzyme activities and mRNA concentrations in rat pancreas by food content. *Biochem. Biophys. Res. Commun.* 127: 937-942, (1985)

3. Goke, B., K. Fenchel, S. Knobloch, R. Arnold, and G. Adler, Increased CCK-response to proteinase inhibitor feeding after induction of pancreatic hypertrophy in rats. *Pancreas* 3: 576-579, 1988.

4. Grossmann, M. I., H. Greengard, and A. C. Ivy. The effect of dietary composition on pancreatic enzymes. *Am. J. Physiol.* 138: 676-682, 1943.

5. Iwai. K., T. Fushiki, S. Fukuoka. Pancreatic enzyme secretion mediated by a novel peptide: monitor peptide hypothesis. *Pancreas* 3: 720-728, 1988.

6. Rosewicz, S., L. D. Lewis, X. Y. Wang, R. A. Liddle, and C. D. Logsdon. Pancreatic digestive enzyme gene expression: effects of CCK and soybean trypsin inhibitor. *Am. J. Physiol.* 256: G733-G738, 1989.

7. Tsuzuki, S., T. Fushiki, A. Kondo, H. Murayama, and E. Sugimoto. Effect of a high-protein diet on the gene expression of a trypsin-sensitive, cholecystokinin-releasing peptide (monitor peptide) in the pancreas. *Eur. J. Biochem.* 199: 245-252, 1991.

8. Tsuzuki, S., A. Kondo, T. Fushiki, and E. Sugimoto. Monitor peptide gene expression is increased by exogenous CCK in the rat pancreas and in a rat pancreatic acinar cell line (AR4-2J). *FEBS Lett.* 307: 386-388, 1992.

# DETECTION OF MITOCHONDRIAL REGULATORY REGION RNA IN CULTURED CELLS AND DIFFERENTIATED TISSUE CELLS: THE IMPLICATIONS FOR CELLULAR GROWTH CONTROL

NOBORU NAKAMICHI, MAMIKO ITO AND TOSHIHARU MATSUMURA
*Meiji Milk Products Co. Cell Technology Center/Meiji Institute of Health Sciences, and RITE Meinyu Branch Laboratory, Odawara, Japan 250-0862.*

**ABSTRACT** In mitochondria, there are many RNAs for which the function has yet to be fully characterized. In this paper, attention is focused on these poorly characterized RNAs and their possible role in regulating mitochondrial function. Using a cDNA clone made from a 0.2 kb L-strand transcript of mitochondrial regulatory region RNA (mrrL-RNA) as a probe, we detected 1.3 kb RNAs from human tissues and cultured cells, and cloned cDNAs to them. Primary sequence studies suggest that they are transcribed from H-strand mitochondrial regulatory region DNA. Both the 1.3 kb RNA, referred to as mrrH-RNA hereafter, and mrrL-RNA were detected in all human tissues and cultured cells examined.

With few exceptions, the level of 1.3 kb RNA is low in the cultured cells while it is high in tissue cells, and the level of 0.2 kb RNA is high in the cultured cells while it varies in tissue cells. Since one of the functions of mrrL-RNA is speculated to prime H-strand DNA synthesis, we propose that mrrH-RNA might function as an antisense RNA to inhibit mitochondrial H-strand DNA replication.

Key words: mitochondrial DNA, mitochondial RNA, D-loop, cDNA

## 1. Introduction

Mitochondria provide cells with ATP (the source of energy for all cells), and thus play an important role in all production biotechnology using cultured cells. The mitochondrial DNA (mtDNA) is a 16.5 kbp closed circular double-stranded DNA molecule. The mitochondrial regulatory region (mrr) between the $tRNA^{phe}$ and the $tRNA^{pro}$ genes within mtDNA contains the D-loop region, a number of regulatory sequences including an L-strand promoter (LSP), two H-strand promoters, i.e., H1 for rRNA transcription and H2 for whole H-strand transcription unit, a TAS sequence motif, mtTF1 binding sites, and three conserved sequence blocks CSB1, CSB2 and CSB3, and are essential for mtDNA replication and RNA transcription (1-3).

Short polyadenylated L-strand transcripts (mrrL-RNAs) of ca. 0.2 kb have been obtained from mammals in several laboratories including ours (4-6). We presented decisive evidence that they are transcribed from mrr (6). Their function may be to prime H-strand DNA replication (7), and they may also play additional regulatory roles during replication and transcription (5, 8).

In this paper we have extended the results of our previous study by using cDNA corresponding to the 0.2 kb mrrL-RNA as a probe (mrr/HeLa-1, 6), and successfully detected 1.3 kb mitochondrial RNA from a number of human tissues and tissue culture cells. Structural analyses showed that most of the cDNAs synthesized from this mRNA correspond to the mrr region. In the literature we found similar RNAs from rats and mice (8, 9).

Strikingly, mrrH-RNA distributes very differently among tissue cells and cultured cells, suggesting that mrrH-RNAs are associated with the control of growth and differentiation of cells.

## 2. Materials and Methods

### 2.1. CELLS, RNA EXTRACTION AND NORTHERN ANALYSES

Hela S3 human cervical carcinoma cells, HepG2 hepatocellular carcinoma cells, IMR90 normal human fibroblasts, KYM rhabdomyosarcoma cells, AT(5BIVA) Ataxia telangiectasia fibroblasts, NB-1 neuroblastoma cells, and KatoIII stomach cancer cells were each grown in chemically defined media supplemented with 10 to 15 % fetal bovine serum. The details of the culture conditions will be described separately. These human cell lines were grown at 37 °C in a humidified atmosphere containing 5 % $CO_2$. The total RNA was isolated from these cell lines using RNAzol B reagent (Biotex Laboratories, Houston, Texas), and finally the poly ($A^+$) RNAs were separated. All total and poly ($A^+$) RNA samples from human tissues used in this study were purchased from CLONTECH Laboratories (Palo Alto, CA). RNA samples were then subjected to Northern analyses using $^{32}P$-labelled mrr/HeLa-1 cDNA as the probe as described previously (6).

### 2.2. cDNA LIBRARY CONSTRUCTION AND SCREENING

To construct the cDNA library, human skeletal muscle poly ($A^+$) RNAs were used for the template. The double stranded cDNAs were inserted into the λgt10 vector and packaged using Giga pack II Gold (STRATAGENE, La Jolla, CA). Approximately $5 \times 10^5$ phages were screened using a $^{32}P$-labeled mrr/HeLa-1 cDNA probe and positive plaques were purified by performing two additional plaque hybridizations. Further details will be described separately.

The cloned-cDNAs in the plaque-purified positive phage DNAs were PCR-amplified using λL- primer (5'CTGCTTCTCATAGAGTCTTG3') and λR-primer (5'ATACATATA CGGTTCTCTCCAGAG3') which anneal to the two arms of the cloning vector respectively. The sizes of the amplified cDNA inserts were estimated by electrophoresis in 0.7 % agarose gels followed by blotting onto a Hybond $N^+$ nylon membrane (Amersham) using a $^{32}P$-end labeled synthetic oligonucleotide corresponding to the mrr sequence (5'AAGATAAAATTTGAAATC 3') as a probe (Ref. to details in below).

## 3. Results

### 3.1. DETECTION OF 1.3 AND 0.2 KB RNA SPECIES

The mrr/HeLa-1 probe, which consists of a cDNA corresponding to one of the 0.2 kb mrrL-RNA, detected several bands including 1.3 and 0.2 kb bands in the RNA preparations obtained from all eight lines of human cultured cells as well as 13 samples of various human tissues as described in the footnote to Table 1.

In general, the level of 1.3 kb RNA was low in the cultured cells while it was high in tissue cells, and the level of 0.2 kb RNA was high in the cultured cells while it varied in tissue cells (Table 1).

Table 1. Detection of RNAs in total RNAs from various cells and tissues

| detected band | cultured cells | | | | | | | | differentiated tissues | | | | | | | | | | | | |
|---|---|---|---|---|---|---|---|---|---|---|---|---|---|---|---|---|---|---|---|---|---|
| sizes (kb) | 1 | 2 | 3 | 4 | 5 | 6 | 7 | 8 | 9 | 10 | 11 | 12 | 13 | 14 | 15 | 16 | 17 | 18 | 19 | 20 | 21** |
| 1.3 | ± | ± | ± | ± | ± | + | ± | + | + | + | + | + | + | + | ± | + | + | + | + | + | + |
| 0.2 | + | + | + | + | + | + | + | + | – | + | ± | + | + | + | ± | + | ± | + | + | + | + |

*:RNA samples were electrophoresed in 1.5 % agarose gels containing formaldehyde, transferred to Hybond $N^+$ nylon membrane (Amersham), and hybridized with $^{32}P$-labeled mrr/HeLa-1 cDNA. Expression levels are shown as "+" for high level, "±" for low, and "-" for negligible. **: 1. Hela, 2. KYM, 3. A T, 4. NB-1, 5. KATOIII, 6. HepG2, 7. IMR90, 8. HL60, 9. Skeletal muscle, 10. Heart, 11. Uterus, 12. Fetal brain, 13. Adult brain, 14. Salivary gland, 15. Mammary gland, 16. Lung, 17. Fetal liver, 18. Adult liver, 19. Pancreas, 20. Kidney, 21. Placenta.

In contrast, HepG2 cells and HL60 cells contained relatively high levels, and mammary gland tissue relatively low level of 1.3 kb RNA. Although these results are too complex to draw simple conclusions, it appears that cultured cells contain more uniform of 1.3 kb RNA and 0.2 kb RNA than tissue cells. The variation in the levels of these RNAs in tissue cells provides an attractive subject for future study.

## 3.2. SCREENING OF COMPLEMENTARY DNAs FOR 1.3 KB RNA

A cDNA library was constructed from skeletal muscle tissue RNA and screened using a mrr/HeLa-1 cDNA as the probe (6). Next, a combination of PCR and southern blot analysis methods was used to estimate the size of the cDNA of 45 positive phage clones using a $^{32}$P-end labeled synthetic oligo-probe (Fig. 1).

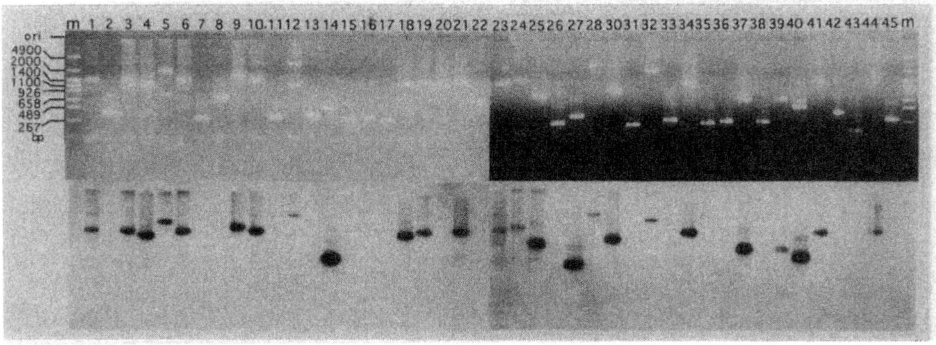

Figure 1. Southern blot analyses of PCR amplified cDNA inserts. The cDNAs cloned in each of 45 positive phages were amplified using λL- and λR-primers as shown in the materials and method section. Amplified cDNAs were resolved electrophoretically in a 0.7 % agarose gel (upper panel). The gels were subjected for southern blot analyses as described in the materials and method section to select clones containing ca.1.3 kb cDNA (lower panel). 1-45: positive clone numbers. m: PHY molecular weight marker (Takara Shuzo, Shiga, Japan).

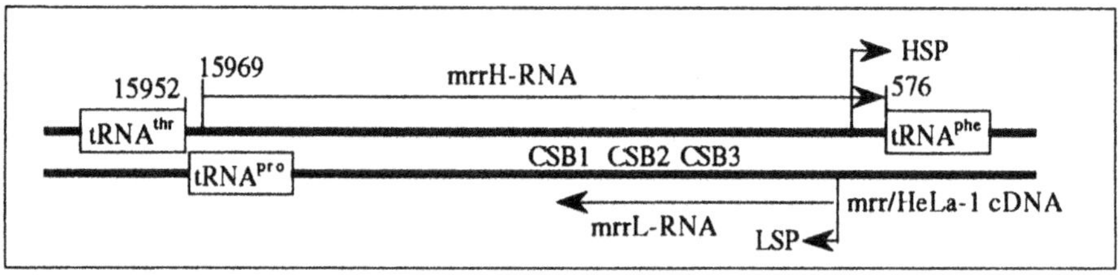


Figure 2. cDNA mapping. Nucleotide sequences of cDNAs were determined by the dideoxy chain termination method (ABI Automated DNA sequencer, Perkin Elmer). A portion of human mtDNA and sequenced cDNA to mrrH-RNA were aligned using two nucleotide sequence analysis softwares (Genetyx Software Development, Tokyo; Hitachi Software co., Tokyo). H:heavy strand, L:light strand, CSB:conserved sequence blocks, HSP: heavy strand promoter, LSP:light strand promoter.

Ten clones containing the ca. 1.3 kb cDNA insert were used for the nucleotide sequence analyses. Among these, five clones contained the longest cDNA insert and the 5' end maps to 17 nucleotides downstream from the $tRNA^{thr}$ gene and the 3' end maps to immediately upstream from the $tRNA^{phe}$ gene (Fig. 2). These clones were determined to be independent clones because the numbers of polyadenyl residues at the 3' end of the cDNAs were all different from one another.

## 4. Discussion

In this study, we detected ca. 1.3 kb polyadenylated RNAs in all human tissue cells and cultured cells which were examined. Sequence analyses indicate that the 5' end of all these 1.3 kb mrrH-RNAs maps to nucleotide 15953, immediately downstream from the $tRNA^{thr}$ gene, and the 3' end to nucleotide 577, immediately upstream from the $tRNA^{phe}$ gene. These 1.3 kb mrrRNAs are therefore homologues of previously described RNAs from the rat and the mouse (8, 9) and suggested to be transcribed from the H-strand of the mitochondrial regulatory region (mrr).

Since the whole unit of H-strand RNA is transcribed from HSP (3), these results suggest that the mrrH-RNA corresponds to the terminal portion of the whole transcription unit. Although the entire terminal portion of the H-strand transcription unit has yet to be completely determined, these results indicate that it extends to at least nucleotide 577.

The present results show that growing cultured cells contain low levels of 1.3 kb H-strand RNA and high levels of 0.2 kb L-strand transcript, while tissue cells contain high levels of 1.3 kb RNA and varying amounts of 0.2 kb RNA. We speculate that the difference in the abundancy of mrrH-RNA between cultured cells and tissue cells may be related to the difference in growth states. A previous observation that H-strand RNA species accumulate by 10-20 fold during in vitro induced myogenesis (9) appears to support this speculation.

It should be noted here that a close correlation has been found between the content of the triple stranded D-loop structure in mitochondria and the cellular growth rate and/or the oxidative capacity of the tissues (10, 11). It should also be recalled here that one function of mrrL-RNAs is thought to be to prime H-strand DNA synthesis (7, 12), and that free RNA relevant to mrrL-RNA associates with a plasmid DNA template containing the mouse mtDNA D-loop region to form an R-loop in the absence of protein factors in vitro (13). Our results together with these previous findings lead us to speculate that the levels of L-strand DNA replication are regulated by 1.3 kb polyadenylated H-strand transcripts as antisense RNAs, similar to how antisense RNA regulation occurs in the rat mtDNA system (8).

Current production biotechnology using cultured cells is not yet able to control cellular energy supply. We believe that these results lay the foundation for a new approach to controlling the cellular energy supply by using the 1.3 kb mrrH-RNA.

***Acknowledgements***

This work was supported in part by the New Energy Industrial Technology Development Organization (NEDO).

## 5. References

1. Doda, J. N., Wright, C. T., and Clayton, D. A. (1981) Elongation of displacement-loop strands in human and mouse mitochondrial DNA is arrested near specific template sequences, *Proc. Natl. Acad. Sci. USA* **78**, 6116-6120.
2. Roberti, M., Musicco, C., Polosa, P. L., Milella, F., Gadaleta, M. N., and Cantatore, P. (1998) Multiple protein-binding sites in the TAS-region of the human and rat mitochondrial DNA, *Biochem. Biophysic. Res. Commun.* **243**, 36-40.
3. Attardi, G. (1993) The human mitochondrial genetic system , in S. DiMauro and D. C. Wallace (eds.), *Mitochondrial DNA in Human Pathology*, Raven Press, Ltd., New York, pp. 9-25.
4. Ojala, D., Crews. S., Montoya, J., Gelfand, R., and Attardi, G. (1981) A small polyadenylated RNA (7S RNA), containing a putative ribosome attachment site, maps near the origin of human mitochondrial DNA replication, *J. Mol. Biol.* **150**, 303-314.
5. Noda, A. (1992) Gene expression in Ataxia Telangiectasia cells as perturbed by bleomycin treatment, *Somatic Cell Mol. Genet.* **18**, 113-122
6. Nakamichi, N., Rhoads, D.D., Hayashi, J.-I., Kagawa, Y., and Matsumura, T. (1998) Detection, localization, and sequence analyses of mitochondrial regulatory region RNAs in several mammalian species, *J. Biochem.* **123**, 392-398
7. Chang, D. D., Hauswirth, W. W. and Clayton, D. A. (1985) Replication priming and transcription initiate from precisely the same site in mouse mitochondrial DNA, *EMBO J.* **4**, 1559-1567
8. Sbisa, E., Nardelli, M., Tanzariello, F., Tullo, A. and Saccone, C. (1990) The complete and symmetric transcription of the main non coding region of the rat mitochondrial genome: in vivo mapping of heavy and light transcripts, *Curr. Genet.* **17**, 247-253
9. Vijayasarathy, C., Zheng, Y.-M., Mullick, J., Basu, A., and Avadhani, N.G. (1995) Identification of a stable RNA encoded by the H-strand of the mouse mitochondrial D-loop region and a conserved sequence motif immediately upstream of its polyadenylation site, *Gene Expression* **4**, 125-141
10. King, T. C., and Low, R. L. (1987) Mitochondrial DNA displacement loop structure depends on growth state in bovine cells, *J. Biol. Chem.* **262**, 6214-6220
11. Annex, B. H., and Williams, R. S. (1990) Mitochondrial DNA structure and expression in specialized subtypes of mammalian striated muscle, *Mol. Cell. Biol.* **10**, 5671-5678
12. Chang, D.D. and Clayton, D.A. (1985) Priming of human mitochondrial DNA replication occurs at the light-strand promoter, *Proc. Natl. Acad. Sci. USA* **82**, 351-355
13. Lee, D. Y., and Clayton, D. A. (1996) Properties of a primer RNA-DNA hybrid at the mouse mitochondrial DNA leading-strand origin of replication, *J. Biol. Chem.* **271**, 24262-24269

**EFFECT OF OVARIAN STEROIDS AND OXYTOCIN ON LOCAL PRODUCTION OF PROSTAGLANDIN $E_2$, PROSTAGLANDIN $F_{2\alpha}$ AND ENDOTHELIN-1 IN COW OVIDUCTAL EPITHELIAL CELL MONOLAYERS.**

M. P. B. WIJAYAGUNAWARDANE[a], Y. H. CHOI[b], A. MIYAMOTO[c], H. KAMISHITA[a], A. Y. KOJIMA[a], M. TAKAGI[a], K. SATO[a]

[a]*Laboratory of Theriogenology,* [b]*Research Center for Protozoan Molecular Immunology, and* [c]*Department of Animal Science, Obihiro University of Agriculture and Veterinary Medicine, Obihiro 080-8555, Japan*

**Abstract**

The cyclic physiological and anatomical changes in the oviduct is mediated by the local countercurrent transfer of the ovarian products. Thus, cow oviductal epithelial cells (COEC) culture were utilized to investigate the effect of ovarian products such as progesterone ($P_4$), estradiol 17β ($E_2$) and oxytocin (OT) on the oviductal prostaglandin $E_2$ ($PGE_2$), $F_{2\alpha}$ ($PGF_{2\alpha}$) and endothelin -1 (ET-1) production. COEC were collected from non-pregnant Holstein cows during the follicular phase and cultured in M199 under standard culture conditions until monolayer formation. Cells in the first passage were incubated with $P_4$ (500 ng/ml), $E_2$ (1 ng/ml), OT ($10^{-9}$ M) and combination of $E_2$+$P_4$ for 24 and 48 h. Administration of $E_2$ significantly increased the production of $PGE_2$, $PGF_{2\alpha}$ and ET-1. However, simultaneous administration of high level of $P_4$ blocked the effect of $E_2$. OT showed no effect on oviductal production of neither PG nor ET-1. These results may imply that locally transferred $E_2$ at high concentration from the developing follicle is important stimulus for the oviductal $PGE_2$, $PGF_{2\alpha}$ and ET-1 production, which in turn may regulate the oviductal contraction to ensure an optimal embryo transport during the peri-ovulatory period.

*Key words*: cow, oviduct, epithelial cell culture, progesterone, estradiol 17 β, prostaglandin $E_2$, prostaglandin $F_{2\alpha}$ and Endothelin-1.

## 1. Introduction

The optimum microenvironment in the oviducts for gamete transport, fertilization and early embryonic development is mediated by the local countercurrent transfer of ovarian products. It has been observed that the oviduct ipsilateral to the corpus luteum (CL) and developing follicle bearing ovary contained higher levels of progesterone (P4) and estradiol 17β (E2) respectively than that of the contralateral side [1]. Moreover, oxytocin (OT) is also locally recirculated to the oviduct in higher concentration than that supplied in the peripheral circulation [2], and involved in the motility of the female genital tract [3]. Recently, we found that the oviductal content of prostaglandin E2 (PGE2), F2α (PGF2α) and endothelin -1 (ET-1) are higher in the side ipsilateral to the developing follicle than in the contralateral side [1]. These data suggest that there are a close relationship in local concentration between ovarian and oviductal products. Thus, in this study cow oviductal epithelial cells (COEC) culture was utilized to investigate the effect of P4, E2 and OT on the production of constrictive substances such as PG and ET-1.

## 2. Material and Methods

Oviducts from non-pregnant Holstein cows during the follicular phase were collected and transported to the laboratory in ice cold Hank's balanced salt solution (HBSS; Nissui pharmaceutical Co., Ltd., Tokyo, Japan). The isolation and culture of COEC was conducted as described by Rosselli et al., [4] with some modifications.

Subconfluency cells were washed twice with serum free M 199 and cultured for 24 and 48 h in 1 ml of serum free M199 with P4 (500 ng/ml; Sigma Chemical Co., St.Louis, MO), E2 (1 ng/ml; Sigma Chemical Co., St.Louis, MO), OT ($10^{-9}$ M ; Peptide Institute Inc., Osaka, Japan ) or combination of E2 and P4 at the same concentration. After the desired period of culture, the cells were dispersed by repeated freezing and thawing at -80°C. Then the medium was collected and centrifuged for 10 min at 1200 g at 4°C. The supernatant was stored at -20°C until hormonal assay. PG was measured directly using EIA. For the purpose of concentration of peptide, the sample was further purified using Sep-Pak $C_{18}$ cartridges as described by Miyamoto et al., [5], resulting in 40 times concentration. The protein concentration in the medium was estimated by Bio-Rad $D_C$ protein assay kit (Bio-Rad Laboratories, Hercules, CA) using bovine serum albumin as a standard. The concentrations of PGE2, PGF2α and ET-1 were measured using second antibody EIAs as described previously [1].

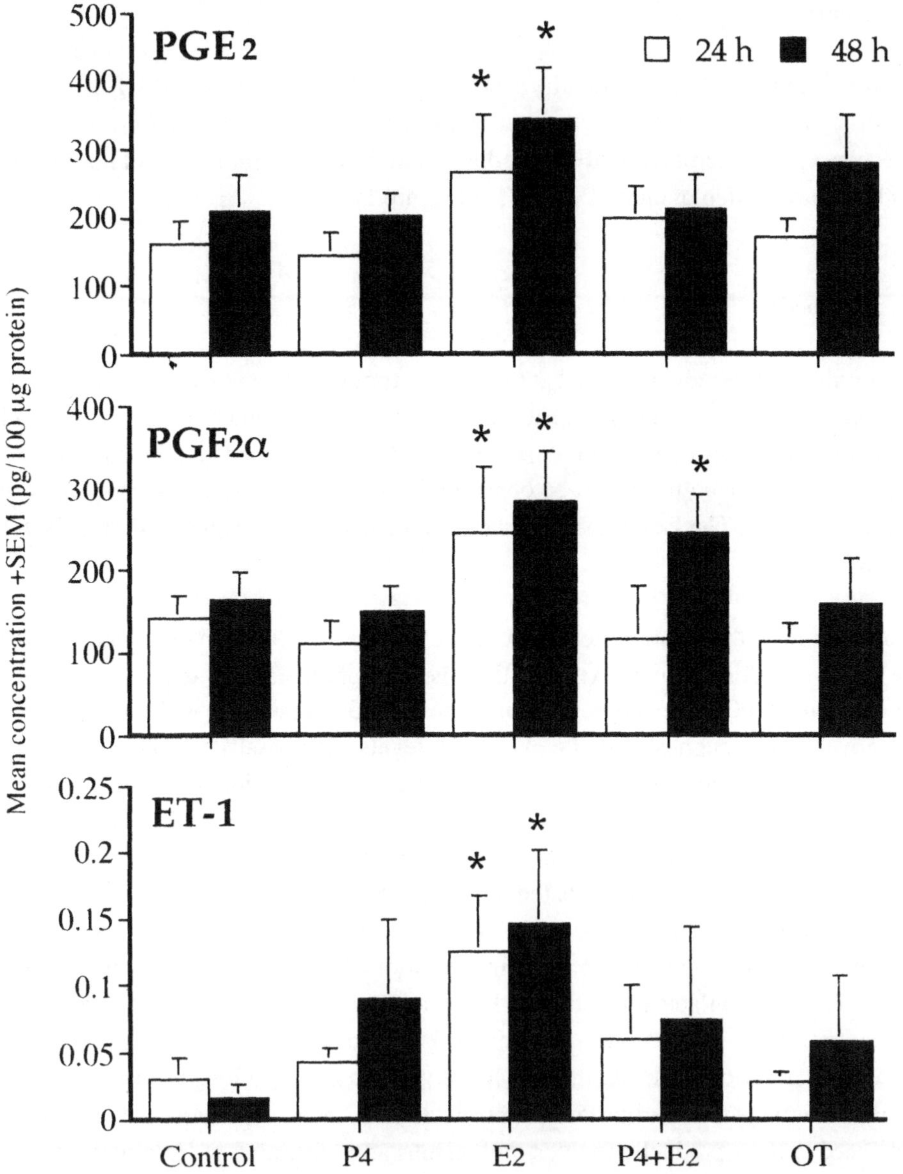


Figure 1 Effect of progesterone (P4), estrogen (E2), P4+E2 and oxytocin (OT) on prostaglandin E2 (PGE2), prostaglandin F2α (PGF2α) and endothelin-1 (ET-1) production after 24 and 48h of culture (n=8).

* Significantly different from controls $p<0.05$.

## 3. Results

$E_2$ alone significantly increased the production of $PGE_2$, $PGF_{2\alpha}$ and ET-1 both at 24 and 48 h of culture, while simultaneous administration of $E_2$ and $P_4$ did not show any effect except the increased $PGF_{2\alpha}$ production at 48 h of culture. OT did not affect the oviductal production of either PG or ET-1 (Figure 1).

## 4. Discussion

The results of this study indicate that the administration of $E_2$ increases the production of oviductal PG and ET-1 by the COEC, and simultaneous administration of high level of $P_4$ blocks the effect of $E_2$. The doses of $E_2$, $P_4$ and OT used in this study were the physiological concentrations to be observed in the oviductal tissue during the normal estrous cycle [1]. The $P_4$ dose was the high concentration that expected during the luteal phase.

It is well known that ovarian steroids regulate the local oviductal PG production through the effect on phospholipase $A_2$, and that the phospholipase $A_2$ activity and the local production of PG were highest during the $E_2$ dominant peri-ovulatory period [6]. Moreover, it was suggested that elevated $E_2$ levels with basal $P_4$ at the peri-ovulatory period may trigger the independent process of oviductal secretion of PG and ET-1 [1]. In the present cell culture study, a simulatenous administration of high $P_4$ with $E_2$ blocked the stimulatory action of $E_2$ alone on $PGE_2$ and ET-1 production. This supports our prevous data that the oviductal content of PG and ET-1 at luteal phase (in the presence of high $P_4$) was lower than those in the oviduct ipsilateral to ovulatory follicle (in the presence of high $E_2$ plus low $P_4$) [1]. In the uterus, $P_4$ makes the tissue less responsive to $E_2$ [7] and modulates the action of $E_2$ by a rapid reduction of the estrogen receptor concentration [8]. Furthermore, $P_4$ reduces PG production by lowering phospholipase $A_2$ activity in the oviductal epithelium [6]. It was reported that long term administration of $P_4$ stimulates the uterine $PGF_{2\alpha}$ production in ewes [9]. Thus, the significantly increased $PGF_{2\alpha}$ production with simultaneous administration of $E_2$ plus $P_4$ might be due to the chronic stimulatory effect of $P_4$. The observation made in this study that OT had no effect on oviductal PG or ET-1 production is similar to a previous report [10] in which OT did not stimulated $PGF_{2\alpha}$ or $PGE_2$ output from the guinea-pig endometrium maintained in tissue culture.

The results of the study may imply that locally transferred $E_2$ at high concentration from the developing follicle is important stimulus for the oviductal $PGE_2$, $PGF_{2\alpha}$ and ET-1 production, which in turn may regulate the oviductal contraction to ensure an optimal embryo transport during the peri-ovulatory period.

## 5. References

1. Wijayagunawardane, M.P.B., Miyamoto, A., Cerbito, W.A., Acosta T.J., Takagi, M., Sato, K., 1998, Local distributions of oviductal estradiol, progesterone, prostaglandins, oxytocin and endothelin-1 in the cyclic cow. *Theriogenology* 49, 607-618.
2. Schramm, W., Einer-Jensen, N., Schramm, G., McCracken, J.A., 1986, Local exchange of oxytocin from the ovarian vein to arteries in sheep. *Biol. Reprod.* 34, 671-680.
3. Maggi M, Genazzani AD, Gianninni S, Torrisi C, Baldi E, Tomasso M DI, Munson PJ, Rodbard D and Serio M., 1988, Vasopressin and oxytocin receptors in vagina, myometrium and oviducts of rabbits. *Endocrinology* , 122, 2970-2980.
4. Rosselli, M., Imthurn, B., Macas, E., Keller, P.J., 1994. Endothelin production by bovine oviductal epithelial cells. *J. Reprod. Fertil.* 101, 27-30.
5. Miyamoto, A., Kobayashi, S., Arata, S., Ohtani, M., Fukui, Y., Schams, D., 1997, Prostaglandin $F_{2\alpha}$ promotes the inhibitory action of endothelin-1 on the bovine corpus luteum in vitro. *J. Endocrinol.* 152, R7-11.
6. Morishita, T., Nozaki, M., Sano, M., Yokoyama, M., Nakamura, M., Nakano, H. 1993. Changes in phospholipase $A_2$ activity of rabbit oviduct ampullary epithelium by ovarian steroids. *Prostag. Leukotr. Ess.* 48, 315-318.
7. Henricks, D.M., Harris, Jr. R.B., 1978. Cytoplasmic estrogen receptors and estrogen concentration in bovine uterine endometrium. *Endocrinology* 103, 176-185.
8. Evans, R.W., Chen, T.J., Hendry III, W.J., Leavitt, W.W., 1980. Progesterone regulation of estrogen receptor in the hamster uterus during the estrous cycle. *Endocrinology* 107, 383-390.
9. Vincent, D.L., Inskeep, E.K.. 1986. Role of progesterone in regulating utero ovarian venous concentrations of $PGF_{2\alpha}$ and $PGE_2$ during the estrous cycle and early pregnancy in ewes. *Prostaglandins* 31, 715-733.
10. Riley, S.C., Poyser, N.L., 1989. Effects of oestradiol, progesterone, hydrocortisone and oxytocin on prostaglandin output from the guinea-pig endometrium maintained in tissue culture. *Prostaglandins* 34, 535-551.

# TGF-β INDUCED CELLULAR SENESCENCE IN CANCER CELLS IS REVERSIBLE AND OPERATES VIA TWO SEPARATE AND INDEPENDENT PATHWAYS

T. MIURA, Y. KATAKURA, E. NAKATA, N. UEHARA, AND S. SHIRAHATA

*Graduate School of Genetic Resources Technology, Kyushu University*
*6-10-1Hakozaki, Higashi-ku, Fukuoka 812-8581, JAPAN*

## 1. Abstract

Normal, somatic cells do not divide indefinitely, because they do not express telomerase, which maintains the length of telomere. When the length of telomere is shortened to a critical level, most of cells stop proliferation by a process termed replicative senescence (1, 2). Recently several groups reported a cellular senescence mechanism independent of telomere shortening, which is termed premature senescence (3). Premature senescence is commonly characterized by a flat enlarged morphology, growth arrest and activity for SA-β-gal within a week after addition of signals. Therefore, these findings indicate that an immediate cellular senescence was caused independently of telomere shortening, which is indistinguishable from replicative senescence dependent upon telomere shortening.
Transforming growth factor-β (TGF-β) is a family of multifunctional proteins that act on many different types of cells. TGF-β has many biological effects including regulation of cell differentiation, stimulation of extracellular matrix formation, and modulation of the immune response. TGF-β also is a potent growth inhibitor for most cell types (4). TGF-β inhibits growth by causing cell-cycle arrest as a result of up-regulation of the CDK inhibitor p15 and/or repression of the Cdc25A expression. However, A549 cells are not inhibited for its growth by TGF-β (5). In our studies, a growth rate of A549 cells became slightly lower in the presence of TGF-β, but TGF-β-treated A549 cells showed no induction of p15 and downregulation of Cdc25A. These results indicate that TGF-β had minimal effects on cell cycle regulation in A549 cells. From the viewpoint of cellular senescence, however, TGF-β was found to cause cellular senescence in A549 cells via two separate and independent pathways. These results indicate that TGF-β can trigger a novel senescence program leading to tumor suppression.

## 2. Materials and Methods

### 2.1 CELL CULTURE

Human lung adenocarcinoma cell lines (A549 cells) were maintained in ERDF medium supplemented with 5% fetal bovine serum (FBS) at 37 °C in 5% $CO_2$. TGF-β was added at a final concentration of 10 ng/ml.

### 2.2 SENESCENCE-ASSOCIATED β-GALACTOSIDASE ACTIVITY

SA-β gal activity was measured by the method described by Dimri *et al.* (6). Cells were washed with PBS, fixed in 3% formaldehyde for 5min, and again washed with PBS. Cells were stained with SA-β-gal staining solution (1 mg/ml 5-bromo-4-chloro-3-indolyl-β-D-galactosidase (X-gal), 5 mM $K_3Fe[CN]_6$, 5 mM $K_4Fe[CN]_6$, 150 mM NaCl, 2 mM $MgCl_2$ in PBS at pH 6.0) overnight at 37°C.

### 2.3 ASSAY FOR TELOMERASE ACTIVITY

1 x $10^6$ cells were pelleted by centrifugation and lysated in 200 μl of ice-cold buffer, which was composed of 10 mM Tris-HCl (pH 7.5), 1 mM $MgCl_2$, 1 mM EGTA, 0.5% CHAPS, 10% glycerol, 0.1 mM AEBSF and 5 mM β-mercaptoethanol, for 30 min on ice. The lysate was centrifuged for 20 min at 13,000 x g, and the supernatant was then examined for telomerase activity by a PCR-based assay termed telomeric repeat amplification protocol (TRAP). Protein concentration was measured using the Protein Assay kit (Bio-Rad Laboratories, Inc., Hercules, CA). Cell extracts was added to the TRAP reaction mixture, which was composed of 50 mM dNTP, 1 x PCR buffer (Boehringer Mannheim, Indianapolis, IN), 0.1 μg TS primer (5'-AATCCGTCGAGCAGAGTT-3'), 1.0 μg *T4* gene 32 protein (Wako, Kyoto, Japan) and 2 units Taq polymerase (Boehringer Mannheim). The reaction mixture was incubated at 20°C for 30min, heated at 94°C for 3 min. During this heating, 0.1 μg of CS primer (5'-$[CCCTTA]_3$CCCTTAA-3')was added to each samples and gently mixed. Then, the samples were subjected to 29 cycles of PCR (45 s at 94°C, 45 s at 50°C, and 60 s at 72°C). PCR products were electrophoresed onto a 10% polyacrylamide gel, and visualized by SYBR GREEN I (Takara, Shiga, Japan) staining.

### 2.4 RT-PCR ANALYSIS

Total RNA was isolated from each subconfluent culture using TRIzol reagent. Reverse transcription was performed using the SuperScript II RNase $H^-$ reverse transcriptase (Gibco BRL, Gaithersburg, MD). PCR amplifications carried out 26 cycles (94°C for 45s, 60°C for 45 s and 72°C for 90s) for hTERT and TEP1. The primer sets used were as followed: for hTERT, LT5 (5'-CGGAAGAGTGTCTGGAGCAA-3') and LT6 (5'-GGATGAAGCGGAGTCTGGA-3') ; for TEP1, TEP1.1 (5'-TCAAGCCAAACCTGAATCTGAG-3') and TEP1.2 (5'-

CCCGAGTGAATCTTTCTACGC-3'). The resulting PCR products were electrophoresed onto 8% non-denaturing polyacrylamide gel and stained with SYBR GREEN I.

### 2.5 DNA EXTRACTION AND ANALYSIS

Genomic DNA was prepared using the DNA Extractor WB kit (Wako) according to the manufacture protocol. Genomic southern hybridization, and telomere-length estimation were carried out as previously described by Katakura *et al.* (7).

## 3. Results and Discussion

### 3.1 TGF-β INDUCED PREMATURE SENESCENCE IN CANCER CELLS

To test whether TGF-β can induce premature senescence in A549 cells, A549 cells were cultured in the presence of 10 ng/ml TGF-β. A549 cells cultured with TGF-β rapidly became flat and enlarged morphology. This morphological change characterizes a cellular senescence. Subsequently, we examined the expression of senescence-associated marker of SA-β-gal at pH 6.0. As shown in Figure 1*A* and *B*, A549 cells cultured with TGF-β for 7 days were SA-β-gal positive. These results demonstrate that TGF-β induces a premature senescence in A549 which is independent of the telomere shortening. In recent studies, Serrano *et al.* showed that expression of oncogenic *ras* in primary human and/or rodent cells induces a premature senescence associated with accumulation of p53 and p16 (3). Xu *et al.* demonstrate that re-expression of functional pRB in RB-defective tumor cells induces cellular senescence within 4-5 days (8). Furthermore, Sugrue *et al.* demonstrate that p53 overexpression can activate the rapid onset of senescence in tumor cells (9). Our findings also may provide that TGF-β can induce an immediate cellular senescence in cancer cells which is indistinguishable from replicative senescence.

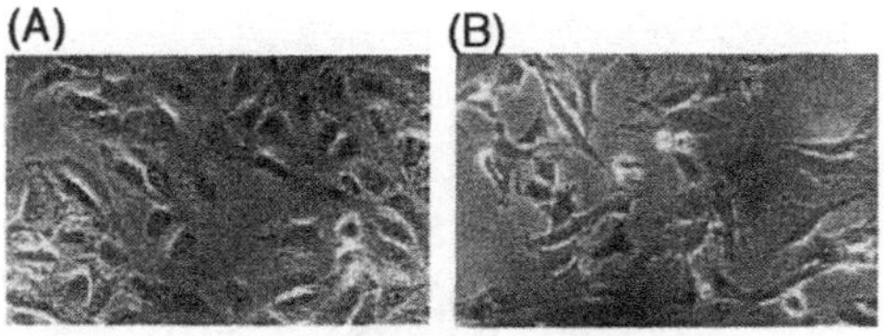


**Figure. 1 TGF-β induces morphrogical changes and senescence-associated β-galactosidase (SA-β-gal) in A549 cells.**

### 3.2 TGF-β TREAT A549 CELLS LOST TELOMERASE ACTIVITY AND SHORTENED TELOMERE

We evaluated the effect of TGF-β on telomerase activity in A549 cells. Telomerase activity was found to be correlated with hTERT (human telomerase transcriptase) expression , but not with TEP1 (human telomerase-associated protein) expression. Expression of hTERT and TEP1 was determined by RT-PCR (Fig. 2*A*). hTERT expression was repressed at day 2 after the addition of TGF-β. Furthermore, telomerase activity was repressed at day 25 after the addition of TGF-β (Fig. 2*B*). These results suggest that repression of telomerase activity was caused via the down-regulation of hTERT expression in TGF-β-treated A549 cells. To analyze the change in telomere length accompanied with the repression of telomerase activity, A549 cells were cultured in the presence of TGF-β for a long term. As shown in Fig. 3, TRF length shortened to 2 kb at day 175 after the addition of TGF-β. This result indicates that A549 cells treated TGF-β entered into the replicative senescence state.

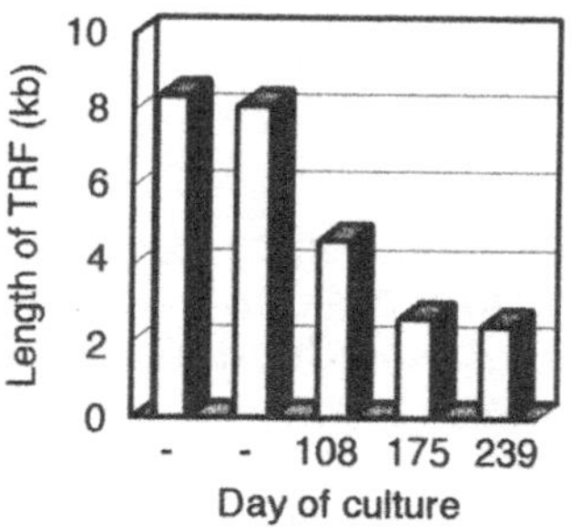


Figure.3 Telomere length in TGF-β-treated A549 cells.

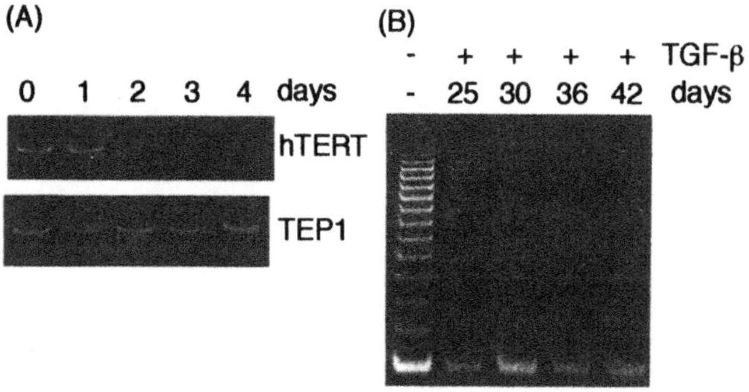


Figure. 2 Inhibitoin of hTERT expression and telomerase activity in A549 cells by TGF-β.

## 3.3 TGF-β-INDUCED CELLULAR SENESCENCE IS REVERSIBLE IN A549 CELLS

To investigate the reversibility of cellular senescence induced by TGF-β, A549 cells treated TGF-β for 78 days were maintained in TGF-β-free medium. Senescent A549 cells came to show the same morphology as original cancer cells, and became SA-β-gal negative at day 39 after removal of TGF-β from the cell culture medium (Fig. 4*B* and *C*). Furthermore, we observed reactivation of telomerase activity at day 20 after withdrawal of

TGF-β, and telomerase activity reached to a similar level to non-treated A549 cells at day 32 after withdrawal of TGF-β (Fig. 4*A*). These results suggested that TGF-β-induced senescence was revesibly regulated in A549 cells.

In this report we provide a novel correlation between TGF-β and a cellular senescence. TGF-β can rapidly induce a senescence, stringently repress hTERT expression and telomerase activity in cancer cells. However, it is not clear whether TGF-β signals directly link to a cellular senescence induction. We presently investigate to clarify a novel senescence mechanism via the TGF-β signal pathway.

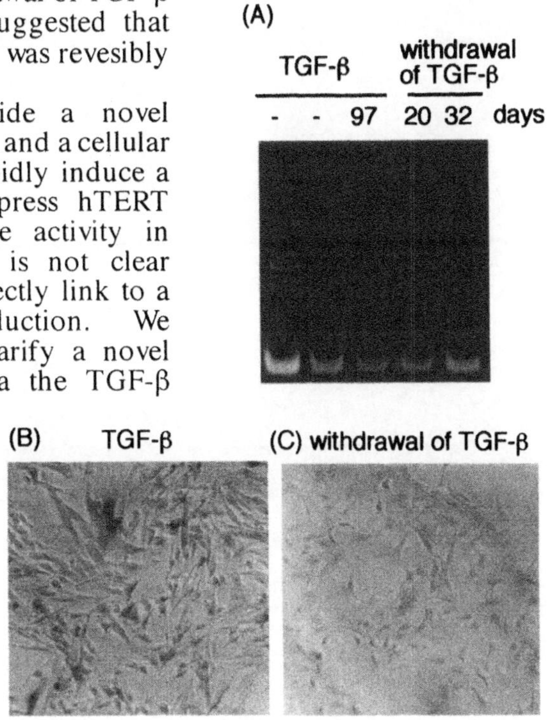


Figure.4 TGF-β induced cellular senescence in A549 cells is reversible.

## References


1. Campisi, J. 1996. Replicative senescence : an old lives' tale? *Cell* 84 : 497-500
2. Allsopp, R. C. and Harley, C. B. 1995. Evidence for a critical telomere length in senescent human fibrobrasts. *Exp. Cell. Res.* 219 : 130-136
3. Serrano, M., Lin, A.W., McCurrach, M. E., Beach, D. and Lowe, S. W. 1997. Oncogenic *ras* provokes premature cell senescence associated with accumulation of p53 and $p16^{INK4a}$. *Cell* 88 : 593-602
4. Heldlin, C. H., Miyazono, K., and Dijke, P. 1997. TGF-β signaling from cell membrane to nucleus through SMAD proteins. *Nature* 390 : 465-471
5. Iavarone, A., and Massague, J. 1997. Repression of the CDK activator Cdc25A and cell-cycle arrest by cytokine TGF-β in cells lacking the CDK inhibitor p15. *Nature* 387 : 417-422
6. Dimri, G. P., Lee, X., Basile, G., Meileen, A., Scott, G., Roskelley, C., Medrano, E. E., Linskens, M., Rubelj, I., Pereira-Smith, O. Peacocke, M., Campisi, J. 1995. A biomarker that identifies senescent human cells in culture and in aging skin in vivo. *Proc. Natl. Acad. Sci.* USA 92 : 9363-9367
7. Katakura, Y., Yamamoto, K., Miyake, O., Yasuda, T., Uehara, N., Nakata, E., Kawamoto, S., and Shirahata, S. 1997. Bidirectional regulation of telomerase activity in a subline derived from human lung adenocarsinoma. *Biochem. Biophys. Res. Commun.* 237 : 313-317
8. Xu, H. J., Zhou, Y., Ji, W., Perng, G. S., Kruzelock, R., Kong, C. T., Bast, R. C., Mills, G. B., Li, J., and Hu, S. X. 1997. Reexpression of the retinoblastoma protein in tumor cells induces senescence and telomerase inhibition. *Oncogene* 15 : 2589-2596
9. Sugrue. M. M., Shin, D. Y., Lee, S. W., and Aaronson, S. A. 1997. Wild-type p53 triggers a rapid senescence program in human tumor cells lacking functional p53. *Proc. Natl. Acad. Sci.* USA 94 : 9648-9653

# MOLECULAR CLONING AND CHARACTERIZATION OF SPECIFIC GENES INVOLVED IN CELLULAR SENESCENCE

N. UEHARA, Y. KATAKURA, T. MIURA, AND S. SHIRAHATA

*Graduate School of Genetic Resources Technology, Kyushu University, Hakozaki 6-10-1, Higashi-ku, Fukuoka 812-8581, Japan*

**INTRODUCTION** Telomerase and telomere shortening are one of the factor which regulate cellular senescence. Telomerase is a ribonucleoprotein enzyme that adds hexameric TTAGGG repeats onto the telomeres to compensate for progressive loss of telomere which occurs with each round of DNA replication (end replication problem). Telomerase activity has been detected in germ line cells, immortal cells, but not in somatic cells (1). Repression of telomerase caused telomere shortening, which signals cellular senescence (2). However, it is not well known about the molecules that are associated with cellular senescence, telomerase regulation and telomere maintenance.

To identify the genes involved in cellular senescence, telomere maintenance and telomerase regulation, we performed a PCR-Select subtraction method by using cDNA libraries prepared from human lung adenocarcinoma derived cell lines. We have established A5DC7, CK and AST-9 cells from the human lung adenocarcinoma cell line A549. A5DC7 cells were isolated from IFN-γ -treated A549 cells and showed nomal senescent cell phenotypes with restored ability of contact inhibition and anchorage-dependent growth as well as lost tumorigenicity (3). We also established CK cells from senescent A5DC7 cells by raising serum concentration from 5% to 10%. CK cells regained telomerase activity and telomere maintenance mechanism as well as resumed proliferation (4). Furthermore, we established AST-9 cells from A549 cells treated with hydrogen peroxide. Although AST-9 cells express almost the same level of telomerase activity as A549 cells, AST-9 cells showed the passage number-dependent telomere shortening. These cell lines are thought to be useful for screening the genes involved in induction and/or maintenance of cellular senescence, maintenance of telomere length and telomerase regulation.

## MATERIALS AND METHODS

*Cell culture.* The A5DC7, CK and AST-9 cell lines were established from the human lung adenocarcinoma cell line A549. AST-9 cells was obtained by treating with 200 μM of hydrogen peroxide ($H_2O_2$) for 4 days, and cloned by standard limiting dilution. All cell lines were grown at 37°C in 5% $CO_2$ atmosphere. A549, A5DC7 and AST-9 cells were cultured in ERDF medium (Kyokuto Pharmaceutical, Tokyo, Japan) supplemented with 5% fetal bovine serum (FBS; Whittaker Bioproducts, Inc., Walkersville, MD, USA). CK cells were cultured in ERDF medium supplemented with 10% FBS (4).

*Assay for telomerase activity.* Telomerase activity was measured by TRAP (Telomeric Repeat Amplification Protocol) assay with some modifications (1, 5). Briefly, $10^6$ cells were pelleted and lysed with lysis buffer. Cell lysates were centrifuged at 15,000 rpm for 20 min., and supernatants were stored at -80°C. TRAP assay products were analyzed by electrophoresis using 10% polyacrylamide nondenaturing gels in 1×TBE buffer. Gels were stained with SYBR Green I (TAKARA, Shiga, Japan).

*DNA extraction and analysis.* Genomic DNA was prepared using the DNA Extractor WB Kit (Wako, Kyoto, Japan) according to the manufacturer's protocol. Length of the Terminal Restriction Fragments (TRF) was determined by southern blot analysis with a telomeric sequence probe and measured the length of telomere according to the procedure of KATAKURA *et al.* (4).

*RNA preparation.* Total RNA was prepared from cell pellets of A549 derived cell lines by TRIzol (GIBCO BRL, Gaithersburg, MD, USA). Poly $(A)^+$ RNA was then selected from total RNA by using oligo d(T) cellulose (Collaborative Biomedical Products, Bedford, MA, USA).

*Construction of the cDNA library and subtractive hybridization.* cDNA libraries were constructed from poly $(A)^+$ RNA of each A549 derived cell lines, and synthesized tester and driver cDNA. We then subtracted driver cDNA from tester cDNA in all combinations of cell lines and amplified the specifically expressed cDNA fragments in tester cDNA pool by using the PCR-Select cDNA Subtraction Kit (CLONTECH Laboratories, Inc., Palo Alto, CA, USA) according to the manufacturer's protocol.

*Dot blot hybridization.* Secondary differential screening was performed with four replicate dot-blots fixed with 1 μg of plasmid DNA representing each cDNA clone from the subtracted cDNA library. Each blot was probed separately with a $[^{32}P]$-labeled cDNA synthesized from total RNA of A549 derived cell lines with oligo dT primer.

*Nucleotide sequence analysis.* Subtractive cDNAs are cloned into the pGEM-T vector (Promega, Madison,WI). Then, sequence analyses were performed with ALF express auto sequencer (Pharmacia, Buckinghamshire, UK) according to the manufacturer's protocol.

## RESULTS AND DISCUSSION

We have established three kinds of cell lines which were derived human lung adenocarcinoma cell line A549, named A5DC7, CK and AST-9. Then we assessed the phenotypes of these cell lines from the viewpoint of cellular senescence, i.e. telomerase activity and telomere shortening as previously reported (Fig. 1). Higher telomerase activity was detected in A549, CK and AST-9 cells, but that of A5DC7 cells was greatly reduced. TRF length of A5DC7, CK and AST-9 cells were analyzed by southern blot hybridization using telomeric sequence (TTAGGG)$_4$ as probe. TRF length of A5DC7 and AST-9 cells shortened, and that of CK cells lengthened dependent upon the number of passagings, although that of A549 remained unchanged. Senescence marker of β-galactosidase activity was observed only in A5DC7 cells. Furthermore, we analyzed the expression of mRNAs for telomerase components (hTERT and TEP1). hTERT is the catalytic subunit of human telomerase, and hTERT mRNA expression is correlated with the telomerase activity. TEP1 is the homologue of the *tetreahymena* telomerase-associated protein p80. Corresponding to the higher telomerase activity detected, hTERT mRNA was highly expressed in A549, CK and AST-9 cells. TEP1 mRNA was constitutively expressed in all kinds of A549 derived cells.

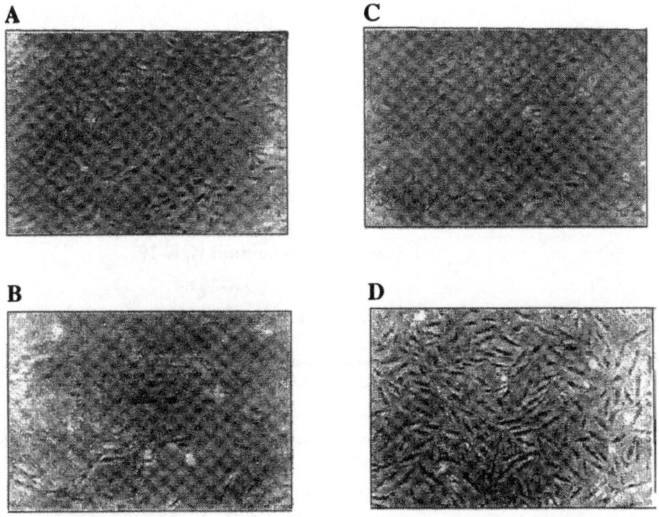


**Fig. 1 Phase-contrast micrographs of A549 derived cell lines. Human adenocarcinoma cell line A549 (A). A5DC7 cells were isolated from IFN-γ-treated A549 cells and showed nomal senescent cell phenotype (B). CK cells were obtained from senescent A5DC7 cells by raising serum concentration from 5% to 10% (C).AST-9 cells were established from A549 cells by treating with hydrogen peroxide (D).**

In this study, we have sought to identify the genes overexpressed and/or specifically expressed in these cells by using PCR-select subtraction method. Consequently, we found 110 genes that specifically expressed in A549, A5DC7, CK and AST-9 cells. We compared expression level of subtracted clones among four A549 derived cell lines by dot blot hybridization in order to select for clones for further analysis. Most of sequences of subtracted clones were demonstrated to match the sequences deposited in the Genbank database (Table 1). Matched sequences are 28S ribosomal RNA gene, glyoxalase I, polycystic kidney disease (PKD1) gene, MHC class III HSP-70-2, Ku protein subunit, Interferon-γ, RasGAP-related protein (IQGAP2), aldolase A, human preferentially expressed antigen of melanoma (PRAME), protein kinase C-binding protein RACK8, human telomerase catalytic subunit (hTERT), poly(A)-binding protein, transcription factor IID, elongation factor 1 α, elongation factor 2 and house keeping proteins such as keratin, α-tublin and actin. However, the rest of the clones do not have any significant homology to the deposited sequences including EST database, suggesting the possibility that these clones codes for novel genes involved in telomerase regulation and cellular senescence.

**Table 1 Sequence of the subtracted cDNA clones**

| Tester | Driver | Sequence (clone number) |
|---|---|---|
| A5DC7 | CK | 28S ribosomal RNA gene (7K-1, 4) |
| | | EST sequence (7K-2, 3, 5, 6) |
| AST-9 | A549 | ACTB mRNA for mutant β-actin (QA-3) |
| | | elongation factor 1 α (QA-5) |
| | | 28S ribosomal RNA gene (QA-11) |
| AST-9 | CK | keratin 8 (QK-2) |
| | | α-tublin (QK-3, QK-11, QK-12) |
| | | glyoxalase I (QK-4) |
| | | polycystic kidney disease (PKD1) gene (QK-5) |
| | | Human homolog of D. melanogaster flightless-I gene (QK-6) |
| | | EST sequence (QK-7, 8, 14, 15, 16, 17, 18, 23) |
| | | MHC class III HSP70-2 (QK-9) |
| | | Ku protein subunit (QK-10) |
| | | interferon-γ (QK-19) |
| | | GARS protein (QK-20) |
| CK | A5DC7 | EST sequence (K7-1, 3, 6, 7, 11, 12, 13, 16,18, 19, 20, 23, 24, 25) |
| | | elongation factor 2 (K7-2) |
| | | preferentially expressed antigen of melanoma(PRAME) (K7-4) |
| | | protein kinase C-binding protein RACK8 (K7-5) |
| | | nonmuscle myosin heavy chain (NMHC) (K7-9) |
| | | cytochrome c oxidase COX subunit IV (K7-15) |
| | | pancreatic tumor-related protein (K7-17) |
| | | prothymosin alpha (K7-21) |
| | | peroxisomal enoyl-CoA hydratase-like prtein (K7-22) |

| Tester | Driver | Sequence (clone number) |
|---|---|---|
| **A549** | **AST-9** | **EST sequence (AQ-1, 3, 6, 7, 9, 10, 13)** |
| | | **telomerase catalytic subunit(hTERT) (AQ-2)** |
| | | **aldolase A (AQ-4)** |
| | | **PM5 protein (AQ-5)** |
| | | **18S rRNA gene (AQ-8)** |
| | | **poly(A)-binding protein (PABP) gene (AQ-11)** |
| **A549** | **A5DC7** | **EST sequence (A7-1, 3, 5, 7, 11, 15, 16)** |
| | | **novel gene (A7-2, 12)** |
| | | **RasGAP-related protein(IQGAP2) (A7-4)** |
| | | **actin-binding protein (A7-6)** |
| | | **keratin 8 (A7-8, 9)** |
| | | **transcription factor TF IID (A7-10, 14)** |
| | | **γ actin (A7-13)** |
| **AST-9** | **A5DC7** | **40S ribosomal protein (Q7-1)** |
| | | **pHL-1 gene (c-myc oncogene containing coxIII sequence) (Q7-2)** |
| | | **EST sequence (Q7-3, 6)** |
| | | **mesothelial keratin K7 (type II) (Q7-5)** |
| | | **novel gene (Q7-7)** |
| | | **elongation factor 2 (EF-2) (Q7-9)** |

## REFERENCES


1. Kim, N. W., Piatyszek, M. A., Prowse, K.R., Harley, C.B., West, M. D., Ho,P. L. C., Coviello, G. M., Wright, W. E., Weinrich, S. L., and Shay, J. W. (1994) *Science* **266,** 2011-2015.
2. Levy, M. Z., Allsopp, R. C., Futcher, A. B., Greider, C. W., Harley, C. B. (1992) *J. Mol. Biol.* **225,** 951-960.
3. Kawamoto, S., Inoue, Y., Shinozaki, Y., Katakura, Y., Tachibana, H., Shirahata, S., Murakami, H. (1995) *Biochem. Biophys. Res. Commun.* **215,** 280-285.
4. Katakura, Y., Yamamoto, K., Miyake, O., Yasuda, T., Uehara, N., Nakata, E., Kawamoto, S. and Shirahata, S. (1997) *Biochem. Biophys. Res. Commun.* **237,** 313-317.
5. Piatyszek, M. A., Kim, N. W.,Weinrich, S. L., Hiyama, K., Hiyama, E., Wright, W. E. and Shey, J. W. (1995) *Meth. Cell Sci.* **17,** 1-15.

# REGULATORY MECHANISMS OF GRANULOSA CELL APOPTOSIS IN OVARIAN FOLLICLE ATRESIA

*Role of Granulosa Cell Apoptosis in Porcine Ovarian Follicle Atresia*

N. MANABE, Y. KIMURA, K. UCHIO, C. TAJIMA, H. MATSUSHITA, M. NAKAYAMA, M. SUGIMOTO, and H. MIYAMOTO
*Unit of Anatomy and Cell Biology, Department of Animal Sciences, Kyoto University Kyoto 606-8502, Japan. manabe@jkans.jkans.kais.kyoto-u.ac.jp*

## Abstract

Porcine ovary samples were prepared for histochemical and ultrastructural analyses. *In situ* analysis of DNA fragmentation was performed on histological sections of follicles using the terminal deoxynucleotidyl transferase-mediated biotinylated deoxyuridine triphosphate nick end-labeling (TUNEL) method. No apoptotic cells were observed in healthy follicles. In atretic follicles, apoptotic TUNEL staining was seen in scattered granulosa cells located on the inner surface of the follicular wall, but not in cumulus cells, internal or external theca cells, or oocytes. Nuclear condensation, a typical feature of apoptosis was seen only in scattered granulosa cells. The neutral $Ca^{2+}/Mg^{2+}$-dependent endonuclease is involved in granulosa cell apoptosis. No endonuclease activity was detected in cumulus cells. An IgM monoclonal antibody (PFG-1) capable of inducing granulosa cell apoptosis was then produced against granulosa cells prepared from healthy antral follicles. Two-dimensional (2D) Western blotting analysis revealed that PFG-1 specifically recognized a cell-membrane protein (PFG-1 antigen, 55 kD, pI 5.9). PFG-1 immunohistochemically reacted with granulosa cells of healthy follicles but not those of atretic follicles. When the isolated granulosa cells prepared from healthy follicles were cultured in medium containing 0.1 $\mu$g/ml of PFG-1, the cells underwent apoptosis. These observations indicated that apoptosis occurs in granulosa cells but not cumulus cells in the atretic follicles and that the PFG-1 antigen, a novel cell death receptor, is different from the apoptosis-mediating receptors Fas antigen or tumor necrosis factor receptor 1 (TNFR-1).

## Follicular Granulosa Cell Apoptosis

In mammalian ovaries, more than 99% of the follicles undergo the degenerative change known as atresia at varying stages of follicle development [1]. A number of studies of the follicular atresia have revealed the morphological and biochemical characteristics of atretic follicles. Recent findings have suggested that apoptosis, originally described in 1972 by Kerr *et al.* [2], is the mechanism underlying ovarian follicular atresia. Apoptotic cell death of granulosa cells of rabbit Graafian follicles with atresia was first observed in 1885 by Flemming [3], who called it "chromatolysis". Recently, Tilly *et al.* [4, 5] reported that the degeneration of atretic follicles in mammalian ovaries can be explained, at least in part, by apoptotic death of granulosa and theca interna cells. However, the degenerative changes in cumulus cells during follicular atresia have not been investigated in detail. We confirmed that apoptosis occurs in granulosa cells but not cumulus cells in the atretic Graafian follicles from porcine ovaries [6-16] (Fig. 1). Briefly, *in situ* analysis of DNA fragmentation was performed on histological sections of follicles using the TUNEL method, and conventional electron microscopic analysis was also performed. Histological and cytological findings confirmed that there were no apoptotic changes in the cumulus cells of the porcine atretic follicles.

**Endonuclease Involved in Granulosa Cell Apoptosis**

As described above, apoptosis was originally defined by morphological criteria [2]. One of the earliest morphological events in apoptotic cells is chromatin condensation which occurs as a consequence of digestion of chromatin DNA. When the cells undergo apoptosis, their chromatin DNA is degraded at internucleosomal sites by endonucleases, giving rise to the characteristic 'ladder' pattern of oligonucleosomal-sized bands corresponding to multiples of about 180-bp. There are multiple pathways for apoptotic nuclear fragmentation, and some types of endonucleases make such DNA breaks during the apoptotic process. Therefore, an understanding of the events that trigger activation of the endonucleases is essential for elucidation of the pathways upstream of apoptosis. However, the endonuclease involved in granulosa cell apoptosis during follicular atresia has not been determined. We designed experiments to characterize the relevant endonucleases (neutral $Ca^{2+}/Mg^{2+}$-dependent endonuclease, neutral $Ca^{2+}$-dependent endonuclease, neutral $Mg^{2+}$-dependent, $Ca^{2+}$-independent endonuclease and acidic cation-independent endonuclease) from apoptotic granulosa cells of pig atretic follicles [17]. When the progesterone/estradiol-17ß ratio of follicular fluid was less than 15, the follicle was classified as healthy. Granulosa and cumulus cells were isolated from healthy and atretic follicles, and then these endonuclease activities were assessed. Extremely high activity of the $Ca^{2+}/Mg^{2+}$-dependent endonuclease was detected only in granulosa cell nuclei prepared from atretic follicles as compared with those from healthy follicles. Extremely low levels of $Ca^{2+}/Mg^{2+}$-dependent endonuclease activity were noted in cumulus cell nuclei prepared from healthy and atretic follicles. No activity of the $Mg^{2+}$-dependent endonuclease or the acidic cation-independent endonuclease was seen in either cumulus or granulosa cell nuclei prepared from healthy or atretic follicles. DNA electrophoresis after endonuclease assay demonstrated internucleosomal DNA cleavage only in the granulosa cell nuclei prepared from atretic follicles. A good correlation was found between the $Ca^{2+}/Mg^{2+}$-dependent endonuclease activity of granulosa cells and the progesterone/estradiol-17ß ratio of follicular fluid in each follicle. However, there was no correlation between the $Ca^{2+}/Mg^{2+}$-dependent endonuclease activity of cumulus cells and the progesterone/estradiol-17ß ratio of follicular fluid. We thus confirmed that the $Ca^{2+}/Mg^{2+}$-dependent endonuclease is involved in granulosa cell apoptosis of the atretic antral follicles in pigs.

**Monoclonal Antibody-Inducible Apoptosis**

Apoptosis is the result of many stimuli which act through many signal transduction pathways culminating in the activation of endonucleases [17]. Apoptotic stimuli and intracellular signal transduction pathways involved in granulosa cell apoptosis remain to be determined. Therefore, understanding of the events that trigger activation of the endonucleases is essential to elucidate the pathways upstream of apoptosis. We produced monoclonal antibodies which recognize cell-surface antigens of granulosa cells [18, 19]. Such antibodies, especially PFG-1 which induces apoptosis of cultured granulosa cells, are good probes for studying the intracellular pathways of granulosa cell apoptosis in porcine atretic follicles. The isolated granulosa cells prepared from healthy antral follicles of pig ovaries were used as antigens for immunization and as target cells in cell-killing activity assay, for observation of nuclear morphology and DNA electrophoresis. According to the standard procedure, female BALB/c mice were immunized with the isolated granulosa cells, and then the spleen cells from immunized mice, which produced anti-granulosa cell antibodies, were fused with Sp2/O-Ag14 mouse myeloma cells. As only IgM antibodies against Fas-antigen had cell-killing activity [20], IgM antibody-producing hybridomas were selected. The hybridoma cells producing IgM antibodies against the granulosa cells were screened by ELISA and immunofluorescent staining techniques. Then, the hybridoma cells which produced antibodies with granulosa cell-killing activity were selected. Finally, one hybridoma clone, named PFG-1, was selected. The antigens on healthy granulosa cells were characterized by 2D Western blotting, and one specific spot (PGF-1 antigen; 55 kD, pI 5.9 ) was observed (Fig. 2).

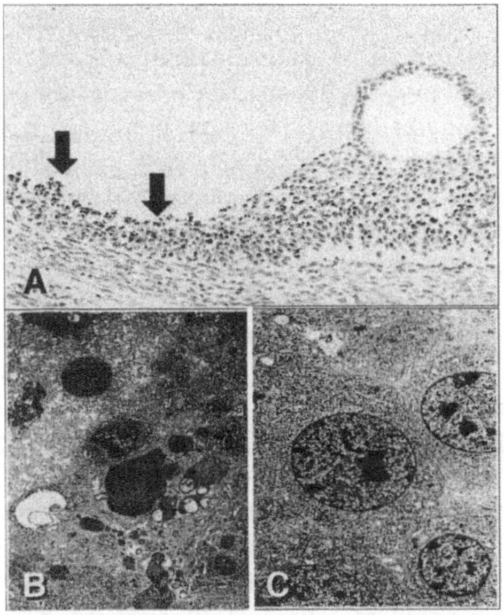


Fig. 1. Atretic follicle sections stained by the TUNEL technique (A) and transmission electron micrographs of granulosa and cumulus cells (B and C). Nuclear DNA fragmentation was observed in granulosa cells scattered on the inner surface of the follicular wall (arrows), but not cumulus cells. Nuclear condensation was observed in granulosa cells but not in cumulus cells (C).

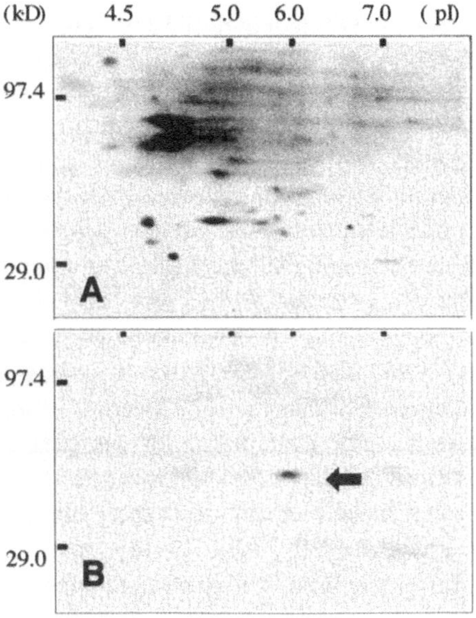


Fig. 2. 2D Western blotting analysis of granulosa cell antigen recognized by PFG-1 (PFG-1 antigen). Cell membrane fractions were prepared from healthy granulosa cells, and separated by 2D-PAGE. Separated proteins were transferred onto nitrocellulose sheets and detected by silver staining (A). PFG-1 antigen (55 kD, pI 5.9; arrow) was visualized immunochemically (B).

Indirect immunofluorescence analysis of cryostat sections of ovaries was used to determine the target specificity of the monoclonal antibodies. PFG-1 was strongly reactive with granulosa cells of healthy follicles. The antibody did not label theca interna or externa cells, basement membrane, or ovarian stroma cells. Moreover, the antibody showed no specific binding to pig stomach, small intestine, large intestine, liver, pancreas, lung, kidney, testis, adrenal grand, heart, spleen or brain, or ovaries of rodents or ruminants. Granulosa cell apoptosis mediated by PFG-1 was determined by assessment of nuclear morphology, DNA electrophoretic analysis and cell cycle analysis using a fluorescence activated cell sorter (FACS). The isolated granulosa cells were co-cultured with PFG-1 at concentrations varying from 0.001 to 100 $\mu$g/ml, for 1 to 48 h at 37 °C, and then the morphology of the nuclei was observed by staining with Hoechst 33258 under a fluorescent microscope. When the isolated granulosa cells were cultured without any additive for 3 h, only round weakly fluorescent healthy granulosa cell nuclei were observed. When the cells were co-cultured with at least 0.1 $\mu$g/ml PFG-1 for 3 h, many small condensed fluorescent signals (representing apoptotic bodies, a morphological hallmark of apoptosis) were observed. After incubation, DNA samples from these isolated granulosa cells were electrophoresed in 2% agarose gels, and the DNA displayed a ladder pattern (biochemical hallmark of apoptosis). However, DNA samples of the cells co-cultured without any additive displayed no such pattern on electrophoresis. The granulosa cells were co-cultured with 0.1 $\mu$g/ml PFG-1 for 3 h, and then the percentages of cells with degraded DNA were determined by FACS analysis. No degraded DNA was observed in cells co-cultured without any additive. However, high percentages of degraded DNA, representing apoptotic bodies, were demonstrated in cells co-cultured with PFG-1.

We confirmed that granulosa cells undergo apoptosis and that no apoptotic cell death occurs in cumulus cells in pig atretic follicles [6-16]. However, it has not been determined which trigger molecules induce granulosa cell apoptosis, or how intercellular apoptotic signals are transmitted in the granulosa cells. A specific monoclonal antibody, which recognizes a cell-surface trigger molecule and induces apoptosis, is essential to define the molecular mechanism of apoptotic signal transmission pathways in granulosa cells. We generated PFG-1 that reacted against a cell-surface protein. In rodents, the Fas antigen, which is a transmembrane glycoprotein that belongs to the TNF/NGF receptor family and mediates apoptosis in a variety of lymphoid and tumor cells [20], can mediate granulosa cell apoptosis in ovarian follicle atresia and luteal cell degeneration [11, 21]. However, this protein has not yet been identified in pig ovaries. Our preliminary comparative studies of the progression of granulosa cell apoptosis in the atretic follicles revealed histochemically that there are species-specific differences in the apoptotic process in granulosa cells [6]. Briefly, apoptosis demonstrated histochemically by TUNEL staining was seen in scattered granulosa cells located on the inner surface of the follicular wall of pig ovaries. In contrast, such apoptotic granulosa cells were demonstrated on the outer surface of the follicular wall of cow ovaries. In rodents, such apoptotic cells were seen randomly throughout the follicular epithelial wall. These observations indicated that local regulation mechanisms of granulosa cell apoptosis are different among mammalian species. TNF also induces apoptosis in a variety of tumor cells, and the TNFR-1 can also mediate apoptosis [22]. The molecular weights of Fas antigen and TNFR are 45 [21] and 65 kD [22], respectively. Fas antigen was histochemically detected in the granulosa cells of both healthy and atretic follicles in the ovaries of rodents [11, 21]. TNFR-1 was not detected in the ovarian follicular cells [22]. The molecular weight of the granulosa cell-surface antigen recognized by PFG-1 is 55 kD, and the antigen visualized histochemically by PFG-1 was only detected in the granulosa cells of healthy follicles. Such differences in localization dependent on the stage of follicle development are important when considering the physiological function of the cell death receptors. We concluded that the antigen recognized by PFG-1 is not Fas antigen or TNFR-1. This antigen was suggested to belong to the TNF/NGF receptor family. This antibody, PFG-1, will be useful as a probe to investigate the cell death receptor on the granulosa cell membrane and its natural ligand, and to define the intercellular pathway of apoptotic signal transmission in granulosa cells of pig ovaries. Biochemical details of this cell death receptor of the pig granulosa cells should be elucidated in future studies. In our laboratory, PFG-1 is currently being used to screen for specific cell death receptor on the granulosa cell membrane.

## References


1. Hirshfield A.N. (1991) Development of follicles in the mammalian ovary. Int. Rev. Cytol., 124: 43-101.
2. Kerr J.F.R. *et al.* (1972) Apoptosis: a basic biological phenomenon with wide ranging implication in tissue kinetics. Br. J. Cancer., 26: 239-257.
3. Flemming W. (1885) Über die Bildung von Richtungsfiguren in Säugethiereiern beim Untergang Graaf'scher Follikel. Arch. Anat. EntGesch., 221-244.
4. Tilly J.L. *et al.* (1991) Involvement of apoptosis in ovarian follicular atresia and postovulatory regression. Endocrinology, 129: 2799-2801.
5. Tilly J.L. (1993) Ovarian follicular atresia: a model to study the mechanisms of physiological cell death. Endocrnol. J., 1: 67-72.
6. Manabe N. *et al.* (1995) Roles and regulation of follicular granulosa cell apoptosis in atretic ovarian follicles. J. Artific. Inseminat., 167: 1-10.
7. Manabe N. *et al.* (1996) Apoptosis occurs in granulosa cells but not cumulus cells in the atretic antral follicles in the pig ovaries. Experientia, 52: 647-651.
8. Manabe N. *et al.* (1996) Signal transmission of granulosa cell apoptosis in the atretic follicles in the pig ovaries. J. Reprod. Dev., 42: 135-141.
9. Manabe N. *et al.* (1997) Apoptosis occurs in granulosa cells but not cumulus cells in the atretic antral follicles in the porcine ovaries. J. Reprod. Dev., 43: 179-180.
10. Manabe N. *et al.* (1997) Apoptosis occurs in granulosa cells but not cumulus cells in the atretic graafian follicles in multiparous pig ovaries. Acta. Histochem. Cytochem., 30: 85-92.
11. Sakamaki K. *et al.* (1997) Involvement of Fas antigen in ovarian follicular atresia and leuteolysis. Mol. Reprod. Dev., 47: 11-18.
12. Kimura Y. *et al.* (1997) Histochemical examination of cellular carbohydrate chains which change during ovarian follicular atresia in the porcine ovary. Jpn. J. Emb. Transfer, 19: 101-110.
13. Kimura Y. *et al.* (1998) Examination of granulosa cell glycoconjugates which change during follicular atresia in the pig ovaries. J. Reprod. Dev., 44: 35-44.
14. Sugimoto M. *et al.* (1998) Ultrastructural changes in granulosa cells in porcine antral follicles undergoing atresia indicate apoptotic cell death. J. Reprod. Dev., 44: 7-14.
15. Manabe N. *et al.* (1998) Regulatory mechanisms of granulosa cell apoptosis in porcine ovarian follicle atresia, in H. Miyamoto and N. Manabe (eds.), *Reproductive biology update*. Nakanishi Pub. Co., Kyoto, Japan, pp. 23-35.
16. Manabe N. *et al.* (1998) Role of granulosa cell apoptosis in ovarian follicle atresia, in T. Yamada and Y. Hashimoyto (eds.), *Apoptosis: Its roles and mechanism*. Academic Societies Book Center, Tokyo, Japan, pp. 97-111.
17. Manabe N. *et al.* (1996) $Ca^{2+}/Mg^{2+}$-dependent endonuclease but not $Ca^{2+}$-dependent, $Mg^{2+}$-dependent or cation-independent endonuclease is involved in granulosa cell apoptosis of pig atretic follicles. J. Reprod. Dev., 42: 247-253.
18. Myoumoto A. *et al.* (1997) Monoclonal antibodies against pig ovarian follicular granulosa cells induce apoptotic cell death in cultured granulosa cells. J. Vet .Med. Sci., 59: 641-649.
19. Manabe N. *et al.* (1997) A Monoclonal antibodiy against pig ovarian follicular granulosa cells recognizes a novel cell death receptor and induces granulosa cell apoptosis, in S. Kawashima *et al.* (eds.) *Advences in comparative endocrinology*, Monduzzi Editore Spa., Bologna, Italy, pp. 1793-1797.
20. Yonehara S. *et al.* (1989) A cell-killing monoclonal antibody to a cell surface antigen co-downregulated with the receptor of tumor necrosis factor. J. Exp. Med., 169: 1747-1756.
21. Hakuno N. *et al.* (1996) Fas/APO-1/CD95 system as a mediator of granulosa cell apoptosis in ovarian follicle atresia. Endocrinology, 137:1938-1948
22. Stauber G.B. *et al.* (1988) Human tumor necrosis factor-α receptor: purification by immunoaffinity chromatography and initial characterization. J. Biol. Chem., 263: 190-198.

# ENHANCEMENT OF FcεRI EXPRESSION BY HYDROCORTISONE INDUCES THE UPREGULATION OF FcεRIγ mRNA IN THE HUMAN LEUKEMIA CELL LINE KU812

Takashi Hara, *Hirofumi Tachibana and *Koji Yamada
*Graduate School of Genetic Resources Technology, Kyushu University,*
**Department of Food Science and Technology, Faculty of Agriculture,*
*Kyushu University, 6-10-1 Hakozaki, Higashi-ku, Fukuoka 812-8581, Japan*

**Abstract** The human leukemia cell line KU812 have been described as an immature pre-basophilic cell line and exhibits a potential to differentiate into mature basophilic cells (1-3, 17). KU812 cells produce modest amount of histamine and express FcεRI, the high affinity IgE receptor, which plays a central role in the allergic responses system. Previously, we observed that hydrocortisone can enhance FcεRI expression in KU812 cells. In this study, we demonstrate that KU812 cells treated with hydrocortisone are able to release histamine in response to stimulation with IgE and anti-IgE antibody, suggesting that the FcεRI induced by hydrocortisone was functional enough to transduce a signal for degranulation. FcεRI is a tetramer consisting of an α chain, a β chain and two γ chains. We measured the mRNA level of each subunit in KU812 cells by RT-PCR followed by Southern blotting. After treatment with 1 μM hydrocortisone for 35 days, the cells showed an increase in the amount of mRNA for FcεRIγ. Almost no significant change in the amount of mRNA for FcεRIα and FcεRIβ was observed. This result suggests that the enhanced expression of FcεRI on the surface of KU812 cells by hydrocortisone was related to the upregulaion of FcεRIγ mRNA.

## INTRODUCTION

The high affinity IgE receptor, FcεRI, which is expressed on the surface of basophils and mast cells, plays a central role in the allergic response system by binding allergen-specific IgE and multivalent allergens, resulting in the production and release of inflammatory mediators. This receptor is a tetrameric structure composed of one α chain, one β chain and two disulfide-linked identical γ chains (4). FcεRIα extends into the extracellular space and contains the entire IgE-binding site (5, 6). FcεRIγ is required for signal transduction and the function of FcεRIβ is most likely to amplify the FcεRIγ-mediated activation signaling (7). Importantly, it is known that the presence of FcεRIγ is indispensable for cell surface expression of FcεRIα (8, 9). The importance of FcεRI in the allergic response has been demonstrated by the observation that FcεRIα-deficient mice completely failed to exhibit IgE-mediated anaphylaxis (10). Therefore, the expression of FcεRI, as well as the synthesis of IgE, is critical for allergic responses.

Using the established human pre-basophilic cell line KU812, we previously found that hydrocortisone can induce an enhancement of FcεRI expression on the cell surface along with increasing in the amount of intracellular histamine. In the present study, we examined whether the FcεRI induced by hydrocortisone is able to trigger intracellular signaling events for histamine release, and whether enhancement of FcεRI expression was related to upregulation of mRNA production.

## MATERIALS AND METHODS

### *Cells*

KU812 cells were obtained from Japan Cancer Research Resources Bank (JCRB). KU812 cells was maintained in RPMI-1640 (Nissui, Tokyo, Japan) supplemented with 10% fetal bovine serum (Intergen, Purchase, NY), 100 U/ml penicillin G, 100 mg/ml streptomycin and 10 mM HEPES buffer. KU812 cells were kept in a humidified atmosphere of 5% $CO_2$ at 37°C and maintained by splitting the cells to a density of 2.5 $\times 10^5$ cells/ml in fresh medium every four days.

### *Treatment with hydrocortisone*

KU812 cells were cultured in the presence of 1 μM hydrocortisone, ethanol alone (0.1%) or in control medium. Hydrocortisone was dissolved in ethanol at 200 μM and this solution or ethanol alone was diluted 1000-fold into medium so that the final ethanol concentration was 0.1%. Cultures were started at a cell density of $2.5\times 10^5$ cells/ml and the spent culture was replaced with fresh medium every 4 days.

### *Determination of high affinity IgE receptor FcεRI expression*

Cells were incubated with mouse anti-human FcεRI monoclonal antibody CRA-1 (Kyokuto, Tokyo, Japan) for 30 minutes at 4°C. Cells were then washed twice in phosphate-buffered saline (PBS; pH 7.5) and labeled with a fluorescein-labeled goat anti-mouse antibody (Biosource, Camarillo, CA) for 30 minutes at 4°C. Cells were again washed twice in PBS and subjected to flow cytometry analysis. As a negative control, instead of mouse anti human FcεRI antibody, mouse control antibody (Vector, Burlingame, CA) was used.

### *Histamine measurement*

Histamine levels were measured by fluorometric assay (11). Total histamine content was assessed by sonicated cell lysates in Tyrode buffer. To achieve histamine release, cells were prepared in Tyrode buffer then stimulated with Ca ionophore A23187 or with anti-human IgE antibody (Biosource, Camarillo, CA) after sensitization with human myeloma IgE (Cosmo Bio, Tokyo, Japan). Histamine in the cell lysates and supernatants was partially extracted, coupled with *o*-phtalaldehyde which gives rise to a fluorescent product that can be measured on spectrofluorometer with the excitation at 360 nm and the emission at 450 nm.

***Analysis of FcεRIα, β and γ mRNA expression by RT-PCR and Southern blotting***

Total RNA was isolated using the TRIZOL reagent (GIBCO RBL). cDNA synthesis was performed using the MMLV-reverse transcriptase (Amersham, UK). The resultant cDNA samples were PCR amplified in the presence of specific sense and antisense primers for FcεRIα, β and γ. As an internal control, glyceraldehyde-3-phosphoryl dehydro-genase (G3PDH) cDNA was amplified. PCR was performed for 16 cycles, each cycle including denaturation (95°C, 30 sec), annealing (60°C for FcεRIα and FcεRIγ or 58°C for FcεRIβ, 30 sec) and polymerization (72°C, 1 min), and a final incubation (72°C, 7 min) after the last cycle. Sequences of PCR primers and sizes of expected products are as follows: for FcεRIα sense 5'-CTTAGGATGTGGGTTCAGAAGT-3', and antisense 5'-GACAGTGGAGAATACAAATGTCA-3' (495 bp); for FcεRIβ: sense 5'-TAATTC-TTCATAAAGACGATCATC(A, T, G or C)GG-3', and antisense 5'-ATATGCCTTTG-TTTTGGAACAATGGTGTG-3' (457 bp); for FcεRIγ: sense 5'-TAGGGCCAGCTGG-TGTTAATGGCA-3', and antisense 5'-GATGATTCCAGCAGTGGTCTTGCT-3' (364 bp); for G3PDH: sense 5'-GCTCAGACACCATGGGGAGGT-3', and antisense 5'-GT-GGTGCAGGAGGCATTGCTA-3' (404 bp). The amplified PCR products were electrophoresed, blotted onto Hybond-N+ membranes (Amersham), hybridized (60°C, 16 h) with probes specific for FcεRIα, β, γ and G3PDH, respectively. The hybridized probes were detected using the Gene Images detection kit (Amersham).

## RESULTS

Flow cytometry analysis revealed that when KU812 cells were treated with 1 μM hydrocortisone for 21 days, FcεRI expression on the cell surface was enhanced (Figure 1). This enhancement was observed at day 3, gradually increased until reaching a plateau at day 7 to 14 and maintained up to 35 days (data not shown). In addition, the amount of intracellular histamine substantially increased after exposure to hydrocortisone. We assessed if hydrocortisone can elicit a grater amount of histamine release from KU812 cells when stimulated with A23187 or IgE/anti-IgE. When cells were cultured in the absence of hydrocortisone, after stimulation with A23187 or IgE/anti-IgE, histamine

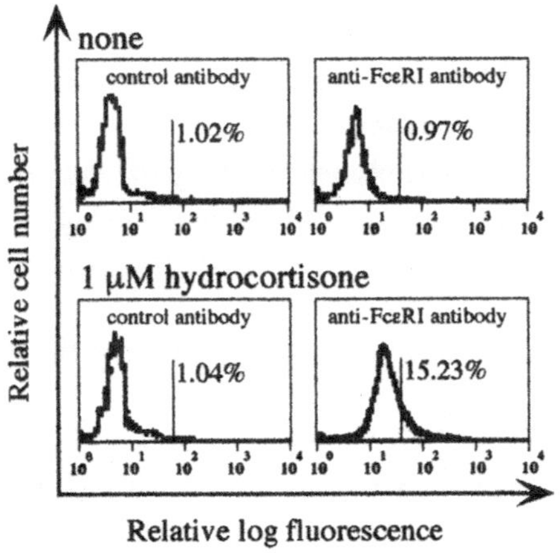


# Figure 1

**Induction of FcεRI expression on KU812 cells by hydrocortisone. After cultivation with 1 μM hydrocortisone for 21 days, cells were incubated with either mouse anti-human FcεRI monoclonal antibody CRA-1 or mouse IgG as subclass-matched negative controls, then stained with FITC-conjugated goat anti-mouse IgG and analyzed by flow cytometry. The typical result obtained from five independent experiments is shown for each sample.**

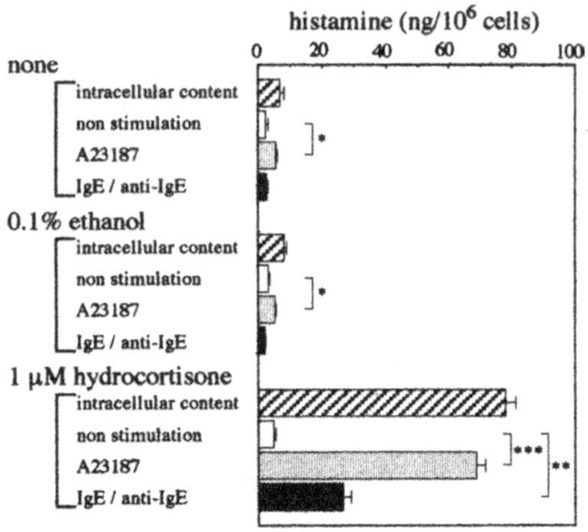


# Figure 2

**Histamine release from KU812 cells. KU812 cells cultivated with or without hydrocortisone (1 μM) for 21 days were stimulated with $Ca^{2+}$ ionophore A23187 or goat anti-human IgE polyclonal antibody after sensitization with human myeloma IgE at 37°C for 20 min. Results are expressed as the mean ± SE (n = 5). *, $P < 0.05$; **, $P < 0.01$; ***, $P < 0.001$ (by Student's t-test).**

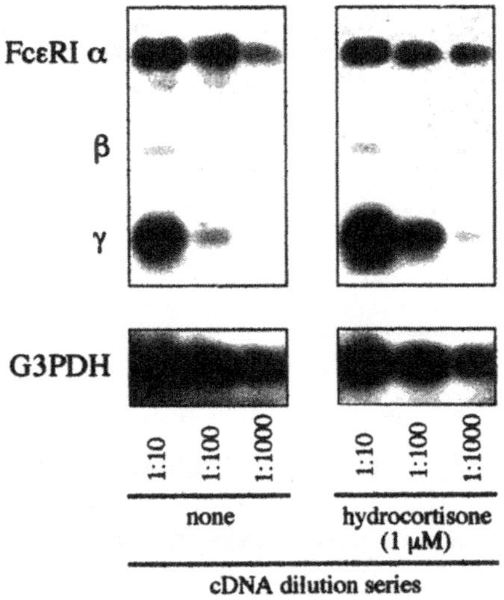


# Figure 3

**Upregulation of FcεRIγ mRNA in KU812 cells treated with 1 μM hydrocortisone for 21 days. mRNA level of FcεRIα, FcεRIβ, FcεRIγ and G3PDH in KU812 cells treated with hydrocortisone. FcεRIα, FcεRIβ, FcεRIγ and G3PDH mRNA were analyzed by RT-PCR. Southern blotting using specific probes for FcεRIα, FcεRIβ, FcεRIγ andG3PDH was performed to confirm the PCR products.**

release were 5.2±0.4 ng/$10^6$ cells and 2.0±0.3 ng /$10^6$ cells, respectively. When cells were cultured with 1 μM hydrocortisone for 21 days, histamine release values after stimulation with A23187 or IgE/anti-IgE were 68.7±2.6 ng /$10^6$ cells and 26.5±2.4 ng /$10^6$ cells, respectively (Figure 2). These results suggest that KU812 cells treated with hydrocortisone became mature enough in function to be able to release histamine. Furthermore, these results show that the hydrocortisone induced FcεRI molecules on the cell surface are able to transduce a sufficient amount of intracellular signals to result in degranulation.

Since flow cytometric analysis revealed that hydrocortisone enhanced the expression of FcεRI at the protein level, we measured the mRNA level of each FcεRI subunit by RT-PCR followed by Southern blotting (Figure 3). Modest amounts of FcεRIα, β and γ mRNA were observed in KU812 cells cultured in medium only. After treatment with 1 μM hydrocortisone for 21 days, the cells showed an increase in the amount of FcεRIγ mRNA produced, but no significant change was observed for FcεRIα and FcεRIβ. This data suggests that the enhanced expression of FcεRI on the surface of KU812 cells by hydrocortisone is related to the upregulation of FcεRIγ mRNA.

## DISCUSSION

Of all the blood cell types, basophils and mast cells are the only cells that can synthesize histamine (12, 13). In addition, the amount of intracellular histamine in basophils and mast cells is reported to increase during their differentiation (14,15). Basophils express FcεRI on the cell surface which are able to bind IgE. Upon antigen challenge, the antigen-specific IgE-FcεRI complexes were cross-linked that is followed by an aggregation of FcεRI, which result in release of mediators and generation of alachidonic acid metabolites. We previously described that hydrocortisone induces a) the enhanced expression of FcεRI, b) the appearance of granulated cells and c) the increase of intracellular histamine in KU812 cells. Furthermore, in this study, the enhanced FcεRI was shown to be functional with regard to transmission of the degranulation signal in response to IgE/anti-IgE antibody stimulation. Thus, it is likely that glucocorticoids including hydrocortisone may contribute to basophilic differentiation and/or maturation, while these same glucocorticoids may have potent immunosuppressive and anti-inflammatory effect on lymphocytes and granulocytes.

Recently, it has been reported that IL-4 induces the expression of FcεRI in human cultured mast cells, which is caused by the increase of FcεRIα mRNA at transcriptional level (16). We previously showed that IL-4 enhanced the expression of FcεRI in KU812 cells, which was related to the upregulation of the expression of mRNA for all three subunits, FcεRIα, β and γ (17). Here we show that treatment of KU812 cells with hydrocortisone resulted in the enhanced expression of FcεRI, notably the increased expression of FcεRIγ mRNA. It is known that FcεRIγ is required for constitutive expression of the FcεRI molecule on the cell surface in basophils and mast cells. Therefor these results suggest that the IL-4 and hydrocortisone may act to induce upregultion of the FcεRI expression via different regulatory mechanisms. More research must be done to fully elucidate these pathways.

## REFERENCES


1 Kishi, K. (1985) Leuk. Res. 9, 381-390.

2 Valent, P., Besemer, J., Kishi, K., Kaltenbrunner, R., Kuhn, B., Maurer, D., Lechner, K., and Bettelheim, P. (1990) J. Immunol. 145, 1885-1889.

3 Nilsson, G., Carlsson, M., Jones, I., Ahlstedt, S., Matsson, P., and Nilsson, K. (1994) Immunology. 81, 73-78.

4 Blank, U., Ra, C., Miller, L., White, K., Metzger, H., and Kinet, J. P. (1989) Nature 337, 187-189.

5 Hakimi, J., Seals, C., Kondas, J. A., Pettine, L., Danho, W., and Kochan, J. P. (1990) J. Biol. Chem. 265, 22079-22085.

6 Blank, U., Ra, C., and Kinet, J. P. (1991) J. Biol. Chem. 266, 2639-2645.

7 Lin, S., Cicala, C., Scharenberg, A. M., and Kinet, J. P. (1996) Cell 85, 985-995.

8 Miller, L, Blank, U., Metzger, H., and Kinet, J. P. (1989) Science 244, 334-337.

9 Maurer, D., Fiebiger, E., Reininger, B., Wolff-Winiski, B., Jouvin, M. H., Kilgus, O., Kinet, J. P. and Stingl, G. (1994) J. Exp. Med. 179, 745-750.

10 Dombrowicz, D., Flamand, V., Brigman, K. K., Koller, B. H., and Kinet, J. P. (1993) Cell 75, 969-976.

11 Shore, P. A., Burkhalter, A., and Cohn, V. H. (1959) J. Pharmacol. Exp. Ther. 127, 182-186.

12 Krauss, S., Gilbert, H. S., and Wassermann, L. R. (1968) Blood 31, 699-709.

13 Maeyama, K., Taguchi, Y., Sasaki, M., Wada, H., Beaven, M. A., Kitamura, Y., and Watanabe, T. (1988) Biochem. Biophys. Res. Commun. 151, 1402-1407.

14 Matuda, H., Kannan, Y., Ushio, H., Kiso, Y., Kanemoto, T., Suzuki, H., and Kitamura, Y. (1991) J. Exp. Med. 174, 7-14.

15 Valent, P., Schmidt, G., Besemer, J., Mayer, P., Kaltenbrunner, R., Hinterberger, B., Lechner, K., Maurer, D., and Bettelheim, P. (1989) Blood 73, 1763-1769.

16 Toru, H., Ra, C., Nonoyama, S., Suzuki, K., Yata, J., and Nakahata, T. (1996) Int. Immunol. 96, 1161-1169.

17 Hara, T., Koji, Y., and Tachibana, H. (1998) Biochem. Biophys. Res. Commun. 247, 542-548.

# TELOMERE SHORTENING IN CANCER CELLS BY ELECTROLYZED-REDUCED WATER

Sanetaka Shirahata, Eri Murakami, Ken-ichi Kusumoto, Makiko Yamashita, Masaaki Oda, Kiichiro Teruya, Shigeru Kabayama, Kazumichi Otsubo,* Shinkatsu Morisawa,* Hidemitsu Hayashi, ** and Yoshinori Katakura

*Graduate School of Genetic Resources Technology, Kyushu University, Hakozaki, Higashi-ku, Fukuoka 812-8581, Japan; *Nihon Trim Co. Ltd., 1-8-34 Oyodonaka, Kita-ku, Osaka 531-0076, Japan; **Water Institute, Nisshin Building 9F, 2-5-10 Shinjuku, Tokyo 160, Japan.*

**ABSTRACT:** Electrolyzed-reduced water (ERW) which is produced near cathode during electrolysis of water scavenges reactive oxygen species and protects DNA from oxidative damage (Shirahata et al., 1997). Most of cancer cells exhibit high telomerase activity to elongate telomere length, insuring their immortality. Here we found that cultivation of human lung adenocarcioma A549, human uterine cancer HeLa and human normal fibroblast TIG-1 cells were growth inhibited in medium containing ERW and drastic morphological changes occurred in A549 and HeLa cancer cells but not in TIG-1 cells. Telomerase activity did not change but telomere length became shorter depending upon cell division in medium containing ERW. Telomere binding activities of telomere binding proteins in cancer cells decreased in medium containing ERW, suggesting that ERW inhibit the binding of telomerase to telomere region via telomere binding proteins, resulting in the shortening of telomere length.

## 1. Introduction

It has long been established that an oxidant/antioxidant imbalance in an organism caused numerous damages to biomolecules and cellular structures that resulted in the development of a variety of pathologic states such as cancer, arteriosclerosis, and aging. Electrolysis of water produces reduced water near cathode which contains a lot of hydrogen molecules and oxidized water near anode which contains a lot of oxygen or hypochloric acid. Electrolyzed-reduced water (ERW) is popular as beneficial drinking water for our health in Japan. We demonstrated that electrolyzed-reduced water scavenges reactive oxygen species (ROS) and protects DNA from oxidative damage (Shirahata *et al.*, 1997). Recently ROS has been noted to have important roles to sustain the phenotypes of cancer cells such as invasion, metastasis, chemotherapy resistance, oncogene activation and genomic instability (Toyokuni *et al.*, 1995). Most of cancer cells exhibit high telomerase activity and no telomere shortening, which guarantee immortality of the cells. Recently it has been reported that ROS is related to telomere regulation mechanisms (von Ziglinicki *et al.*,

1995). Here we report that electrolyzed-reduced water causes the telomere shortening in cancer cells without changing the telomerase activity.

## 2. Materials and Methods

### 2.1 Preparation of electrolyzed reduced water and culture medium

Ultrapure water produced by an ultrapure system (Type Synthesis, Millipore) containing 0.1 g/l of NaCl was electrolyzed by an electrolyzing device (Type TI-7000S, Nihon Trim) and alkaline ERW (pH 11; ORP: - 600~-800 mV) was collected as described previously (Shirahata *et al.*, 1997). MEM/ERW medium was prepared in a usual manner for preparation of culture medium except for using ERW instead of Milli Q water. Conventional MEM medium with Milli Q water was named MEM/Milli Q medium. Since ERW had no buffering effect, alkaline ERW at about pH 11 was easily neutralized by HEPES buffer (pH 7.4) which were previously added. The medium was sterilized by autoclaving. The medium was stocked in a 250 ml plastic T flask at 4 °C.

### 2.2 Cells and cell culture

Human lung adenocarcinoma A549 cells, human uterine cancer HeLa cells and human normal fibroblast TIG-1 cells were cultivated in MEM/ERW medium or usual MEM/Milli Q medium supplemented with 10% fetal bovine serum in the presence or absence of ERW. Medium was exchanged every 2 days.

### 2.3 Determination of telomerase activity and telomere length

Telomerase activity in A549 and HeLa cells were determined by telomeric repeat amplification protocol assay (Katakura *et al.*, 1997). The telomere length was examined by Southern blotting method (Katakura *et al.*, 1997).

### 2.4. Electrophoresis mobility shift assay of telomere binding proteins

Telomere binding proteins were detected by the electrophoresis mobility shift assay .

## 3. Results and Discussion

### 3.1. Growth inhibition and morphological changes of cancer cells by electrolyzed-reduced water

The growth of lung cancer A549 cells and uterine cancer HeLa cells was suppressed in MEM/ERW medium. The growth of normal fibroblast TIG-1 cells was also suppressed and reached to confluence at lower cell density, but the normal cells in MEM/ERW medium seemed to be in a good state (Shirahata et al., 1998). Analysis of signaling pathways using specific inhibitors revealed that ERW inhibited the activation of MAP kinase which is related to the regulation of cell cycle and differentiation (data not shown). Cultivation of cancer cells in MEM/ERW medium for 90 days induced drastic morphological changes in cancer cells. The small size of cancer cells changed to big sizes and stretched shapes. However, the morphology of normal cells did not change. ROS is known to stimulate cell proliferation (Burdon & Gill, 1993; Burdon *et al.*, 1994). Assay of intracellular redox state using fluorescent dye, 2',7'-dichlorofulorescein-diacetate (DCFH-DA) which specifically reacts with intracellular $H_2O_2$ revealed that the amount of intracellular $H_2O_2$ in cancer cells decreased in ERW (data not shown), suggesting that phenotypes of cancer cells might be changed by shifting the oxidized state in cancer cells in MEM/Milli Q medium to more reduced one in MEM/ERW medium.

### 3.2 Telomerase activity and telomere shortening

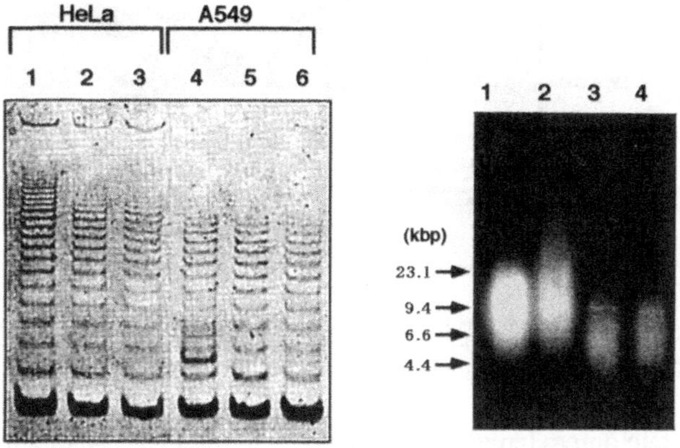


Fig. 1. Telomerase activity in A549 and HeLa cells (A) and telomere shortening in A549 cells (B) in medium containing electrolyzed reduced water. A549 and HeLa cells were cultivated in MEM/ERW medium for 39 days (A) or 153 days (B) and telomerase activities and telomere length were examined by the TRAP assay and electrophoresis mobility shift assay, respectively. A. Lane 1, MEM/MilliQ; lane 2, MEM/HOCl (1 ppm); lane 3, MEM/ERW; lane 4, MEM/MilliQ; lane 5, MEM/HOCl (1 ppm); lane 6, MEM/ERW. B. Telomere length was examined by electrophoretic method. Lane 1, control; lane 2, HOCl ( 1ppm); lane 3, ERW; lane 4, After treatment of the cells for 10 days with MEM/ERW medium, the medium was changed to MEM/MilliQ medium.

It has been known that most of cancer cells exhibit telomerase activity but most somatic cells do not exhibit the activity. Specific telomerase inhibitors have intesively been investigated in or der to develope new types of anticancer reagents which can change immortal cancer cells to mortal cells. When A549 and HeLa cells were cultivated in MEM/ERW medium for 90 days, we found no change in telomerase activity in both cell lines (Fig. 1, left panel). However, RT-PCR analysis for the expression of the hTRT gene also supported the results. However, the telomere lengths in A549 cells cultured in MEM/ERW medium was significantly shortened. ERW was contaminated with 1 ppm of HOCl which was produced near anode, the same concentration of HOCl did not shortened the telomere length. When A549 cells were cultured in MEM/Milli Q medium after cultivation in MEM/ERW medium for 10 days, the telomere length was a little elongated, suggesting the effect of ERW was reversible. Similar phenomenon was also observed in HeLa cells.

### 3.3 Lowered activities of telomere binding proteins in reduced water

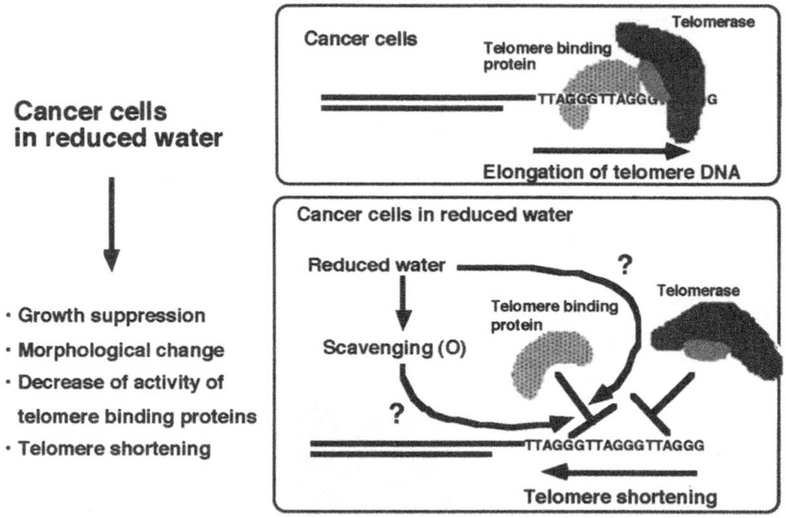


Fig. 2. Putative mechanisms on the number of passage dependent telomere shortening in cancer cells by electrolyzed-reduced water.

The regulation mechanisms of telomere length have not been clarified yet. Telomere binding proteins which bind to single strand or double strand telomere regions is known to regulate the binding of telomerase to telomere regions. If telomere binding proteins do not work well, telomere length will be shortened dependent upon number of cell division, even if telomerase existed. We investigated telomere binding activities of telomere binding proteins using radiolabelled double strand and single strand synthetic DNA probes which had telomere-specific sequence. The double strand DNA probes were mixed with cell lysate and electrophoresesed. The mobility of the probes decreased dependent upon the molecular size of bound proteins. We found the binding activities of specific telomere binding proteins were reversibly decreased by ERW in A549 cells (data not shown). These results suggested that ERW inhibit the binding of telomerase to telomere region by reversibly inhibiting the binding of telomere binding proteins to telomere DNA. The inhibition of telomerase to telomere regions will cause the number of passages-dependent telomere shortening.

In conclusion, cancer cells cultivated in medium containing ERW caused growth suppression, morphological changes, decrease of activities of telomere binding proteins and telomere shortening. Reduced water may inhibit the binding of telomerase to telomere via redox regulation by scavenging ROS in cancer cells or via other mechanisms, resulting in senescence or impaired tumor phenotypes of cancer cells (Fig. 2). Some medical doctors observed that cancer patients were drastically improved by daily intake of reduced water in some cases. We have proposed that active hydrogen in reduced water scavenges ROS and regulate cell functions (Shirahata et al., 1997; Shirahata et al., 1998). Further intensive investigation on the functions of reduced water in immunology, oncology and cell biology is desirable.

## References

1. Burdon, R. H., Alliangana, D., and Gill, V. (1994) Endogeneously generated active oxygen species and cellular glutathione levels in relation to BHK-21 cell proliferation. *Free Rad. Res.*, **21**, 121-133.
2. Burdon, R. H. and Gill, V. (1993) Cellulary generated active oxygen species and Hela cell proliferation. *Free Rad. Res. Comms.*, **19**, 203-213.
3. Katakura Y., Yamamoto, Y., Miyake, O., Yasuda, T., Uehara, N., Nakata, E., Kawamoto, S., and Shirahata, S. (1997) Bidirectional regulation of telomerase activity ina subline derived from human lung adenocarcinoma. Biochem. Biophys. Res. Comms. 237, 313-317.
4. Shirahata, S., Kabayama, S., Kusumoto, K., Gotoh, M., Teruya, K., Otsubo, K., Morisawa, S., Hayashi, H., and Katakura, K. (1998) Electrolyzed reduced water which can scavenge active oxygen species suppresses cell growth and regulates gene expression of animal cells. In: O.-W. Merten et al. (eds.), New Development and New Applications in Animal Cell Technology, pp.93-96, Kluwer Academic Publishers, the Netherlands.
5. Shirahata, S., Kabayama, S., Miura, T., Kusumoto, K., Gotoh, M., Hayashi, H., Otusbo, K., Morisawa, S., and Katakura, Y. (1997) Electrolyzed-reduced water scavenges active oxygen species and protects DNA from oxidative damage. Biochem. Biophys. Res. Commns., 234, 269-274.
6. Toyokuni, S., Okamoto, K., Yodoi, J., and Hiai, H. (1995) Persistent oxidative stress in cancer. *FEBS Letters*, **358**, 1-3.
7. von Zglinicki, T., Saretzki, G., Docke, W., and Lotze, C. (1995) Mild hyperoxia shortens telomeres and inhibits proliferation of fibroblasts: a model for senescence? Exp. Cell Res. 220, 185-193.

# ACETIC ACID SUPPRESSES THE INCREASE OF GLYCOSIDASE ACTIVITY DURING CULTURE OF CACO-2 INTESTINAL EPITHELIAL CELLS

NOBUMASA OGAWA,* HIDEO SATSU, HIROHITO WATANABE,
MASAHIRO FUKAYA,* YOSHINORI TSUKAMOTO,*
YUSEI MIYAMOTO AND MAKOTO SHIMIZU
*Department of Applied Biological Chemistry*
*The University of Tokyo, Tokyo 113-8657, Japan*
**Nakano Vinegar Co. Ltd., Handa, Aichi 475-8585, Japan*

## Abstract

This research was designed to characterize an inhibitory effect of acetic acid on the activity of glycosidases expressed in Caco-2 cells. Cells were cultured in a medium containing 5 mM acetic acid for 15 days, and meantime the activity of sucrase was monitored. On the 3rd day in culture the activity of sucrase was approximately 3 $\mu$ mol/hr/mg protein. This activity was not significantly different from that in cells cultured without acetic acid (control). Thereafter, in control cells, the activity of sucrase increased and reached 12 $\mu$mol/hr/mg protein on the 15th day. However, the activity of sucrase in acetic acid treated cells barely increased and reached 4.7 $\mu$mol/hr/mg protein on the 15th day. The activity of sucrase was suppressed in a dose dependent manner. We measured the activity of glycosidases and other hydrolases after 15-day treatment with 5 mM acetic acid. The activity of sucrase and maltase was suppressed to the level of 40 to 50% of the control and the activity of trehalase and lactase was drastically decreased to less than 10% of the control. However, the activity of alkaline phosphatase (ALP), aminopeptidase-N (AP-N), dipeptidylpeptidase-IV (DPP-IV) and $\gamma$-glutamyltrans peptidase ($\gamma$-GTP) which are also present in the intestinal brush border membrane was not affected by acetic acid. Acetic acid treatment did not affect either cell growth or cell viability. To understand mechanisms underlying this suppression of glycosidase activity by acetic acid, we performed northern and western analyses of sucrase. Neither mRNA nor protein level of sucrase was decreased by treatment with 5 mM acetic acid for 15 days. These results suggest that acetic acid may suppress the increase of glycosidase activity during the growth of Caco-2 cells by blocking a biological step following transcription and translation, such as trafficking of *de novo* synthesized glycosidases and their incorporation into the brush border membrane.

## 1. Introduction

We often use vinegar as a seasoning when we cook and eat meal to give sour taste. A main ingredient of vinegar is acetic acid. Our preliminary experiment has shown that the concentration of acetic acid reaches approximately 5 mM in the intestinal tract after an oral administration of vinegar at 30 mM in chews. Since the concentration of acetic acid becomes such a high level in the intestine, acetic acid may perturb intestinal functions. Previous studies have shown that oral administration of vinegar for several weeks lowers the increase of blood glucose level after meal (Nakajima and Ebihara, 1988). To elucidate how vinegar decreases the glucose level in blood, we studied effects of acetic acid on glucose uptake with a human colonic carcinoma cell line, Caco-2. This cell line is known to show the nature of differentiated epithelial cells of the small intestine and used as a good experimental tool to study physiological functions of the intestine (Blais et al., 1987). Glucose uptake was not inhibited by acetic acid, indicating that lowering of the blood glucose level by acetic acid is not due to inhibition of glucose transport. Thus, we studied effects of acetic acid on the activity of glycosidases. Acetic acid treatment suppressed the increase of glycosidase activity during cell culture. Our finding suggests that disaccharides in food are not hydrolyzed to monosaccharides in the intestine owing to reduced activity of glycosidases by chronic administration of vinegar. This leads to the decrease of the bioavailability of sugar and subsequently lowering of the blood glucose level. In the present study we characterized the suppression of glycosidase activity by acetic acid using Caco-2 cells.

## 2. Methods and Materials

The composition of a medium used was DMEM supplemented with 10% FBS, 1% non essential amino acid, 2% glutamine, 100 U/ml of penicillin and 100 mg/ml of streptomycin. The medium pH was adjusted to 7.4 with $NaHCO_3$. Cells for study were seeded at a density of 0.14 million cells per well in 24-well microplates coated with collagen and cultured with a medium containing acetic acid. Fresh medium was given every two days. The passage of the cells was the 64th-78th.

A buffer containing a substrate was placed on the monolayer of Caco-2 cells, incubated at 37℃ for a predetermined time and collected to measure the activity of hydrolases. The activity of glycosidases was determined by Dahlqvist's method with a minor modification (Dahlqvist A., 1964). The activity of ALP, AP-N, DPP-IV and $\gamma$-GTP was assayed by using p-nitrophenol phosphate, ala-7-amino-4-methylcoumarin, gly-pro-7-amino-4-methylcoumarin and L-glutamic $\gamma$-p-nitroanilide as substrates, respectively. Cell proliferation was estimated by the increase of cell number and cellular protein determined by Bradford method. The viability of cells was assessed by their ability to reduce alamarBlue (Biosource int., CA, USA) and to maintain LDH inside cells (Wako, Osaka, Japan).

Total RNA extracted from Caco-2 cells was electrophorated on formaldehyde agarose gel and blotted on a nylon membrane. A DNA probe for sucrase-isomaltase was synthesized by PCR and labeled by using $^{32}P$-ATP with klenow polymerase. The DNA of $\beta$-actin was used as an internal control of northern analysis.

For western analysis, cells were harvested from two 10 cm dishes and homogenized in the presence of NP-40. Sucrase was immunoprecipitated with a monoclonal antibody generously provided by Dr. Hauri. The precipitant was subjected to SDS-PAGE electrophoresis and sucrase protein was detected by an ECL method.

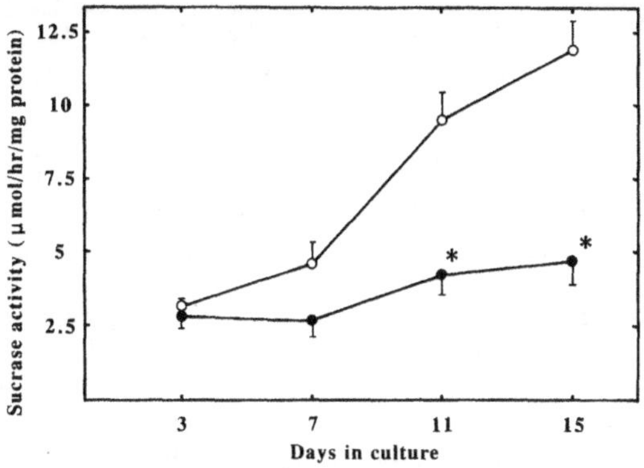


FIGURE 1 Effect of acetic acid on the activity of sucrase in Caco-2 cells. The results are given as means with bars representing S.E.M. for 6 determinations. Control, (○); acetic acid, (●).
* Significantly different from control ($p<0.001$).

## 3. Results and Discussion

Figure 1 shows the increase of sucrase activity in Caco-2 cells during culture. In control cells cultured in acetic acid-free medium, the activity of sucrase was 3.14 ± 0.22 $\mu$mol/hr/mg protein on the 3rd day, increased monotonically and reached 11.90 ± 1.02 $\mu$mol/hr/mg protein on the 15th day. In cells treated with 5 mM acetic acid sucrase activity was 2.80 ± 0.41 $\mu$mol/hr/mg protein on the 3rd dayd. But thereafter the increase of sucrase activity was very slow and barely reached 4.70 ± 0.80 $\mu$mol/hr/mg protein on the 15th day. These data show that acetic acid suppresses the increase of sucrase activity during the growth of Caco-2 cells. When cells were treated with acetic acid of 1, 2.5 and 5 mM for 15 days, sucrase activity decreased to 61 ± 1, 48 ± 2 and 41 ± 1 % of the control, respectively, showing that acetic acid suppresses the activity of sucrase in a dose dependent manner.

We measured the activity of glycosidases and other hydrolases after 15-day treatment with 5 mM acetic acid. As shown in Table, the activity of sucrase and maltase was suppressed to 45 ± 5 and 40 ± 4% of the control, respectively. The activity of trehalase and lactase was drastically decreased to 8 ± 9 and 4 ± 22 % of the control, respectively. However, the activity of ALP, AP-N, DPP-IV and $\gamma$-GTP was not affected by acetic acid. The results indicate that the suppression of increase of glycosidase activity by acetic acid may be specific.

Table Effects of acetic acid on the activity of hydrolases in Caco-2 cells

| Hydrolases | Activity | | |
|---|---|---|---|
| | Control ($\mu$mol/hr/mg) | Acetic acid ($\mu$mol/hr/mg) | % |
| ALP | 0.45 ± 0.04 | 0.51 ± 0.02 | 113 |
| AP-N | 30.23 ± 1.44 | 31.51 ± 1.90 | 104 |
| DPP-IV | 73.67 ± 1.98 | 69.52 ± 2.40 | 94 |
| $\gamma$-GTP | 1.21 ± 0.11 | 1.22 ± 0.13 | 101 |
| Sucrase | 10.20 ± 1.08 | 4.54 ± 0.53 * | 45 |
| Maltase | 55.73 ± 8.10 | 21.40 ± 2.11 * | 40 |
| Lactase | 4.42 ± 0.22 | 0.37 ± 0.39 * | 8 |
| Trehalase | 1.39 ± 0.19 | 0.05 ± 0.30 * | 4 |

Cells were treated with 5 mM acetic acid for 15 days. The data are given as means ± S.E.M. for twelve determinations. * Significantly different from control (P< 0.001).

To examine the effect of acetic acid on the growth of Caco-2 cells, cell number and cellular protein were determined while cells were cultured with 5 mM acetic acid for 15 days. Acetic acid did not give any effect on the increase of cellular protein (protein on the 3rd , 7th, 11th and 15th days: 193 ± 8, 248 ± 7, 291 ± 9 and 291 ± 4 mg/well in control cells, 181 ± 11, 216 ± 6, 270 ± 9 and 290 ± 7 mg/well in acetic acid treated cells, n = 9). The number of cells was 2.7 ± 0.1 and 2.9 ± 0.1 million cells per well (n = 4) in control and acetic acid treated cells, respectively, on the 15th day. Cell viability was estimated by the reduction of alamarBlue and release of LDH after cells were treated with 5 mM acetic acid for 15 days. In cells treated with acetic acid, the value of reduced alamarBlue was 100 ± 6 % (n = 4) when that value of control cells is set at 100%. The percentage of LDH released from over remained in cells was 3.5 ± 0.3 and 2.9 ± 0.4 % (n = 3) in control and acetic acid treated cells, respectively. The data show that acetic acid in the medium does not only disturb the growth of Caco-2 cells but also decrease their viability. The suppression of the increase of glycosidase activity by acetic acid is not due to non-specific damage of cells.

To understand mechanisms underlying the suppression of increase of glycosidase activity in Caco-2 cells cultured with acetic acid, we performed northern and western blot analyses of sucrase. The results are shown in Fig. 2. Cells were cultured with 5 mM acetic acid for 15 days and total RNA was prepared. The message level of sucrase was not decreased in acetic acid treated cells in comparison with control cells (panel A). The panel B shows the results of western analysis. A similar treatment with acetic acid did not decrease the protein level of sucrase expressed in the cells either. These results demonstrate that acetic acid does not inhibit a *de novo* synthesis of glycosidases at either transcription or translation process during cell growth.

Taken together, culturing Caco-2 cells with acetic acid at millimolar concentration causes the suppression of increase of glycosidase activity during cell proliferation. Because glycosidases hydrolyze disaccharides and produce monosaccharides which are efficiently absorbed from the small intestine, this suppression may partially account for lowering of the blood glucose level observed in animals chronically administered with vinegar. In this study we tried to understand mechanisms underlying the suppression of glycosidase activity by acetic acid. Though the data indicates that acetic acid may interfere a biological step following transcription and translation, we fail to point out an exact step and delineate molecular mechanisms of acetic acid inhibition. Hauri et al. (1985) have reported that the transport of glycosidases from the rough surfaced endoplasmic reticulum to the trans-Golgi apparatus was considerably slower than that of peptidases in Caco-2 cells. Therefore, we assume that glycosidases and peptidases utilize different trafficking pathways after protein synthesis and acetic acid may intercept only the pathway for glycosidases. Consequently, glycosidase proteins incorporated into the brush border membrane decrease and their activity becomes small. Further experiments are necessary to shed light upon molecular mechanisms underlying the suppression of glycosidase activity by acetic acid during cell proliferation.

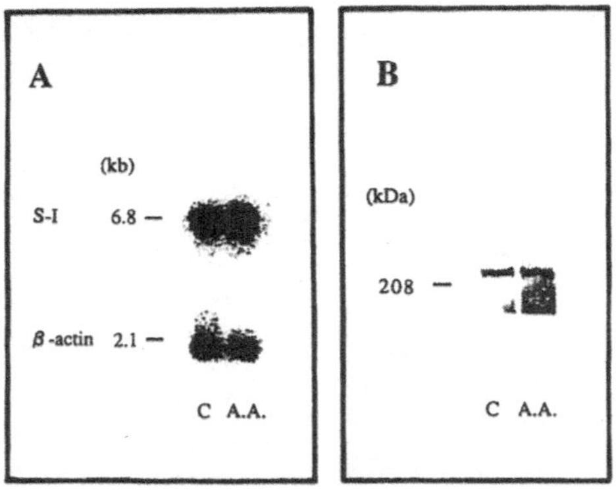


FIGURE 2 Effects of acetic acid on mRNA and protein levels of sucrase. The message for sucrase was detected as mRNA of sucrase-isomaltase complex. C and A.A. represent control and acetic acis, respectively. S-I; sucrase-isomaltase. Panel A, northern blot; panel B, western blot.

## 4. References

Blais, A., Bissonnette, P., & Berteloot, A. (1987) Common characteristics for Na+-dependent sugar transport in Caco-2 cells and human fetal colon. *J. Membrane Biol.* **99**, 113-125.

Dahlqvist, A. (1964) Method for assay of intestinal disaccharidases. *Anal. Biochem.* **7**, 18-25.

Hauri, H-P., Sterchi, E.E., Bienz, D., Fransen, A.M., & Marxer, A. (1985) Expression and intracellular transport of microvillus membrane hydrolases in human intestinal epithelial cells. *J. Cell. Biol.* **101**, 838-851.

Nakajima, A. and Ebihara, K. (1988) Effect of prolonged vinegar feeding on post prandial blood glucose response in rats. *J. JPN. Soc. Nutr. Food Sci.* **41**, 487-489.

# EFFECTS OF DIGESTED SKIM MILK ON THE PROLIFERATION ACTIVITY OF HUMAN ACUTE MYELOID LEUKEMIA (HL-60) CELLS

MOLAY KUMAR ROY, YASUO WATANABE, YOUICHI TAMAI*

*Department of Bioresources, Faculty of Agriculture, Ehime University, Matsuyama, Ehime 790, Japan*

*Corresponding Author
Phone: +81-089-946-9848, Fax: +81-089-977-4364

## 1. Abstract

The bovine skim milk digested with cell free extract of yeast (*Saccharomyces cerevisiae* X2180-1A) was found to exhibit the proliferation inhibition activity toward HL-60 cells. The optimum pH for digestion of skim milk and production of the inhibition factors was pH 4.8. Analysis of this cell proliferation inhibition activity was revealed to be due to the induction of apoptosis which was demonstrated by the formation of apoptotic bodies and fragmentation of DNA in treated cells. The proliferation inhibition factors produced were recovered in the soluble fraction of 92% ethanol, suggesting that the factors are hydrophilic low molecular weight substances derived from skim milk.

## 2. Introduction

Apoptosis, a highly regulated process of cell death, occurs in response to many physiological and biochemical stimuli that activate a series of genetic events and culminate in the death and efficient disposal of a cell (Wyllie, 1993).

Recent research has been showing some food components which may not have any nutritive role but biologically active and can regulate cell growth, differentiation, and apoptosis(Kuo et al., 1996; Sakamoto et al. 1997). Some milk components for example lactoferrin, immunoglobulins, lysozyme, fatty acid, glycoconjugates have been actively investigated over the past two decades specially for antibacterial activity (Hakansson et al., 1995; Fiat et al., 1989; Andersson et al., 1983; Matthews et al., 1976). Recently inhibitory actions of milk components on the proliferation of transformed cell lines have been drawing considerable attention in connection with the prevention of carcinoma. Hakansson et al., 1995 reported multimeric $\alpha$-lactalbumin of human milk induce apoptosis in variety of transformed cell lines where raw bovine milk was reported inactive and no effect on cell viability. However fermented milk has been proved to contain active factors that perform some physiological functions (Tamai et al.,1995; Kuwabara et al., 1995; Yamamoto et al., 1994a). The active factors which are believed to be peptides possibly produced by the proteolytic activity of some fermenting organism. Peptides derived from different milk proteins have also been shown to perform several physiological function. Karaki *et al.* showed tryptic hydrolysate of bovine milk casein exert antihypertensive effects on hypertensive rats. Antibacterial peptides have been identified in pepsin hydrolysed bovine lactoferrin (Dionysius et al. 1997). Previously we studied fermented milk prepared by various lactic acid bacteria and a yeast have proliferation inhibitory activity towards HL-60 cell lines where as a single strain of lactic acid bacteria for example *Bifidobacterium longum* cultured milk was inactive. In this study we investigated bovine skim milk digested with cell free extract of yeast *Saccharomyces cerevesiae* X2180-1A have proliferation inhibitory activity on acute myeloid leukemia (HL- 60) cells.

## 3. Materials and methods

Cell proliferation assay (MTS assay) system was purchased from Promega. RPMI medium, Dulbecco's PBS(-), BL-agar medium, and GAM-broth were purchased from Nissui (Tokyo). Bovine skim milk was from Shikoku Milk Product Co Ltd (Kawauchi, Ehime). Other chemicals used were purest grade available.

### 3.1. PREPARATION OF CELL FREE EXTRACT FROM YEAST AND DIGESTION OF SKIM MILK

Yeast (*S. cerevisiae* X2180 1A) was grown for 36 hours on YEPG (glucose 2%, yeast extracts 1%, peptone 2%) medium at 30°C with shaking. The early stationary phase cells were collected and clear cells (wet weight, 30 gm) were broken by shaking with glass beads (100 gm, diameter 0.3 mm) at 0°C for 20 min with Vibrogen homogenizer. The homogenate was centrifuged at 15000 x g for 10 min and the supernatant was dialyzed at 0°C against distilled water. The resulting precipitates were removed by centrifugation and clear supernatant (protein, 4.5 mg/ml) was used as cell free extract. **Protein content** of the sample was measured by Micro BCA protein assay kit (PIERCE) using bovine serum albumin as a standard. **Skim milk digestion** was conducted through following way: 1% skim milk (5 ml) suspended in distilled water was mixed with lyophilized cell free extracts (4.5 mg) and adjusted pH to 3.6, 4.4, 4.8, 5.2, 5.6, 6.0 with acetic acid and nonadjusted suspension (pH 6.8) and then incubated at 37°C for 3 hours with shaking. After incubation samples were heated at 90°C for 5 min to inactivate enzymes, and readjusted pH at 7.0 with 8N NaOH. The resulting precipitates were removed by centrifugation at 15,000 x g for 10 min and clear supernatant was lyophilized after filtration through a sterilized filter paper (0.45 $\mu$m). The lyophilizate was dissolved in 0.5 ml (1/10 th volume of initial skim milk suspension) RPMI 1640 medium prior to assay cell proliferation activity.

### 3.2. PROLIFERATION INHIBITION ACTIVITY, DNA FRAGMENTATION ASSAY AND MORPHOLOGICAL EXAMINATION OF HL-60 CELLS

Human acute myeloid leukemia (HL-60) cells were obtained from the Japanese Cancer Research Resources Bank (JCRB) and routinely cultured in RPMI-1640 medium as described before (Ito et al., 1998). **Cell proliferation** activity was measured with cell titer 96™ aqueous non-radioactive cell proliferation assay (MTS assay) and by counting viable cells using trypan blue exclusion method with a minimum of 100 cells/field under light microscope. For **DNA fragmentation assay** HL-60 cells (2 x $10^6$) suspended in 9 ml culture medium were added with 1 ml DSM and cultured for 5, 12, 16, 24 and 30 h. The DNA was extracted with ApopLadder Ex™ kit, Takara Biomedicals. Obtained DNA fragments were loaded onto a 1.5% agarose gel in TBE buffer (89 mM Tris, 89 mM boric acid, and 2 mM EDTA) and electrophoresed at constant voltage (100 V). The gel was stained with ethidium bromide and photographed on a UV transilluminator. **Morphological examination** of HL-60 cells incubated for 24 hours in the presence or absence of DSM was observed under a light microscope.

### 3.3. ETHANOL FRACTIONATION OF DIGESTED SKIM MILK

Skim milk (100 ml) digested with lyophilized crude yeast extract (90 mg) at pH 4.8 was heated at 95°C for 5 min then readjusted pH to 7.0 with NaOH. The resulted precipitates were removed by centrifugation. Ice-cooled ethanol was added to the supernatant to make ethanol concentration up to 93% and kept overnight at -20°C. Following centrifugation at 15000 x g for 10 min precipitation (EtOH ppt) was lyophilized and supernatant was evaporated. Obtained materials (EtOH sup) were lyophilized by dissolving it with distilled water. Prior to study both EtOH ppt and EtOH sup were dissolved in 10 ml culture medium.

## 4. Results and Discussions

In the present study we demonstrated skim milk digested with crude yeast enzyme inhibit cell proliferation of HL-60 cells and the morphological changes for example shrinkage, membrane belbing, formation of apoptotic bodies are compatible with the occurrence of apoptosis. Prior to cell proliferation assay we measured optimum digestion condition for active materials production from skim milk with yeast extract by conducting digestion at different pH up to 12 hours at 37°C. During digestion sample was taken at different time intervals and absorbance at 240 nm of 5% TCA soluble fraction was measured and highest value was found when digestion was conducted at pH 4.8 for 3 hours (data not shown). The result is correlated with the activity as the materials formed at pH 4.8 inhibited cell proliferation of HL-60 cells most extensively (fig-1) measured by MTS assay. In a separate study we found crude enzyme itself have no inhibitory effect on cell proliferation where as nondigested skim milk (NDSM) exhibited minor inhibitory effect on proliferation activity (data not shown). The results suggest cell proliferation inhibition factor latently present in skim milk, and yeast enzyme(s) can release the factors most specifically at pH 4.8. These factors are usually considered low molecular weight hydrophilic compounds as the materials were recovered from DSM in 92% ethanol and the insoluble materials were found inactive (data not shown). While investigating cell proliferation by counting viable cells we found number of dead cells linearly increased within 24 hr. incubation period (fig-2) in DSM treated HL-60 cells. Whereas proliferation of human normal dermal fibroblast (Fb) cells was not affected by DSM (data not shown).

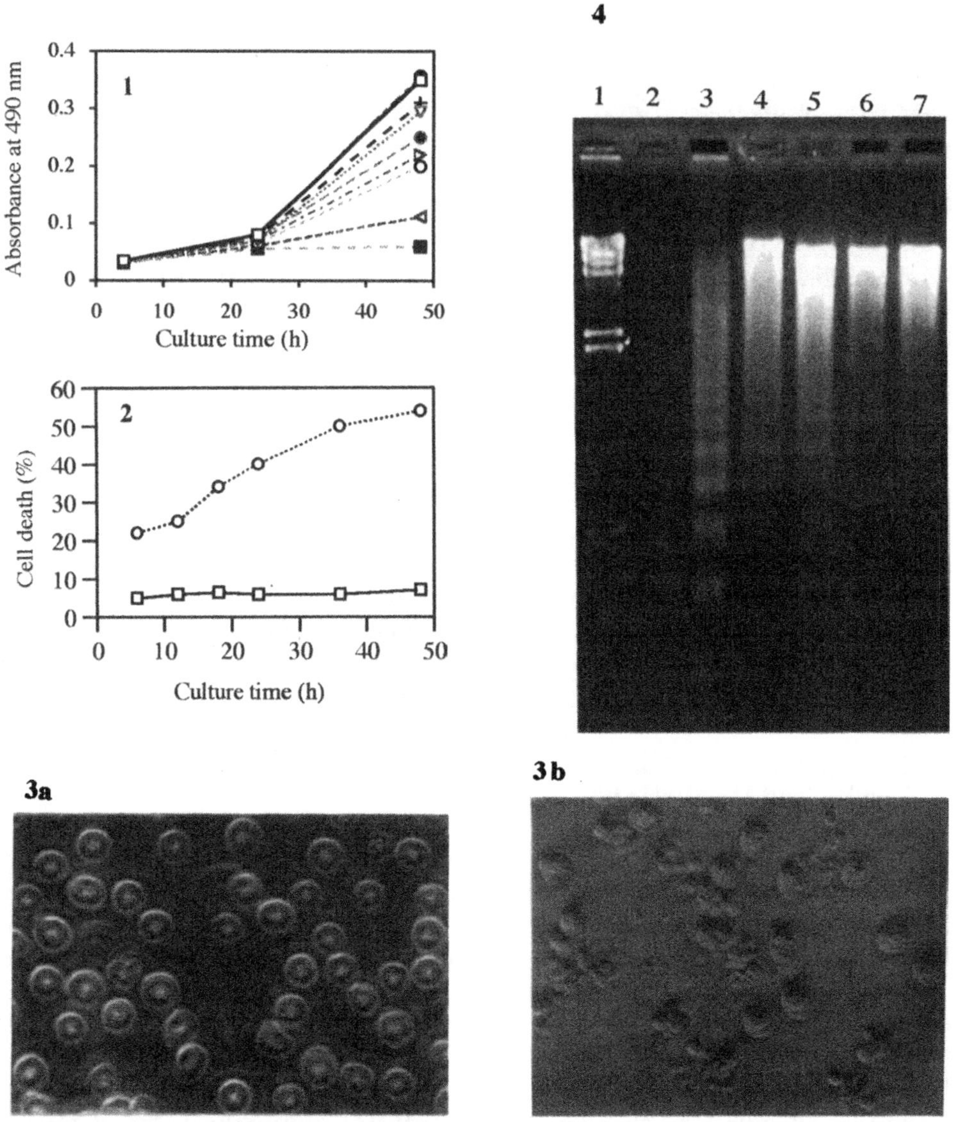


**Fig: Characterization of DSM induced proliferation inhibition in HL-60 cells. 1.Effects of DSM prepared at different pH (□, control; ∇, pH 3.6; O, pH 4.0; ◁, pH 4.4; ■, pH 4.8; ▷, pH 5.2; ● pH 5.6; ⊕, pH 6.8; +, pH** 6.0) in the formation of formazan (MTS assay), 2. Time course analysis of DSM induced cell death (O, in the presence of DSM; □ in the absence of DSM), 3a. Non treated Cells, 3b. DSM treated cells and 4. Agarose Gel electrophoresis of DNA extracted from HL-60 cells (lane 1, molecular size marker; lane 2, non-treated cells cultured for 12 h; lane 3, cells cultured with actionmycin D for 5 h; lane 4-7, cells cultured with DSM for 5 h, 12 h, 16 h and 20 h respectively.

Morphological examination under light microscope of the treated cells (fig-3a) apparently shows typical apoptotic characteristics such as cell shrinkage, membrane belbing whereas nontreated cells maintained their normal morphology (fig-3b) during investigation period. Agarose gel electrophoresis of the DNA of DSM treated HL-60 cells showed a ladder like pattern of DNA fragments consisting multiples of around 200 base pairs (Fig-4). The results provide us to consider DSM contains active materials that can inhibit cell proliferation activity as well as can induce apoptosis in HL-60 cell lines. Whereas NDSM exerts minor effects on cell proliferation of HL-60 cell lines. So the active fragments, derived from skim milk by yeast enzymes are relatively inactive within the structure of precursors molecules of bovine skim milk. During study we observed DNA fragmentation (ladder DNA) was occurred within 5 hours after incubation of HL-60 cells with DSM. But proliferation inhibition was not clearly found within this time. We consider this behavior as DSM earlierly collapse chromatin structure of HL-60 cells and then diminish the activity of dehydrogenase enzymes.

Physiological properties of milk have been elucidating from long before and in most cases' milk proteins or the peptides liberated from were found active. Peptides derived from bovine casein, a main milk protein by hydrolyzing with trypsin or protenase of *Lactobacillus helveticus* has been shown to exert various biological activities (Chiba & Arai 1988, Yamamoto *et al.*, 1994). *S. cerevisiae* is used widely for the production of fermented milk product in food industries. So the interest is in the enzymes of *S. cerevisiae* that can release these factors from bovine milk proteins. Their carbohydrae fermenting activity is well known but proteolytic activity e.g. protein degradation ability during fermentation not yet fully understood. Yeast vacuole, a lysozome like organism contains several kinds of protease. Protease A, an aspartyl endoprotenase active towards azocasein, azocol, hemoglobin and shows optimum activity around pH 5.0 (Yokosawa *et al.*, 1983). Protease B is a serine endoprotenase with a pH optimum near neutrality has molecular weight range from 31000 to 44000 (Fugishiro *et al.*, 1980; Huse *et al.*, 1982). Protease B is also active towards azocasein, azocol, hemoglobin. Several other proteases such as aminopeptidases, carboxypeptidses have also been identified in *S. cerevisiae*. The specificity of these enzymes towards milk proteins not yet investigated properly. We are trying to identify the enzymes particularly involved in the process in addition to isolate the active components from digested skim milk.

## 6. References


Andersson, B., Dahmen, J., Frejd, T., Leffler, H., Mugnusson,G., Noori, G., and Savanborg, C. (1983). *J. Exp. Med.* 158, 559-570.

Chiba, H. and Arai, S. (1988). Kagakuto Seibutsu, **26**, 34-40.

Dionysius, D. A. and Milne, J.M. (1997). *J. Dairy Sci.* **80**, 667-674.

Fujishiro, K., Sanada, Y., Tanaka, H. and Katunuma, N., (1980). *J. Biochem.* **87**, 1321-1326

Fiat, A. M., and Jolles, P. (1989). *Mol. Cell. Biochem.* **87**. 5-30.

Hakansson, A., Zhivotovsky, B., Orrenius, S. and Sabharwl, H. and Svanborg, C. (1995). *Proc. Natl. Acad. Sci. USA*, **92**, 8064-8068.

Huse, K., Kopperschlager, G. and Hofman, E. (1982). Acta.biol. med. germ., **41**, 991-1002.

Ito, M., Roy, M.K., Kurihara, K., Watanabe, Y. and Tamai, Y. (1998). *Food Sci. Technol. Int. Tokyo.* **4**, 125-129.

Karaki, h., Doi, K., Sugano, S., Uchiwa, H., Sugai, R., Murakami, U., Takemoto, S. (1990). *Comp. Biochem. Physiol. C.* 96, 367-371.

Kuo, M.L., Huang, T.S. and Lin, J.K. (1996), *Biochim. Biophys. Acta* **1317**, 95-100.

Kuwabara, Y., Nagai, S., Yoshimitsu, N., Nakagawa, I., Watanabe, Y. and Tamai, Y. (1995). *J. Ferment.. Bioeng.*, **80**, 294-295.

Matthews, T.H.J., Nair, C.D.G., Lawrence, M.K. and Tyrell, D.A.J. (1976). *Lancet* 25, 1387-1389.

Migliore- Samour, D., Floc'h, F. andJolles,P. (1989). *J.DairyRes.***56**,357-362.

Sakamoto, K., Lawson, L.D. and Milner, J.A. (1997), *Nutr. Cancer* **29**, 152-156.

Tamai, Y., Oishi, H., Nakagawa, I., Watanabe, Y., Shinmoto, H., Kuwabara, Y., Yamato, K. and Nagai, S. (1995), *Nipp. Shoku. Ka. Ko. Kai.* **42**, 383-387.

Wyllie, A. H. (1993). *Br. J. Cancer* **67**, 205-208.

Yamamoto, N., Akino, A. and Takano, T. (1994a). *Biosci. Biotech. Biochem.* **58**, 776-778,

Yamamoto, N., Akino, A. and Takano, T. (1994b). *J. Dairy Sci.* **77**, 917-922..

Yokosawa, H. , Ito, H., Murata, S., and Ishii, S. (1983). *Anal. Biochem.* **34**, 210-215.

# DIFFERENTIATION-INDUCING ACTIVITY OF HUMAN LEUKEMIA CELL LINES BY EXTRACTS OF WILD PLANTS, SEAWEED AND MUSHROOMS IN AKITA

KEISHI HATA, KYOKO ISHIKAWA, KAZUYUKI HORI
*Akita Research Institute of Food & Brewing (ARIF)*
*4-26 Sanuki, Araya-machi, Akita 010-1623, Japan*

The differentiation-inducing activity of 38 wild plants, 28 types of seaweed and 6 mushrooms was tested using the human leukemia cell line, HL60. Seven water, 6 hot water, 11 ethanol and 23 methanol extracts were found to differentiate the cells. After incubation with extracts of *Cirsium japonicum*, *Laportea macrostacha*, *Osmunda japonica*, *Symphytum officianale Solieria mollis* and *Undaria pinnatifia*, the differentiation of over 30 % of HL60 cells was induced. Furthermore, HL60 cells differentiated into mature granulocytes following treatment with an aqueous extract of *O. japonica* at 25-50 μg/ml, as demonstrated by light microscopic examination .

## 1. Introduction

The human leukemia cell line, HL60 differentiates into mature granulocytes or monocyte/macrophages following treatment with vitamins, cytokines and other chemical compounds (Collins, 1987). By monitoring cellular properties such as nitroblue tetrazolium (NBT) reduction activity, the potential of test agents to induce cell differentiation can be readily measured. Recently, plant extracts were screened for their cell differentiation activity and the anticancer compounds were isolated from some plants and crude drugs using these cells or other leukemia cells (Mazur etal.,1996, Suh etal.,1995, Umehara etal.,1996).

In Japan, wild plants, seaweed and mushrooms have been utilized as the traditional food material after food-processing. Some plants, such as *Glehnia littoralis*, have been recognized as " Kanpou" in Japan and China. We examined the biological activity in such food materials in Akita. We previously investigated the differentiation-inducing activity of cells induced by extracts of 22 edible wild plants (Hata etal., 1998). We report here, the results of screening for the induction of HL60 cell differentiation by extracts of 72 plants and mushrooms and the effects of extracts from *O. japonica* on the cell differentiation .

## 2. Materials and Methods

### 2.1. MATERIALS

Wild plants and mushrooms from Akita were collected between 1996 and 1998 and the seaweed samples were a generous gift from Akita Fishery Corporative. All-*trans*-retinoic acid and nitroblue tetrazolium were purchased from Wako Pure Chemical Industry, α-naphthol AS-D-chloroacetate and α-naphthylbutyrate were purchased from Sigma Chemical Co., Ltd and RPMI-1640 medium was purchased from Nissui Pharmaceutical Co., Ltd.

### 2.2. CELL CULTURE

The human leukemia cell line, HL60 was provided by Institute for Fermentation, Osaka. The cells were maintained in RPMI-1640 medium supplemented 10 % heat-inactivated fetal bovine serum (EQUITECH-BIO, INC.), 100 units of penicillin /ml and 100 μg of streptomycin /ml in a 5 % $CO_2$ humidified atmosphere at 37°C.

### 2.3. DIFFERENTIATION-INDUCING ASSAY

Phosphate buffered saline (PBS), hot PBS, ethanol and methanol extracts were prepared for 72 plant and mushroom samples. In brief, the samples (10 g) were homogenized in 3 volumes of the solvent and the extract was recovered by centrifugation. NBT assay and esterase activity of the cells were measured as previously described (Hata etal., 1998).

### 2.4. EFFECT OF AQUEOUS MATERIALS OF *O. JAPONICA* ON HL60 CELLS

*O. japonica* (500 g) was homogenized in 1 L of distilled water and the supernatant was collected by centrifugation. The water soluble materials were lyophilized (9.5 g). HL60 cells were incubated with the materials for 72 h at 10 - 100 μg/ ml. After the incubation period, the morphology, NBT reduction activity and esterase activity of the cells were examined.

## 3. Result and Discussion

The effects of 72 plant and mushroom materials (x 4 solvents) on HL60 cells

TABLE 1. Summary of differntiation-inducing activity of HL60 cells by extracts of wild plants and seaweed

| Sientific name (Japanese name) | Percentages of NBT-positive cells Water | Hot water | EtOH | MeOH |
|---|---|---|---|---|
| **Wild plants** | | | | |
| *Anemone flaccia* (Nirinsou) | - | - | 13.1 | 17.4 |
| *Angelia ursina* (Ezonyuu) | - | - | - | 15.3 |
| *Anthricus sylvestris* (Shyaku) | - | - | - | - |
| *Allium grayi* (Nobiru) | - | - | - | 25.9 |
| *Aralia cordata* (Udo) | - | - | 21.3 | - |
| *Aralia elta* (Taranoki) | - | - | 23.0 | - |
| *Asiasarum sieboldii* (Usubasaishin) | 21.5 | 11.2 | - | 25.3 |
| *Astilbe thunbergii* (Toriashishyouma) | - | - | - | 29.2 |
| *Brasenia shreberi* (Jyunsai) | - | - | - | 11.6 |
| *Cirsium japonicum* (Noazami) | - | - | 30.0 | 13.4 |
| *Cirsium purpuratum* (Fujiazami) | - | - | 19.9 | - |
| *Equisetum arvense* (Sugina) | - | - | 26.7 | 19.6 |
| *Eutrema wasabi* (Mizuwasabi) | - | - | 20.8 | 11.4 |
| *Hemerocallis fulva* (Yabukanzou) | - | - | 21.7 | 12.1 |
| *Laportea macrostacha* (Miyamairakusa) | 28.7 | 10.3 | 31.4 | 16.5 |
| *Matteuccia struthiopteris* (Kusasotetsu) | - | - | 25.0 | - |
| *Osumunda japonica* (Zenmai) | 34.0 | 12.7 | 28.7 | 31.4 |
| *Pilea hamaoi* (Mizu) | - | - | 25.3 | - |
| *Pueraria lobata* (Kuzu) | 14.2 | 11.3 | - | 11.3 |
| *Sambucus sieboldiana* (Niwatoko) | 26.0 | - | 22.7 | - |
| *Smilax riparia* (Shiode) | - | - | 30.7 | - |
| *Symphytum officianale* (Konfrii) | - | 11.2 | 14.6 | 30.0 |
| **Seaweed** | | | | |
| *Ceramium kondai* (Igisu) | - | - | - | 23.0 |
| *Chondrus ocellatus* (Tsunomata) | - | - | - | 14.6 |
| *Codium fragile* (Miru) | - | - | - | 11.7 |
| *Gelidium amansii* (Makusa) | 13.2 | - | 13.0 | 23.3 |
| *Gracilaria verrucosa* (Ogonori) | - | - | - | 18.5 |
| *Grateloupia okamurai* (Kyounohimo) | - | - | - | 11.0 |
| *Gymnogongrus flabelliformis* (Okitsunori) | - | - | 12.3 | - |
| *Lomentaria hakodatensis* (Fushitsunagi) | - | - | 10.0 | 18.1 |
| *Solieria mollis* (Hosobamirin) | 45.5 | 39.5 | - | 33.6 |
| *Undaria pinnatifida* (Wakame) | - | - | - | 38.3 |

differentiation were examined. The results are summarized in TABLE 1.

NBT reduction activity of HL60 cells was induced by treatment with 7 water and 5 hot water extracts. The induction was decreased or inactivated by heat treatment except with the extract of *Solieria mollis*. The findings suggested that the active substances were unstable under this condition.

After incubation with various concentrations of the aqueous materials of *O. japonica* for 72 h, the morphological and histochemical properties of HL60 cells were investigated. HL60 cells were mainly composed of promyelocytes (Figure 1. left) and some mature granulocytes. The cells were induced to differentiate in vitro to a several cell types. The induced cells exhibited the appearance of mature granulocytes as well as cells treated by retinoic acid (Figure 1, right).

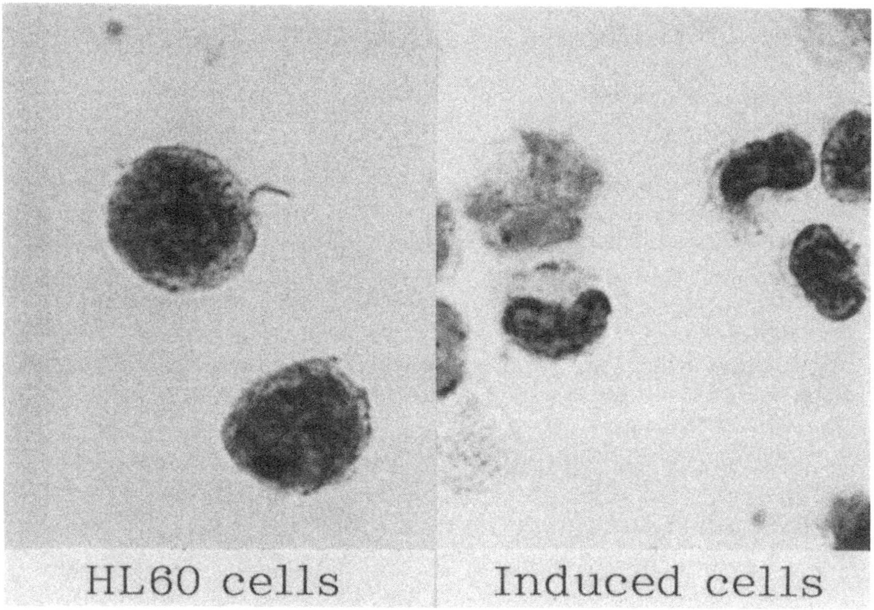


*Figure 1.* Morphological change of HL60 cells treated with aqueous materials of *O. japonica*. Control HL60 cells (left) and induced cells incubated with the materials at 50 μg/ ml for 72 h (right).

Concerning NBT reduction ability, 36 % of HL60 cells were induced by treatment for 72 h at 50 μg/ml. The percentage of the induced cells at this concentration was lower than that by retinoic acid (> 70 %). Over 50 μg/ml, the ability could not be detected by cytotoxic effects (Figure 2).

To confirm the types of the induced cells, the esterase activities expressed in the cells were examined. Usually, α-naphthol AS-D chloroacetate esterase is expressed in

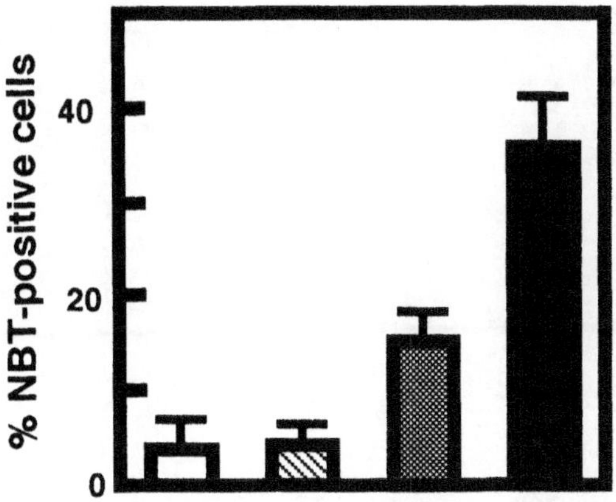


*Figure 2.* NBT reduction activity of HL60 cells treated with the water soluble materials of *O. japonica*. HL60 cells were incubated with none ( ), 10 μg/ml ( ) , 25μg/ml ( ) or 50 μg/ml ( ) for 72 h.

HL60 cells and nonspecific esterase activity can not be detected. Following treatment with the samples, the induction of nonspecific esterase was not observed. The result indicated that the aqueous extracts differentiated HL60 cells into mature granulocytes.

## 4. References

Collins, S.J. (1987) The HL-60 promyelocytic leukemia cell line: Proliferation, differentiation, and cellular oncogene expression, *Blood* **70**, 1233 -1244.

Hata, K., Ishikawa, K. and Hori, K. (1998) Differentiation-inducing activities of human leukemia cell line (HL60) by extracts of edible wild plants in Akita, *Nat. Med.* **52**, 269 -272.

Mazur, X., Becker, U., Anke, T. and Sterner, O. (1996) Two new bioactive diterpenes from Lepista scordida, *Phytochemistry* **41**, 405 -407.

Suh, N., Luyenjgi, L., Fong, H.H.S., Kinghorn, A.D. and Pezzuto, J.M. (1995) Discovery of natural product chemopreventive agents utilizing HL-60 cell differentiation, *Anticancer Res.* **15**, 233 -239.

Umehara, K., Nakamura, M, Miyase, T., Kuroyanagi, M. and Ueno, A (1996) Study on differentiation inducers VI. Lignan derivative from *Arcium fructus*, *Chem. Pharm. Bull* **44**, 2300 -2304.

## ANTI-TUMORIGENIC PROTEIN FROM *ARALIA ELATA*

*The Induction of Apoptosis in Transformed Cells*

MAKOTO TOMATSU, KYOKO ISHIKAWA& NORIO SHIBAMOTO
*Akita Research Institute of Food and Brewing*
*4-26, Sanuki, Arayamachi, Akita 010-1623, Japan*

We purified a novel anti-tumorigenic protein, araliatumoricin (ATM), from shoot buds of *Aralia elata*. ATM is a tetrameric protein that consists of two A-chains (33kDa) and two B-chains (36kDa) and is selectively cytotoxic against human transformed cells. In particular, the *N*-terminal sequence (25 amino acid residues) of the B-chain showed 52% similarity with that of the B-chain of ricin, a toxic lectin isolated from caster beans known to induce apoptosis. ATM also exhibited lectin activity, which was inhibited by lactose and galactose. Due to the presence of ATM, the TUNEL reaction was positive in VA13 cells and DNA fragmentation was observed in HL60 cells, indicating that ATM also induces apoptosis particularly on SV40-transformed cells.

### 1. Introduction

Many growth-inhibitors of human tumor cells have been isolated from plants. Among them, diallyl disulfide which was isolated from garlic oil increases intracellular free calcium levels and Ca-ATPase activity and simultaneously induces apoptosis of human tumor cells (Sundaram& Milner, 1996). These compounds were found to be effective against several tumor cells, although the susceptibility to normal cells was not examined.

In northeast Asian countries, plants belonging to the Araliaceae family are widely cultivated as ingredients of medical herbs. Particularly in Japan and China, bark and root cortex of *Aralia elata* have been traditionally collected for medicines with tonic, anti-arthritic, and anti-diabetic activities. In addition, shoot buds of *Aralia elata* (called "Taranome" in Japan), are a popular garnish especially in the spring in Japan. Various compounds of relatively-low molecular weight with interesting biological activities have been isolated from this family: two diterpenes from *Aralia cordate* with anti-tumor activities (Ryu *et al.*, 1996) and saponins and elatosides from *Aralia elata* with blood-sugar lowering and hypoglycemic activities (Yoshikawa *et al.*, 1995), respectively.

In this paper, we present the physical and biological properties of the novel anti-tumorigenic protein ATM with particular reference to induction of apoptosis in transformed cells.

## 2. Materials and Methods

### 2.1. Cell culture

WI38 (human lung) and VA13 (SV40-transformed WI38) were obtained from the RIKEN cell bank; Flow2000 (human lung), Hs68 (human foreskin), and HL60 (human lymphocyte) were obtained from the Institute for Fermentation, Osaka; and MKN45 (human stomach cancer) and MIA PaCa2 (human pancreatic cancer) were obtained from the Health Science Research Resources Bank. All cell lines except for HL60 were cultured in Eagle's minimal essential medium (MEM) supplemented with 10% fetal bovine serum (FBS), penicillin (100μg/ml), and streptomycin (100μg/ml) at 37℃ under 5% $CO_2$ in a humidified atmosphere. HL60 cells were cultured in RPMI-1640 in place of MEM under the same conditions described above.

### 2.2. Assay of selective cytotoxicity

Selective cytotoxicity was estimated by observing viable cells of VA13 and WI38 cells in the presence and absence of test samples using the WST-1 assay. Specifically, using 96-well plates, cell suspensions (100μl) were pre-incubated under the conditions described above for 24 hours at a concentration of $1.0 \times 10^4$ cells/ml. After the addition of the sample solution (10μl), incubation was performed for 44 hours. Then, the WST-1 solution (10μl) was added, and incubation was continued for an additional 4 hours. Subsequently, the absorbance at 450nm was measured with that at 655 nm as a reference, and the viability index (VI) was calculated by the following equation;

$$\text{Viability index (\%)} = (A - B) / A \times 100$$

where A and B were the absorbance in the absence and presence of samples, respectively. In the primary screening, we selected samples whose VIs were less than 10 on VA13 cells and greater than 50 on WI38 cells.

2.3. Purification of ATM

All the procedures described below were performed at 4℃. Shoot buds of *Aralia elata* were homogenized with five volumes of buffer A [0.1M Tris-HCl(pH7.4), 1mM EDTA] in a MAXIM homogenizer (Nihon Seiki Co., Tokyo). The homogenate was centrifuged at 8,700 × g for 30 min, and ammonium sulfate was slowly added onto the supernatant with stirring to bring it to 80% saturation. The mixture was allowed to stand for 60 min and centrifuged again at 8,700 × g for 30 min. The precipitate was dissolved in a small volume of buffer B [50mM Tris-HCl (pH 7.4), 1mM EDTA], and dialysis was performed against the same buffer using a Stirred Cell 8400 and a YM10 membrane (Amicon Co.). The dialyzed protein solution was loaded onto a QAE-Toyopearl 550C (Tosoh Co.) column (4 × 33 cm), previously equilibrated with buffer B. After thorough washing with the same buffer, an elution was made with a linear gradient of NaCl (2,000 ml) from 0 to 500 mM at 1ml/min. Subsequently, the collected active fractions were pooled and concentrated using the Stirred Cell 8400 (YM10 membrane), followed by dialysis against buffer B. The resulting protein solution was analyzed by native PAGE, and ATM was electroeluted from the separated band with a Maxyield-NP (Atto Co.). The eluted protein solution was concentrated and loaded onto a TSK-GEL G3000 SWXL column (Tosoh Co.), previously equilibrated with buffer C [0.1M Phosphate buffer (pH7.4), 0.1M $Na_2SO_4$] and eluted with the same buffer at 0.5ml/min. Finally, the active fractions were pooled and dialyzed against 10mM Tris-HCl (pH7.4) .

2.4. Molecular weight determination

The purified protein sample was analyzed by 5-20% gradient gel electrophoresis for 10 hours with a NPG520L gel (Atto Co.) to determine the native molecular weight using molecular markers (Pharmacia). The sample was also analyzed by SDS-PAGE (12.5% gel) to determine the size of the subunits.

2.5. Amino acid sequence analysis

The purified ATM was transferred onto a PVDF membrane, and the *N*-terminal amino acid sequence of A- and B-chains was analyzed with a Shimazu PPSQ-10 protein sequencer. The protein sequence was examined using databases of FASTA software and the AFFRC server (http://www. dna.affrc.go.jp).

2.6. Other analytical methods

Apoptosis was determined by the DNA fragmentation assay using the Apoptosis Ladder Detection Kit (Wako Co.) and TUNEL reaction using the Apoptosis *in situ*

Detection Kit (Wako Co.). The amount of protein was determined by the method of Lowry with bovine serum albumin as a standard. Lectin activity was measured by the observation of cytoagglutination of red blood cells.

## 3. Results and Discussion

During the course of screening, we found strong selective cytotoxicity in the water-extracted phase of shoot buds of *Aralia elata*. After a series of chromatographic purifying procedures detailed in the preceding section, we obtained 0.5 mg of ATM from approximately 500 g of the shoot buds. Gradient gel electrophoresis revealed that ATM is 140 kDa in molecular weight, and SDS-PAGE analysis demonstrated that the ATM is made up of two subunits: A-chain (33kDa) and B-chain (36kDa). Simultaneously, density analysis of these two bands on SDS-PAGE indicated that the ratio of A- to B-chains was almost 1:1. ATM, therefore, is a tetrameric protein with two A-chains and two B-chains.

The *N*-terminal sequences of the B-chain showed 52 % similarity towards that of the B-chain of ricin, a toxic lectin isolated from caster beans (*Ricinus communis*), in the first 25 amino acid residues. Ricin is a hetero-dimeric protein with A- and B-chains. However, no distinct homology was observed between A-chains of ATM and ricin in their *N*-termini.

The 50% growth inhibitory values ($LD_{50}$) of ATM on VA13 and WI38 were about 1 and 10ng/ml, respectively. The selective cytotoxicity of ATM was evaluated using tumor cell lines such as MKN45 and MIA PaCa2 and normal cell lines such as Flow2000 and Hs68 cells. Significant growth inhibition was observed in only the two tumor cell lines. The $LD_{50}$ values on MKN45 and MIA PaCa2 were 7 and 15ng/ml, respectively, while the values on both the two normal cell lines were over 100ng/ml. These results indicated that ATM is selectively toxic to tumor cells.

During the cytotoxic studies, we observed that in the presence of ATM, 24-hour cultivated VA13 cells formed membrane blebbing, which was indicative of apoptosis. The effect of ATM on apoptosis, therefore, was studied by the TUNEL reaction in VA13 cells. As shown in Figure 1, the number of positive cells increased with increasing concentrations of ATM. In addition, DNA fragmentation was induced by ATM treatment in HL60 cells. These results strongly suggested that ATM induced apoptosis in human tumor cells.

As described above, the *N*-terminus of the ATM B-chain was homologous to the B-chain of ricin. Since ricin is a representative lectin, we also studied lectin activity on ATM. As a result, we confirmed that ATM caused cytoagglutination of red blood cells and that the reaction was inhibited by lactose and galactose. ATM, therefore, belongs to the lectin family and is recognized by lactose and galactose. It should be noted that ricin was reported to induce apoptosis on tumor cells (Kim *et al.*,1993; Qda *et al.*, 1997).

In conclusion, ATM is the first anti-tumorigenic protein isolated from the shoots of *Aralia elata*, although several bioactive compounds of relatively-low molecular weight have already been isolated from this plant family. ATM belongs to the lectin family, and structurally, ATM is a 140kDa tetrameric protein with two A-chains and two B-chains. Of the lectins, ATM shares a strong structural homology with ricin in their N-terminal domains of their B-chains. Lectin activities of ATM and ricin recognize lactose and galactose, and furthermore, the two lectins induce apoptosis. The correlation between ATM and ricin in their detailed structures and bioactivities should be investigated in the future.

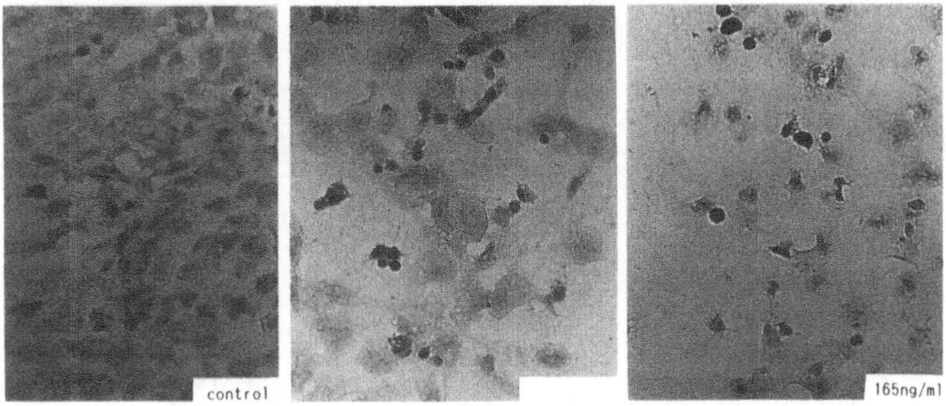


*Figure 1.* Apoptosis detected in ATM-treated VA13 cells by TUNEL-reaction.

## 4. References

Kim,M., Rao,M.V., Tweardy,D.J., Prakash,M., Galili,U., and Gorelik,E.(1993) Lectin-induced apoptosis of tumor cells, *Glycobiology* **3**, 447-453.

Oda,T., Komatsu,N. and Muramatsu,T. (1997) Cell lysis induced by ricin D and ricin E in various cell lines, *Biosci. Biotech. Biochem.* **61**, 291-297.

Ryu,S.Y., Ahn,J.W., Han,Y.N., Han,B.H., and Kim,H. (1996) *In vitro* antitumor activity of diterpenes from Aralia cordate, *Arch. Pharm. Res.* **19**, 77-78.

Sundaram,S.G., and Milner,J.A.(1996) Diallyl disulfide induces apoptosis of human colon tumor cells, *Carcinogenesis* **17**, 669-673.

Yoshikawa,M., Yoshizumi,S., Ueno,T., Matsuda,H., Murakami,T., Yamahara,J., and Murakami,N.(1995) Medicinal foodstuffs. I . Hypoglycemic constituents from a garnish foodstuff "Taranome", shoot buds of *Aralia elata* Seem:elatosides G, H, I, J, and K, *Chem. Pharm.Bull.* **43**, 1878-1882.

# SUPPRESSION OF UV DAMAGE-INDUCED APOPTOSIS OF HUMAN MELANOMA CELLS BY A FERMENTED MILK, KEFIR

Junko Narisawa[1], Tsutomu Nagira[1], Ken-ichi Kusumoto[1], Kiichiro Teruya[1], Yoshinori Katakura[1], David W. Barnes[2], Sennosuke Tokumaru[3], Sanetaka Shirahata[1]
[1]*Laboratory of Cellular Regulation Technology, Graduate School of Genetic Resources Technology, Kyushu University, Hakozaki, Higashi-ku, Fukuoka 812-8581, Japan;* [2]*American Type Culture Collection, Manassas VA 20110, USA;* [3]*Nihon Kefir Co. Ltd., Asahi-cho, Fujisawa 251-0054, Japan.*

## ABSTRACT

Kefran-Kefir powder was suspended in distilled water and centrifuged to remove insoluble fraction. The supernatant was sterilized by filtration, and named Kefir extract. Human melanoma HMV-1 cells were UV-irradiated, and the DNA damage-induced apoptosis and thymine dimer formation were examined. The Kefir extract remarkably suppressed UV irradiation-induced apoptosis. This effect was observed in all cases of Kefir-treatment that was done before, during, and after UV irradiation. Furthermore, we examined the effect of Kefir on UV-induced thymine dimer formation, which is the main cause of cell death by UV irradiation. UV-induced thymine dimer formation was suppressed by Kefir-treatment before and during UV irradiation. Kefir-treatment after UV irradiation resulted in rapid decrease of the amount of thymine dimer. In conclusion, these results demonstrate that Kefir components enhance repair of UV-damaged DNA and protect DNA from UV-damage.

## 1. Introduction

It has been known that ultraviolet ray (UV) produces free radicals or reactive oxygen species in a cell, which damage chromosomal DNA. This damage results in genetic mutation, cell death or cellular carcinogenesis. For human skin, UV-irradiation may cause cell death or precancerous lesions. Kefir is the fermented milk originating from the Caucasus mountains and is known to have numerous benefits. It has been demonstrated that sphingomyeline derived from Kefir augmented the interferon-β production in cells in which virus infection is mimicked by stimulation with poly I: poly C, demonstrating that Kefir-derived components can activate the immune system to avoid virus infection [1, 2]. Here we report that Kefir can suppress apoptosis of human melanoma cells caused by UV damage.

## 2. Materials and Methods

### 2.1. Cells

A human melanoma cell line HMV-1 cell was provided by Japanese Cancer Research Resources Bank (JCRB). This cell line is deficient to produce melanin pigment and relatively resistant to irradiation.

### 2.2. Preparation of a Kefir extract sample

Kefran-Kefir powder was obtained from Nihon Kefir Co. Ltd. The power was suspended in distilled water and centrifuged to remove insoluble fraction. The supernatant was sterilized by filtration, then mixed with F12 medium. We used it as a Kefran–Kefir extract sample.

### 2.3. Kefir extract treatment

Treatment of human melanoma HMV-1 cells with the Kefir extract before UV irradiation; after incubation HMV-1 cells with culture medium including Kefir for 5 hours, we aspirated medium, washed cells, and charged medium to fresh one. UV irradiation was performed by UVC at the dose of $600J/m^2$ for a minute to HMV-1 cells. Treatment of HMV-1 with Kefir during UV; Kefir was added to medium only during UV irradiation. Treatment HMV-1 cells with Kefir after UV; after UV irradiation, HMV-1 cells were incubated with medium including Kefir for 5 hours.

### 2.4. Detection of apoptosis

We used Apoptosis Detection System, Fluorescein Kit (Promega) to detect apoptosis. Apoptic cell was detected by TUNEL method, which detects 3'OH end of chromosomal DNA breakage by apoptosis [3].

2.5. Detection of thymine dimer

We detected thymine dimer formed on chromosomal DNA by mouse monoclonal anti-thymine dimer antibody (KYOWA) [4, 5]. Fixed cells were treated with the antibody, and labeled with FITC-conjugated anti-mouse IgG antibody.

## 3. Results and Discussion

3.1. Suppressive effect against apoptosis

Kefir remarkably suppressed UV irradiation-induced apoptosis. This effect was observed in all cases of Kefir-treatment that was done before, during, and after UV irradiation. (Fig.1)

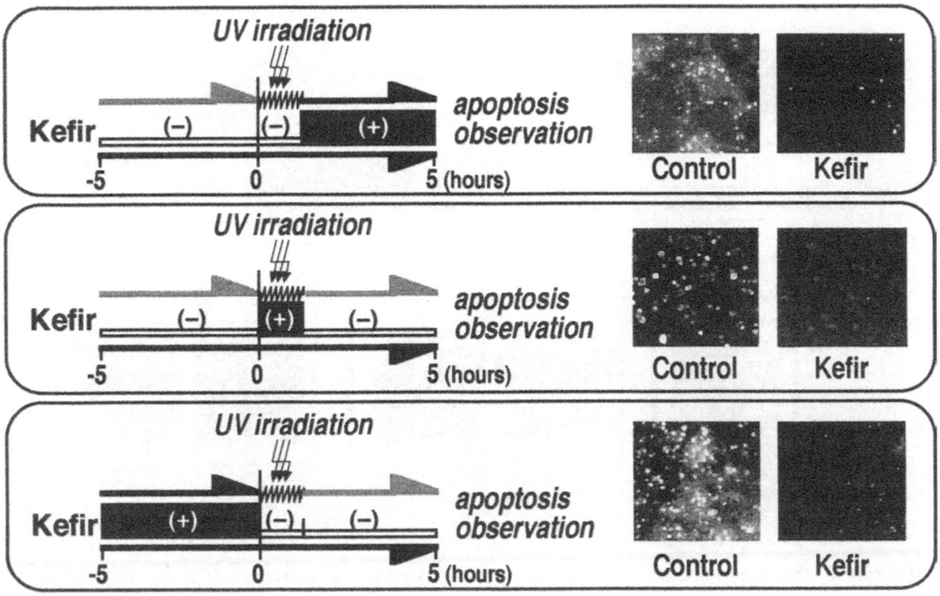


Figure 1. Suppressive effect against apoptosis Top figure shows the effect of Kefir treatment before UV irradiation against cell apoptosis by UV damage. Center figure shows the effect of Kefir treatment during UV irradiation against cell apoptosis by UV damage. The bottom figure shows the effect of Kefir treatment after UV irradiation against cell apoptosis by UV damage.

3.2. Suppressive effect of Kefir against UV-induced thymine dimer formation

Furthermore, we examined the effect of Kefir on UV-induced thymine dimer formation, which is the main cause of cell death by UV irradiation. Pretreatment of human melanoma HMV-1 cells with Kefir extract before UV irradiation resulted in suppression of the thymine dimer formation (Fig. 1). The results from Kefir pretreatment suggest that Kefir gives cells the resistance against UV damages and/or stimulates the repair of UV-damaged chromosomal DNA. Treatment of HMV-1 cells with Kefir extract only during UV irradiation suppressed the apoptosis and thymine dimer formation (Fig.2). These results suggest that Kefir has protection effect from UV irradiation. Treatment of HMV-1 cells with Kefir extract after UV irradiation caused that the number of thymine dimer in the cell rapidly decrease. These results suggest that Kefir stimulate the repair of UV-damaged chromosomal DNA.

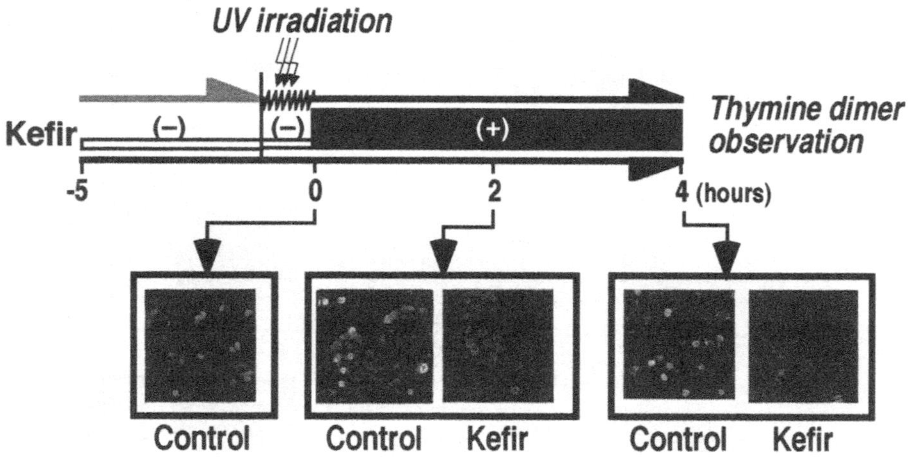


Figure 2. Suppressive effect of Kefir against thymine dimer formation (Kefir treatment after UV irradiation)

## 4. References


1. Osada, K., Nagira, K., and Murakami, H. (1994) Enhancement of interferon-β production with sphingomyelin from fermented milk. *Biotherapy* **7**, 115-123.
2. Kabayama, S., Osada, K., Tachibana, H., Katakura, Y., and Shirahata, S. (1997) Enhancing effects of food components on the production of interferon-β from animal cells suppressed by stress hormones. *Cytotechnology* **23**, 119-125.
3. Gavrieli, Y., Sherman, Y., Ben-Sasson, S. A. (1992) Identification of programmed cell death in situ via specific labeling of nuclear DNA fragmentation, *J. Cell. Biol.* **119**, 493.
4. Koji, T. and Nakane, P. K. (1990) Localization in situ of specific mRNA using thymine-thymine dimerized DNA probes. Sensitive and reliable non-radioactive in situ hybridization. *Acta Pathol. JPN*, **40**, 783-807.
5. Razzaque, M. S., Koji, T., Kawano, H., Harada, T., Nakane, P. K., Taguchi, T. (1994) Glomerular expression of type III and type IV collagens in benign nephrosclerosis: immunohistochemical and in situ hybridization study. *Pathol. Res. Pract.* **190**, 493-499.

# RECOGNITION SYSTEM FOR DIETARY FATTY ACIDS IN THE RAT SMALL INTESTINAL CELLS AND TASTE BUDS

TOHRU FUSHIKI[1], TSUTOMU FUKUWATARI[2], TERUO KAWADA[1], MIHO TSURUTA[1], TAKENORI HIRAOKA[1], TOSHIHIKO IWANAGA[3] AND ETSURO SUGIMOTO[2]

[1]Division of Applied Sciences, Graduate School of Agriculture, Kyoto University, Kyoto 606-8502, Japan
[2]School of Human Culture, University of Shiga Prefecture, 2500 Hassaka-cho, Hikone, Shiga 522-8533, Japan
[3]Graduate School of Veterinary Medicine, Hokkaido University, Kita18 Nishi9, Kita-ku, Sapparo 060-0818, Japan

Fatty acid transporter (FAT) protein and its mRNA, originally expressed in adipose tissue, were found in the tongue of rats. Northern blot analysis showed a significant expression of FAT mRNA in the epithelial layer of circumvallate papillae. Immunohistochemical staining revealed that immunoreactivity for FAT is specifically localized in the apical part of taste bud cells, possibly gustatory cells, in the circumvallate papillae.

---

Fat in food is not only a source of essential nutrients, but also plays an important role in taste sensation. It is generally known that some laboratory animals, such as rats and mice, have a preference for high fat diets [1]. Recently, several studies strongly suggest that dietary fat is recognized in the oral cavity and that long chain fatty acid (LCFA) may be a potential taste stimulus [2, 3]. In mammals, five kinds of proteins have been identified as LCFA-binding proteins present on the cell membrane [4-9]. Fatty acid transporter (FAT) is one of the above proteins and 85% similar to human CD36 glycoprotein [8, 9]. Previously we showed the selective expression of FAT in the brush border of jejunal epithelial cells which participates in the fat-sensory mechanism of intestinal epithelial cells [10]. FAT mRNA in the small intestine is increased by a LCFA-rich diet, suggesting the participation of FAT in the sensation or absorption of dietary fat [11]. We considered that if FAT or related proteins are involved in the sensation of dietary fat in the oral cavity, they must be expressed in taste organ. In this study, we examined the expression and distribution of FAT mRNA and protein in taste buds of circumvallate papillae in rats.

## MATERIALS AND METHODS

### *Northern blot analysis*

The tongue and epididymal fat pad for control were dissected out from twenty male Wistar rats weighing about 180g. The lingual epithelium was isolated from the tongue according to Striem et al. [12], and the epithelial layer were divided into three parts: the circumvallate papillae, the epithelium covering the posterior one third of the tongue except circumvallate papillae (tongue nonsensory epithelium), and the epithelium covering the anterior two thirds of the tongue (anterior tongue epithelium). RNA after fractionation on agarose gel was transferred onto nylon membrane, and hybridized with randomly primed $^{32}$P-labeled rat FAT cDNA probe or rat b-actin cDNA probe at 42°C. After washing, it was subjected to autoradiography at -70°C using Fuji AIF X-ray film.

### *Immunohistochemical staining*

Another five rats were perfused via the aorta with a physiological saline, followed by 4% paraformaldehyde in 0.1M phosphate buffer, pH 7.3. Lingual tissues containing the circumvallate papillae were dissected out and immersed in the same fixative for additional 6 hr. After dipping in 30% sucrose overnight, the tissues were quickly frozen in liquid nitrogen. Cryostat sections, about 15 mm in thickness, were prepared and processed to avidin-biotin complex (ABC) method using the anti-FAT serum. The sections were incubated with the antiserum diluted in 1 : 3,000 overnight. The antigen-antibody reaction was detected using a streptavidin-biotin kit (Histfine, Nichirei, Tokyo).

## RESULTS AND DISCUSSION

To examine whether FAT gene is expressed in lingual epithelium, Northern blot analysis using FAT cDNA was performed for RNAs obtained from circumvallate papillae, surrounding nonsensory epithelium which lacks taste buds (tongue nonsensory epithelium), and anterior tongue epithelium, where fungiform papillae with a few taste buds are scattered. Adipose tissue RNA from the epididymis was also used as an internal control. In circumvallate papillae, a small but significant amount of FAT mRNA was expressed (Fig. 1, lane 3). However, no FAT mRNA was detected in the tongue nonsensory epithelium and the anterior tongue epithelium (Fig. 1, lane 1, 2).

Since FAT mRNA was found to be expressed specifically in the epithelium of circumvallate papillae, immunohistochemical analysis of the tongue was performed using anti-FAT antibody to identify the expression site of FAT in circumvallate papillae. Immunoreactivity for FAT was localized exclusively in taste buds, in which immunoreactive cells and immunonegative cells were intermingled (Fig. 2). In immunoreactive taste bud cells, the apical part of cells were intensely labeled (Fig. 2). Western blot analysis using the same anti-FAT serum failed to detect the band corresponding to FAT in circumvallate papillae as well as in the tongue nonsensory epithelium and the anterior tongue epithelium (data not shown), indicating that a very little amount of FAT protein may be expressed in circumvallae papillae.

In the rat tongue, mRNA of FAT was specifically expressed in the epithelium of circumvallate papillae, and the immunoreactivity for FAT was restricted to taste buds, mainly in the apical part of taste bud cells. These finding strongly suggest the involvement of FAT in taste sensation of the circumvallate papillae. FAT, which was isolated and purified from adipocytes, increases the uptake of fatty acid when expressed in the Ob17PY fibroblast, and combines with LCFA reversibly [8, 14]. Since FAT and CD36 differ in structure from most of the membrane carriers with many transmembrane regions, they are considered to function as receptors for LCFA rather than LCFA transporters. The fact that FAT has a region which interacts with Src kinase at the C-

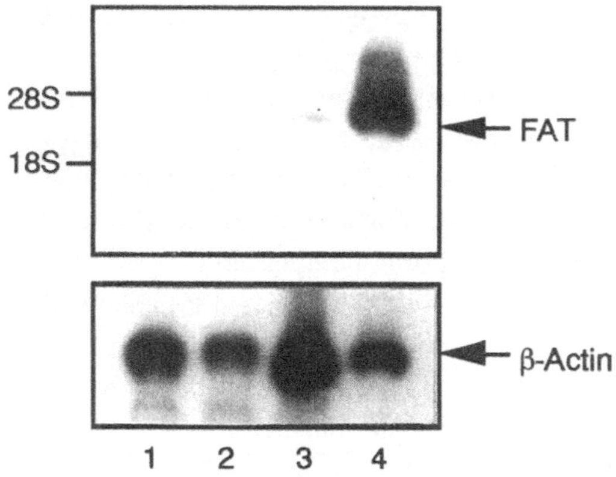


Fig. 1. Northern blot analysis of FAT in tongue epithelium and adipose tissue of rat. Electrophoresis was obtained using 7 mg total RNA obtained from lingual epithelium separated into anterior tongue fungiform (lane 1), tongue nonsensory epithelium (lane 2), and circumvallate papillae (lane 3). Lane 4 shows RNA from adipose tissue.

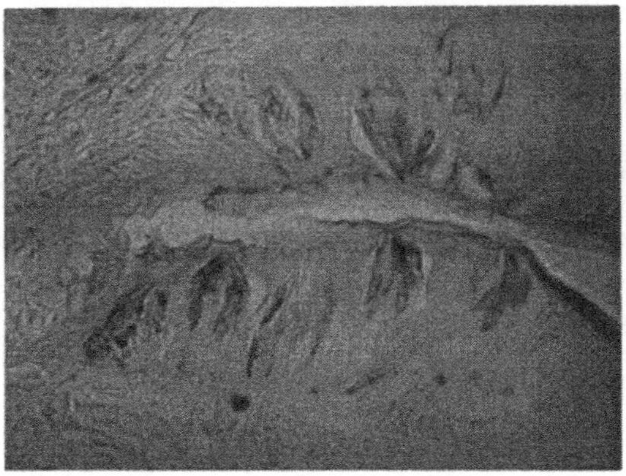

Fig. 2. Immunohistochemical staining of rat circumvallate papillae with anti-rat FAT. Some cells in each taste bud are stained positively with more intense labeling in the apical part of cells.

terminal suggests a role in the signal-transferring process initiated by binding with LCFA [9, 15, 16]. Therefore, FAT may play a role in cellular signal transmission by binding diet-derived LCFA at the apical cell surface of taste buds rather than responding at the basolateral cell surface.

For detection of fat by taste buds, fat must be digested into LCFA in the oral cavity. The circumvallate papillae are associated with the proper salivary glands called Ebner's glands. The excretory ducts of the glands are open to the bottom of circular furrow. All taste buds lined along the furrow are exposed directly to the fluid secreted from Ebner's gland. Several studies revealed rich existence of lipases in the secretions from the Ebner's gland [17-19]. Therefore, LCFA generated from triglyceride may be carried to taste cells by lipophilic carrier molecules, and sensed as the presence of fat in the diet.

The intake of food is known to be rapidly regulated through chemoperception of diet-derived molecules before absorption in the small intestine [20]. α-gustducin, which is the α-subunit of trimeric GTP-binding protein complex expressed in taste buds [21], is also found in the intestinal epithelium [22]. Since α-gustducin participates in intracellular signal transfer of bitterness and sweetness by acting with the taste receptors on cell membrane [23], the chemoperception mechanism for the diet-derived molecules in taste cells is considered similar to that in the digestive tract. Previously, we suggested the participation of receptor protein in the fat-sensory mechanism of intestinal epithelial cells [24, 25]. And, selective expression of FAT in the brush border of jejunal epithelial cells suggests that FAT participates in sensation or absorption of dietary fat [10, 11]. In addition, we showed that the rat's preference for fat is quite similar to the recognition of fat in small intestinal epithelial cells [26]. Thus, we conclude that some dietary fat detecting system similar to that in the intestinal epithelium exists in taste cells, and that FAT participates in the sensory mechanism. These findings provide clues to the long-sought molecular and cellular basis for chemoreception in the oral cavity.

## REFERENCES


1. Hamilton, C.L. (1964) *J. Comp. Physiol. Psychol.* **58**, 459-460.
2. Mattes, R.D. (1996) *Am. J. Clin. Nutr.* **63**, 911-917.
3. Gilbertson, T.A., Fontenot, D.T., Liu, L., Zhang, H. and Monroe, W.T. (1997) *Am. J. Physiol.* **272**, C1203-C1210.
4. Trigatti, B.L., Mangroo, D. and Gerber, G.E. (1991) *J. Biol. Chem.* **266**, 22621-22625.
5. Stremmel, W., Strohmeyer, G., Borchard, F., Kochwa, S. and Berk, P.D. (1985) *Proc. Natl. Acad. Sci. USA* **82**, 4-8.
6. Fujii, S., Kawaguchi, H. and Yasuda, H. (1987) *Lipids* **22**, 544-546.
7. Schaffer, J.E. and Lodish, H.F. (1994) *Cell* **79**, 427-436.
8. Harmon, C.M., Luce, P., Beth, A.H. and Abumrad, N.A. (1991) *J. Membr. Biol.* **121**, 261-268
9. Abumrad, N.A., el-Maghrabi, M.R., Amri, E.Z., Lopez, E. and Grimaldi, P.A. (1993) *J. Biol. Chem.* **268**, 17665-17668.
10. Murota. K, Kawada, T., Iwanaga, T. and Fushiki, T. (1998) in preparation
11. Poirier, H., Degrace, P., Niot, I., Bernard, A. and Besnard, P. (1996) *Eur. J. Biochem.* **238**, 368-373.
12. Striem, B.J., Naim, M. and Lindemann, B. (1991) *Cell Physiol. Biochem.* **1**, 46-54.
13. Chomczynski, P. and Sacci, N. (1987) *Anal. Biochem.* **162**, 156-159.
14. Baillie, A.G.S., Coburn, C.T. and Abumrad, N.A. (1996) *J. Membr. Biol.* **153**, 75-81.
15. Huang, M.M., Bolen, J.B., Barnell, J.W., Shattil, S.J. and Brugge, J.S. (1991) *Proc. Natl. Acad. Sci. USA* **88**, 7844-7848.

16. Schuepp, B.J., Pfister, H., Clemetson, K.J., Silverstein, R.L. and Jungi, T.W. (1991) *Biochem. Biophys. Res. Commun.* **175**, 263-270.
17. Field, R.B., Spielman, A.I. and Hand, A.R. (1989) *J. Dent. Res.* **68**, 139-145.
18. Hamosh, M. and Hand, A.R. (1977) *Lab. Invest.* **37**, 603-608.
19. Schmale, H., Holtgreve-Grez, H. and Christiansen, H. (1990) *Nature* **343**, 366-369.
20. Mei, N. (1985) *Physiol. Rev.* **65**, 211-237.
21. McLaughlin, S.K., McKinnon, P.J. and Margolskee, R.F. (1992) *Nature* **357**, 563-569.
22. Hofer, D., Puschel, B. and Drenckhahn, D. (1996) *Proc. Natl. Acad. Sci. USA* **93**, 6631-6634.
23. Wong, G.T., Gannon, K.S. and Margolskee, R.F. (1996) *Nature* **381**, 737-738.
24. Shintani, T., Takahashi, N., Fushiki, T. and Sugimoto, E. (1995) *Biosci. Biotech. Biochem.* **59**, 479-481.
25. Shintani, T., Takahashi, N., Fushiki, T. and Sugimoto, E. (1995) *Biosci. Biotech. Biochem.* **59**, 1428-1432.
26. Tsuruta, M., Kawada, T., Fukuwatari, T. and Fushiki, T. (1998) *Phisiol. behav.* in press

# EFFECT OF ROSE BENGAL ON IMMUNOGLOBULIN PRODUCTION BY MOUSE B LYMPHOMA, WEHI-279 CELLS

Yuichiro KURAMOTO[1,2], Yoshiyuki MIYAZAKI[2], Shinichiro INOUE[2], Akiko FUJISE[2], Yuichiro YAMAGUCHI[2], Hiroharu KAWAHARA[1], Hirofumi TACHIBANA[2], Michihiro SUGANO[3] and Koji YAMADA[2]

*[1]National Research Institute of Vegetables, Ornamental Plants and Tea, 2769 Kanaya, Shizuoka 428-8501, [2]Faculty of Agriculture, Kyushu University, 6-10-1 Hakozaki, Higashi-ku, Fukuoka 812-8581, and [3]Faculty of Human Life Sciences, Prefectural University of Kumamoto, Kumamoto 862-8502, JAPAN*

***Abstract.*** Effect of Rose Bengal (RB) on immunoglobulin (Ig) production by mouse B lymphoma WEHI-279 cells was examined. At 100 μM, RB increased IgE concentration in the culture medium but decreased the levels of IgG and IgM. To examine whether RB affect to the secretion of IgE from WEHI-279 cells or its synthesis, intracellular and extracellular Ig levels were compared. Cytoplasmic IgE level in the cells treated with 100 μM RB was higher than in the cells cultured without RB. Reverse transcriptase-polymerase chain reaction (RT-PCR) experiment showed that the level of productive ε transcripts of RB-added group was higher than that of control but RB did not affect germ-line ε transcripts. These results suggested that RB enhancement of IgE was due to the increase of the Ig synthesis by WEHI-279 cells, neither to the increase of secretion activity of Ig from the cells nor to the enhancement of class-switching to IgE producing cell.

## 1. Introduction

Food additives are used for coloring, preservating, flavoring, and so on for our benefit. But many adverse reactions are reported when ingested. For instance, sulfites causes asthmatic attack and Tartrazine reveal urticaria (Tarlo and Sussman, 1993; Weber, 1993; Wüthrich, 1993). However, only a few of these agents are currently known to play a role in promoting allergiclike reactions and few information about the mechanism of allergic reaction to the additives are known. Therefore, we focused on the effect of food additives on immunoglobulin production. Allergies against food or environmental allergens are mainly induced by the reaction classified as type I allergy, in which the induction of allergen-specific IgE plays an essential role (Metcalfe, 1991). In this paper, we report the effect of Rose Bengal on Ig production by WEHI-279, mouse B lymphoma cell.

## 2. Materials and Methods

### 2.1 REAGENTS

Rose Bengal was purchased from Wako Pure Chemicals (Osaka, Japan). Mouse IgG, M antibody, goat anti-mouse IgG and HRP-conjugated goat anti-mouse IgG were provided from Zymed Lab. (San Francisco, CA) and mouse IgE from Seikagaku Co. (Tokyo, Japan). Rat anti-mouse IgE was purchased from Experimental Immunology (Brussels, Belgium) and goat anti-mouse IgM and HRP-conjugated goat anti-mouse IgM from Cappel (West Chester, PA). Biotin-conjugated rat anti-mouse IgE was provided from Experimental Immunology and HRP-conjugated avidin from Dakopatts (Glostrap, Denmark).

### 2.2 CELLS AND CELL CULTURE

Mouse B lymphoma, WEHI-279 cells were kindly gifted from Medical Institute of Bioregulation, Kyushu University (Fukuoka, Japan). The cells ($1 \times 10^5$ cells/ml) were cultured with RB or PBS in RPMI1640 medium supplemented with 10% fetal bovine serum at 37°C in 5% $CO_2$ circumstance. Then culture supernatants were harvested for Ig measurement. To examine the cytoplasmic Ig level, the cell homogenate prepared by Handy Sonic (model UR-20P; Tomy Seiko co., Ltd.; Tokyo, Japan) was centrifuged at 2000 x g for 25 min and the supernatants were recovered. Ig concentration in the culture supernatants or the cell lysates was measured by ELISA, as reported previously (Yamada et al., 1993). Data were analyzed by a one-way analysis of variance followed by Duncan's new multiple-range test to evaluate significant differences (Duncan, 1955).

TABLE 1. Primers sequences used in this experiment.

| Amplified RNA | Primer sequence (Position) | Size of amplified product |
|---|---|---|
| IgE mRNA | sense: 5'-TGGACTACTGGGGTCAAGG-3' (no numbered)<br>antisense: 5'-AGCGATGAATGGAGTAGC-3' (991-1008) | 365 bp |
| ε germ-line transcripts | sense: 5'-ACTAGAGATTCACAACG-3' (771-778)<br>antisense: 5'-AGCGATGAATGGAGTAGC-3' (991-1008) | 423 bp |
| β-Actin | sense: 5'-TGGAATCCTGTGGCATCCATGAAAC-3'<br>antisense: 5'-TAAAACGCAGCTCAGTAACAGTCCG-3' | 348 bp |

### 2.3 AMPLIFICATION OF mRNA USING RT-PCR

After the culture, mRNA extracted from cells were amplified by RT-PCR. The primers used in this experiment was summerized in Table 1. The temperatures and reaction periods used during amplification were 94°C for 30 sec for DNA denaturation, 55°C for 1 min for primer annealing and 72°C for 1 min for primer extension and the reaction in the RT-PCR was performed in 35 cycles. The PCR products were separated by agarose gel electrophoresis and visualized by staining by ethidium bromide to take electronical pictures using digital devices (Eastman Kodak Co., Rochester, NY).

## 3. Results and Discussion

### 3.1 EFFECT OF ROSE BENGAL ON IMMUNOGLOBULIN PRODUCTION BY WEHI-279 CELLS

We have already reported that RB increased IgE production and inhibited IgG and IgM production by rat spleen lymphocytes (Kuramoto et al., 1997). To investigate the mechanism regulated by RB, we used WEHI-279 and excluded the possibility of T cell commitment. At first, we examined dose-dependent effect of RB on Ig production by WEHI-279 cells.

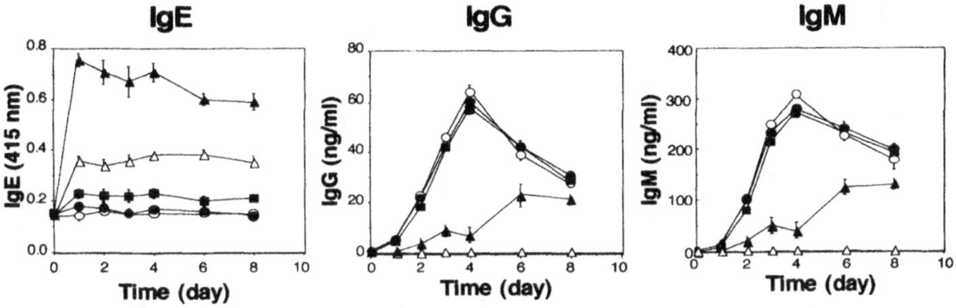


Fig. 1. Time-courses of Ig production by WEHI-279 cells. Cells were cultured with PBS (as vehicle of RB; ○), 1 μM of RB (●), 10 μM of RB (■), 100 μM of RB (▲), or 1 mM of RB (△) for various times and the culture supernatant was collected to measure the Ig concentrations by ELISA (n=3). Both IgG and IgM concentrations were indicated as ng/ml and IgE as optical density (A415). The results indicated here were the mean±SE.

As shown in Fig. 1, Ig concentration in the culture supernatant increased to 60 ng/ml for IgG and 300 ng/ml for IgM when cultured for 4 days with 0-10 μM RB. When cells were cultured with 100 μM RB, these Ig levels increased gradually until 8 days. In both cases, the cell viability was over 90% (data not shown). On the other hand, IgE level in the culture supernatant increased to 0.8 in O.D. when cultured for 1 days with 100 μM RB and then the level decreased gradually. When cells were cultured with 10 μM RB, the IgE level increased more gradually than that with 100 μM RB, while 0-1 μM RB did not affect IgE level strongly.

## 3.2 EFFECT OF ROSE BENGAL ON INTRACELLULAR AND EXTRACELLULAR IMMUNOGLOBULIN LEVELS

To clarify whether RB affect to production or secretion of these Igs, WEHI-279 cells were cultured for 24 hr in the presence of RB and intracellular and extracellular Ig levels were determined. WEHI-279 cells were cultured with or without various concentrations of RB for 24 hr and culture supernatant was collected to apply to ELISA. On the same time, the cells were homogenized to measure cytoplasmic Ig concentration.

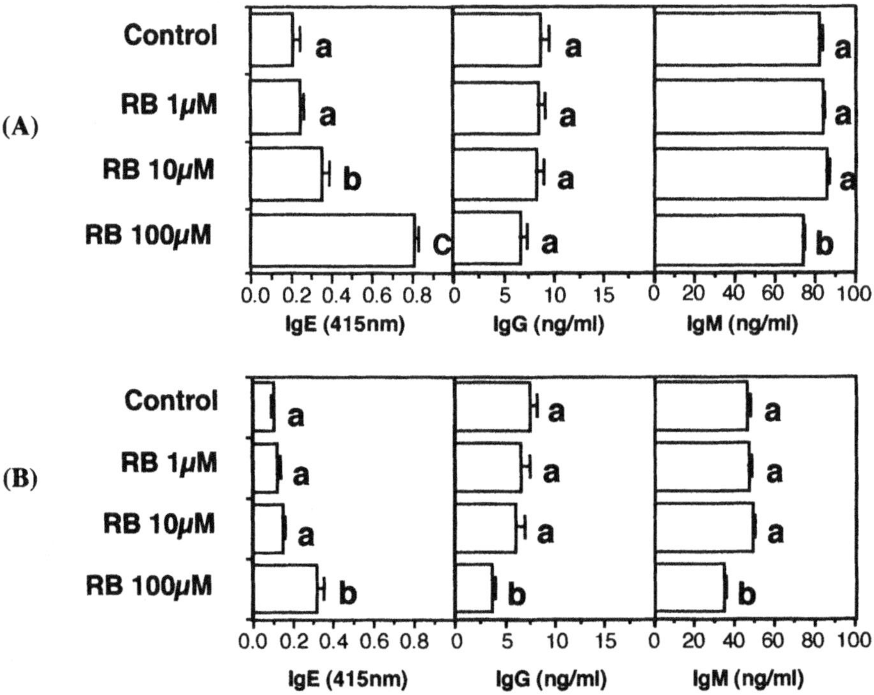


Fig. 2. (A) Effect of RB on Ig production by WEHI-279 cells and (B) effect of RB on cytoplasmic Ig concentration in WEHI-279 cells. Cells were cultured for 24 h in the absence or presence of RB and the Ig concentration in the culture supernatant were determined (for result of A) or were lysed by sonication to determine the Ig concentration in the cells (for result of B) by ELISA (n=5). Both IgG and IgM concentrations were indicated as ng/ml and IgE as optical density at 415 nm. [a-c]Values in the same group without a common letter are significantly different at $p < 0.05$.

As shown in Fig. 2, RB elevated both intra- and extracellular IgE level at 100 μM and weakly at 10 μM. On the contrary, RB did not affect both intra- and extracellular IgG level at 1-10 μM, however 100 μM RB suppressed intracellular IgG level weakly. The effects were also seen in the case of both intra- and extracellular IgM level. These results suggest that RB enhancement of IgE at 100 μM was due to the increase of the Ig production by WEHI-279 cells, not to the increase of secretion activity of Ig from the cells.

## 3.3 THE EXPRESSION OF ε TRANSCRIPTS FROM WEHI-279 CELLS

WEHI-279 cells were cultured for 72h with or without RB to clarify the effect of RB on transcription of IgE gene. In addition to IgE mRNA, the level of germ-line ε transcripts, which is expressed in the course of class-switching to IgE producing cell, is also determined. As shown in Fig. 3, the IgE mRNA level was enhanced in the presence of RB, while germ-line ε transcripts were expressed irrespective of the presence of RB. Thus, it is suggested that RB enhances IgE production of WEHI-279 cells through the increase of IgE mRNA expression and not through the stimulation of class-switching to IgE producing cell.

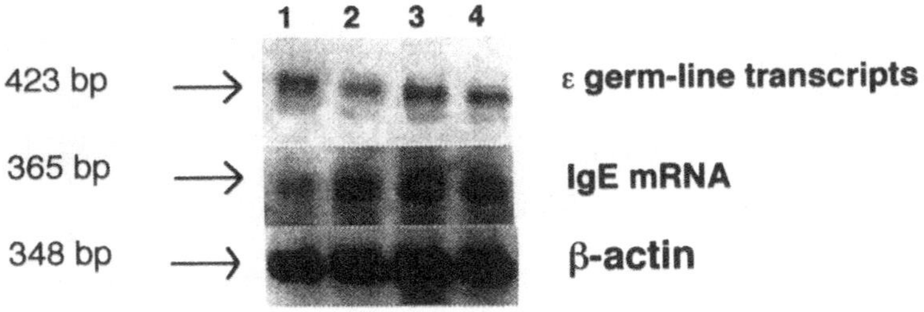


Fig. 3. Effect of RB on the expression of Pε and Gε of WEHI-279 cell. Cells were cultured for 72 h in the presence or absence of RB at each concentration and RNA was extracted from the cells to apply to RT-PCR. Each lane indicates; Lane 1: Control (PBS as a vehicle of RB); Lane 2: 1 μM of RB; Lane 3: 10 μM of RB; Lane 4: 100 μM of RB.

## 4. Refernces

Duncan, D. B. (1955) Multiple range and multiple F test. *Biometrics*, **11**, 1-42.

Kuramoto, Y., Yamada, K., Lim, B. O., Sugano, M. (1997) Stimulating effect of xanthene dyes on immunoglobulin produced *in vitro* by rat spleen lymphocytes. *Biosci. Biotech. Biochem.* , **61**, 723-725.

Metcalfe, D. D. (1991) Food allergy. *Current Opinion Immunol.* , **3**, 881-886.

Tarlo, S. M., Sussman, G. L. (1993) Asthma and anaphylactoid reactions to food additives. *Can. Fam. Physician* , **39**, 1119-1123.

Weber, R, W. (1993) Food additives and allergy. *Ann. Allergy*, **70**, 183-192.

Wüthrich, B. (1993) Adverse reactions to food additives. *Ann. Allergy* , **71**, 379-384.

Yamada, K.; Lim, B. O.; Sugano, M. Suppression of immunoglobulin production of rat lymphocytes by bile acids. *In Vitro Cell. Develop. Biol.* **1993**, 29, 840-841.

# CLASS-SPECIFIC REGULATION OF IMMUNOGLOBULIN PRODUCTION BY ROSE BENGAL IN HUMAN B CELL LINES AND PERIPHERAL BLOOD LYMPHOCYTES

Yoshiyuki Miyazaki, Yuichiro Kuramoto, Hirotaka Haruta, Hirofumi Tachibana, Michihiro Sugano* and Koji Yamada
*Laboratory of Food Science, Department of Food Science and Technology, Faculty of Agriculture, Kyushu University, Fukuoka 812-8581, Japan.*
**Faculty of Human Life Science, Prefectural University of Kumamoto, Kumamoto 862-0920, Japan.*

**Abstract** We have previously reported that Rose Bengal (RB) can enhance IgE production in rat spleen lymphocytes without increasing the production of IgG and IgM. To examine the effect of RB in human B cell lines, U-266 (IgE producer), HMy-2 (IgG producer) and NAT-30 (IgM producer) cells were cultured with RB, and the medium immunoglobulin (Ig) level was measured by ELISA. Though RB enhanced IgE production in rat lymphocytes, it inhibited IgE production in U-266 and IgG production in HMy-2 cells in dose dependent manner. In NAT-30 cells, RB slightly inhibited IgM production at 0.1 μM, while it enhanced production at 10 μM. These results suggest that RB directly modulates Ig production in B cells, and that the modulatory effect is different between rat lymphocytes and human cell lines. Therefore, we examined the effect of RB on Ig production in human peripheral blood lymphocytes. We found that RB did not enhance IgE production in lymphocytes as well as human cell lines.

## Introduction

Allergic reactions are usually classified into 4 types and type I allergy plays an important role in the expression of allergies against food allergens. In this type of allergy, IgE triggers the allergic response by stimulating mast cells and basophils, which release inflammatory chemical mediators. Conversely, IgG inhibit allergic reactions through the competition with IgE. It has been reported that some food components, such as bile acids (Lim *et al.*, 1994) and unsaturated fatty acids (Yamada *et al.*, 1996), enhance IgE secretion and suppress the production of IgG and IgM in rat lymphocytes. In addition, it has been reported that some food additives exert an adverse effect on allergic diseases (Weber, 1993; Tarlo *et al.*, 1993; Fuglsang *et al.*, 1994). Thus, it is important to clarify the effect of food components on the regulation of allergic responses.

Among food additives, Rose Bengal (RB) has been shown to exert strong regulatory control on the production of Ig in rat lymphocytes (Kuramoto *et al.*, 1997), but the mechanism by which RB regulates Ig production has not been clarified. In this study, we compared the effect of RB on Ig production in human established cell lines and peripheral blood lymphocytes to measure the influence of RB on human Ig production.

## Materials and Methods

RB, 4, 5, 6, 7-tetrachloro-2', 4', 5', 7'-tetraiodofluorescein ($C_{20}H_2Cl_4I_4Na_2O_5$, M.W. 1017.64) was purchased from Wako Pure Chemical Industries (Osaka, Japan) and dissolved in phosphate buffered saline, pH 7.4.

We used the following human cell lines to examine the regulatory control of RB on Ig production; the IgE producing myeloma cell line U-266, the IgG producing B lymphoblastoid cell line HMy-2 and the IgM producing Burkitt's lymphoma cell line NAT-30. These cells were cultured in ERDF medium (Kyokuto Pharmaceutical, Tokyo, Japan) supplemented with 5% fetal bovine serum (FBS; Intergen Co., NY) at 37°C under humidified 5% $CO_2$-95% air. Cells were inoculated at $1x10^5$ cells/ml and cultured with or without RB for 144 hr. Human peripheral blood lymphocytes (PBL) were isolated from two volunteers, as previously mentioned (Murakami *et al.*, 1997). PBL were inoculated at $1x10^6$ cells/ml and cultured in 10% FBS-ERDF medium containing 0 to 100 μM RB for 72 hr.

Cell number was counted using a hematocytometer and cell viability was determined using the trypan blue staining method. Ig content in culture medium was determined by ELISA, according to the method described previously (Yamada *et al.*, 1993).

## Results

### *Effect of RB on proliferation and Ig production in various human cell lines*

First, we examined the effect of RB on proliferation and IgE production in U-266 cells. As shown in Fig. 1, a decrease in cell viability was not seen in the cells cultured with either 0.1 or 1 μM of RB. On the other hand, cell proliferation was inhibited when cultured with 10 μM RB and cell viability declined to 87% after a 144-hr continuous cultivation. RB did not affected the level of IgE in the culture medium during the first 72 hr, but dose-dependently inhibited the production there after.

Next we examined the effect of RB on proliferation and IgG production in HMy-2 cells. As shown in Fig. 1, the cells exponentially proliferated during the first 72 hr and then the cell number ceased to increase. With the cell number reaching a plateau, a linear decrease in cell viability was observed irrespective of the concentration of RB. Medium IgG level increased linearly after a 24-hr lag period and inhibition of IgG accumulation became significant at the 96 hr point after inoculation in the presence of 1 or 10 μM RB.

When NAT-30 cells were cultured with RB, inhibition of cell proliferation was observed only at a concentration of 10 μM. Interestingly, RB slightly inhibited IgM production at 0.1 μM, while production was enhanced at 10 μM (Fig. 1).

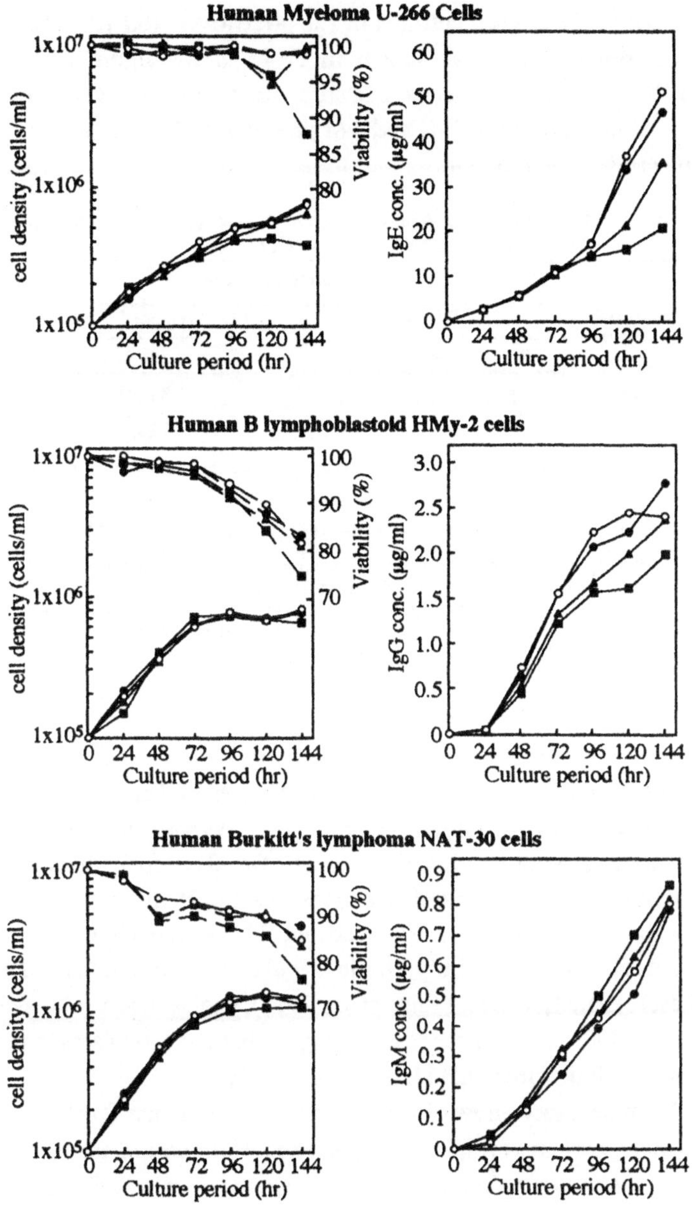


**Fig 1. Effect of Rose Bengal on Proliferation and Immunoglobulin Production of Human B Cell Lines**
Cells were inoculated at $1x10^5$ cells/ml in 5% FBS-ERDF medium supplemented with 0 (○), 0.1 (●), 1 (▲) and 10 (■) μM of Rose Bengal and cultured for 144 hr. In the left figures, solid lines show cell density and broken lines show cell viability.

***Effect of RB on Ig production by human PBL***

To investigate the effect of RB on Ig production in human primary cultured cells, PBLs were cultured in 10% FBS-ERDF medium with or without RB for 72 h. Cell viability was maintained above 90% throughout the entire culture time period when cultured without RB, but decreased to 70% when cultured in the presence of 100 μM RB after the 72-hr cultivation (data not shown). As shown in Table 1, RB significantly decreased IgE production in PBL isolated from donor 1 at all concentrations used here, but not in those from donor 2. When measuring IgG and IgM production, the inhibitory effect of RB was observed at 100 μM in both donors.

**Table 1. Dose-dependent Effect of Rose Bengal on Immunoglobulin Production by Human Peripheral Blood Lymphocytes**

| Rose Bengal conc. (μM) | Ig concentration (ng/ml) | | | | | |
|---|---|---|---|---|---|---|
| | Donor 1 | | | Donor 2 | | |
| | IgE | IgG | IgM | IgE | IgG | IgM |
| None | 11 ± 1 | 53 ± 2 | 23 ± 2 | 3 ± 1 | 33 ± 4 | 26 ± 2 |
| 0.1 | 6 ± 2*** | 51 ± 3 | 28 ± 2* | 2 ± 1 | 32 ± 2 | 22 ± 2 |
| 1 | 5 ± 1*** | 50 ± 2 | 22 ± 1 | 2 ± 1 | 31 ± 1 | 23 ± 3 |
| 10 | 6 ± 2*** | 45 ± 3** | 22 ± 2 | 3 ± 1 | 30 ± 1 | 21 ± 2* |
| 100 | 5 ± 2*** | 25 ± 4*** | 5 ± 1*** | 3 ± 1 | 16 ± 2*** | 6 ± 1*** |

Human PBL were inoculated at $1x10^6$ cells/ml in 10% FBS-ERDF medium supplemented with none or various concentrations of Rose Bengal and cultured for 72 hr. Then, the amount of IgE, IgG and IgM in each culture medium was measured by ELISA. Data are mean±SD (n=4) and significantly different from none at *$p < 0.05$, **$p < 0.01$ and ***$p < 0.005$.

**Discussion**

In the previous study, Kuramoto *et al.* (1997) reported that RB class-specifically regulates Ig production in rat spleen lymphocytes. Because various types of cells are present in splenocytes, identification of target cells for RB is difficult using lymphocytes. Thus, we used three different human B cell lines each producing only a single Ig class to investigate the direct effect of RB. We found that RB does module the production of Ig in all three of these cell lines. Recently, similar regulatory control of Ig production has been reported in diesel exhaust particles (Tsien *et al.*, 1997), enolase and alcohol dehydrogenase-I (Sugahara *et al.*, 1998, 97). These results suggest that RB modulates Ig production in B lymphocytes through its direct interaction with B cells.

However, the modulatory effect of RB observed in the human B cell lines used here is different from that reported in rat spleen lymphocytes (Kuramoto *et al..* 1997). RB inhibited IgE production in human myeloma U-266 cells, whereas it enhanced IgE production in rat spleen lymphocytes. Furthermore, RB concentration thresholds needed to modulate Ig production were much lower in human cell lines (0.1 to 100 μM) than in rat lymphocytes (50 to 1000 μM). When assaying human lymphocytes, RB suppressed IgE production in PBLs from donor 1 at relatively low concentrations between 0.1 to 100 μM. It is known that sensitivity to physiological active components is different between animal species (Veronesi et al., 1993). These results suggest that RB regulates Ig production in human B cells through a different manner than in rat B cells.

**References**

Fuglsang G, Madsen G, Halken S, Jorgensen S, Ostergaard PA and Osterballe O (1994) Adverse reactions to food additives in children with atopic symptoms. Allergy 49: 31-37.

Kuramoto Y, Yamada K, Lim BO and Sugano M (1997) Stimulating effect of xanthene dyes on immunoglobulin produced in vitro by rat spleen lymphocytes. Bioscience, Biotechnology and Biochemistry 61: 723-725.

Lim BO, Yamada K and Sugano M (1994) Effects of bile acids and lectins on immunoglobulin production in rat mesenteric lymph node lymphocytes. In Vitro Cellular & Developmental Biology. Animal. 30: 407-13

Murakami F, Sasaki T and Sugahara T (1997) Lysozyme stimulates immunoglobulin production by human-human hybridoma and human peripheral blood lymphocytes. Cytotechnology 24: 177-182

Sugahara T, Furutani H and Sasaki T (1997) Alcohol dehydrogenase-I from horse liver stimulates immunoglobulin production by human hybridoma and human peripheral blood lymphocytes. Molecular and Cellular Biochemistry 173: 113-119.

Sugahara T, Shimizu S, Abiru M, Matsuoka S and Sasaki T (1998) A novel function of enolase from rabbit muscle; an immunoglobulin production stimulating factor. Biochimica et Biophysica Acta 1380:163-76.

Tarlo SM and Sussman GL (1993) Asthma and anaphylactoid reactions to food additives. Canadian Family Physician 39: 1119-1123.

Tsien A, Diaz-Sanchez D, Ma J and Saxon A (1997) The organic component of diesel exhaust particles and phenanthrene, a major polyaromatic hydrocarbon constituent, enhances IgE production by IgE-secreting EBV-transformed human B cells in vitro. Toxicology and Applied Pharmacology 142: 256-263.

Veronesi B and Ehrich M (1993) Differential cytotoxic sensitivity in mouse and human cell lines exposed to organophosphate insecticides. Toxicology and Applied Pharmacology 120: 240-246.

Weber RW (1993) Food additives and allergy. Annals of Allergy 70: 183-190.

Yamada K, Lim BO and Sugano M (1993) Suppression of immunoglobulin production of rat lymphocytes by bile acids. In Vitro Cellular & Developmental Biology. Animal. 29: 840-1

Yamada K, Hung P, Yoshimura K, Taniguchi S, Lim BO and Sugano M (1996) Effect of unsaturated fatty acids and antioxidants on immunoglobulin production by mesenteric lymph node lymphocytes of Sprague-Dawley rats. Journal of Biochemistry. 120: 138-44

# INHIBITORY ACTION OF CAROTENOIDS IN CRAB SHELL ON THE INVASION OF HEPATOMA CELLS CO-CULTURED WITH MESOTHELIAL CELLS

S. NAKAHARA, Y. MIURA and K. YAGASAKI
*Department of Applied Biological Science, Tokyo Noko University, Fuchu, Tokyo 183-8509, Japan*

## ABSTRACT

The effects of crab shell extract and related carotenoids on the invasion of hepatoma (AH109A) cells were studied in a co-culture system of AH109A with mesentery-derived mesothelial cells. Extracts from both raw and boiled crab shell showed significant inhibition of the AH109A invasion. Both astaxantin and β-carotene, carotenoids with antioxidative activity and identified in crab shell extract, inhibited the invasion of AH109A at a concentation of 5 $\mu$M without affecting the proliferation of the hepatoma cells. The loss of anti-invasive activity due to boiling was more prominent in astaxanthin than in β-carotene. This supports an assumption that compounds with antioxidative activity would generally suppress the invasion of cancer cells.

## 1. Introduction

Cancer cells possess two biological characteristics, namely, endless proliferation and metastasis. Among the complicated and sequential steps of metastasis, the invasion of cancer cells is regarded as an important and characteristic step [1]. The inhibition of these biological events is expected to prolong the life span of a host with cancer. We have recently reported that extracts of two popular beverages, tea [2] and coffee [3], and sera obtained from rats orally given tea [4] or coffee [5] extract suppress the invasion of hepatoma cells *in vitro*. Both tea and coffee contain polyphenolic components with antioxidative activity. From the results of these and other works, we assumed that compounds with antioxidative activity would generally suppress the invasion of cancer cells. One means to verify this assumption is to investigate the effects of antioxidants structurally unrelated with polyphenols. Astaxanthin is a carotenoid with antioxidative activity [6] and is contained in crustaceans [7]. In the present study, the effects of crab shell extract and related carotenoids on the invasion of hepatoma cells were studied in a co-culture system.

## 2. Materials and Methods

*Cell culture and invasion assay.* As model cancer cells, an ascites hepatoma cell line of AH109A was used [8,9]. AH109A cells were maintained in DM-160 medium containing 10% calf serum (10% CS/DM-160). The effects of crab shell extract and related carotenoids on the proliferation of hepatoma cells were measured by WST-1 method [4]. Their effects on the invasion of hepatoma cells were measured by co-culture system of AH109A with mesentery-derived mesothelial cells (M cells) according to the procedure of Akedo *et al.* [10] with slight modifications as described previously [2]. Briefly, primarily cultured M cells from mesentery of rats were cultured in 10% CS/DM-160 for 7-10 days to attain a confluent state. AH109A cells were then seeded on the monolayers and cultured for 48 hr. Invading AH109A cells and colonies underneath M cells were counted with a phase-contrast microscope. Usually at least 10 areas were counted and the invasive activity of AH109A was indicated by the number of invading cells and colonies/$cm^2$.

*Preparation of crab shell extract and experimental medium.* Crabs were washed with water and cut into two pieces. One piece was heated in boiling water (boiled crab), while the other was not (raw crab). Shells were removed from raw and boiled crabs, frozen in liquid nitrogen, and ground up. From the resultant powder of the crab shell, carotenoids were successively extracted with methanol and ethyl acetate. The pooled solvents were evaporated to dryness and dissolved in dimethyl sulfoxide (DMSO). The DMSO solution of raw crab extract was heated for 5 min in boiling water if necessary. Astaxanthin (Sigma) and β-carotene (Sigma) were also dissolved in DMSO. The DMSO solutions of the carotenoids were allowed to stand for 5 min in boiling water when necessary. The DMSO solution of crab shell extract and carotenoids was added to 10% CS/DM-160 at a final DMSO concentration of 0.5%. After being sterilized by membrane filtration, the effects of samples on the proliferation and invasion of AH109A were tested *in vitro*. The pooled extract solvents were also subjected to analysis by thin-layer chromatography using a Silica gel $60F_{254}$ TLC plate (Merck). Corotenoids were developed in a solvent of hexane-ethyl acetate (1:2, by volume) and visualized with ultraviolet lamp (2536Å).

*Statistical analysis.* Data were expressed as means ± standard errors. Statistical analysis was done using one way analysis of variance followed by Tukey's Q test.

## 3. Results and Discussion

Examination of the effects of extracts from both raw and boiled crab shell showed significant inhibition of the AH109A invasion by 47.3% and 40.4%, respectively (Exp. 1A in Table 1). Crab extract boiled after extraction from raw shell also significantly inhibited the invasion of hepatoma cells (Exp. 1B in Table 1). In contrast, they had little effect on the AH109A proliferation at the same concentration (data not shown).

Table 1. Effects of crab shell extract on the invasion of AH109A cells.

| Treatment | Boiled before extraction* | Boiled after extraction** |
|---|---|---|
| | Exp. 1A | Exp. 1B |
| | % of control | |
| Vehicle (control) | 100.0± 9.1$^{a}$ | 100.0± 5.3a |
| Raw crab shell extract | 52.7±10.2$^{b}$ | 51.5± 7.8$^{b}$ |
| Boiled crab shell extract*,** | 59.6±11.8$^{b}$ | 46.5±13.0$^{b}$ |

Each value represents the mean ± SEM for 10 areas. Raw or boiled crab shell extract was added to experimental medium at a concentration of 52 $\mu$g/ml. Control values are 2633 ± 239 (Exp. 1A) and 5323 ± 285 (Exp. 1B) (mean ± SEM, number of invading cells and colonies/$cm^2$). Values not sharing a common letter are significantly different at $p < 0.05$ by Tukey's Q test.

These results indicate that a thermostable component(s) with anti-invasive activity might be contained in the crab shell extract. To identify such component(s), the extract was chromatographed with a Silica gel $60F_{254}$ TLC plate. At least two main bands were detected under a UV lamp; their Rf values were 0.43-0.45 and 0.90-0.93, being equivalent to those of astaxanthin and β-carotene used as markers, respectively. Thus, we next examined the effects of astaxanthin and β-carotene on hepatoma cells.

Table 2. Effects of astaxanthin and β-carotene on the invasion of AH109A cells

| Treatment | Astaxanthin Exp. 2A | β-Carotene Exp. 2B |
|---|---|---|
| | % of control | |
| Vehicle (control) | 100.0± 9.1$^{a}$ | 100.0± 9.1a |
| Carotenoid (without boiling) | 60.0± 7.7$^{b}$ | 59.6±11.8$^{b}$ |
| Carotenoid (with boiling) | 86.1±15.1$^{a}$ | 70.2±10.0$^{b}$ |

Each value represents the mean ± SEM for 10 areas. Carotenoids with or without boiling were added to experimental medium at the concentration of 5 $\mu$M. Control values are 5418 ±491 (Exp. 2A) and 2633 ± 239 (Exp. 2B) (mean ± SEM, number of invading cells and colonies/$cm^2$). Values not sharing a common letter are significantly different at $p < 0.05$ by Tukey's Q test.

Astaxanthin commenced to suppress the AH109A invasion at 2.5 $\mu$M and maintained the suppressive effect up to 20 $\mu$M, whereas it exerted no influence on the AH109A proliferation at the same concentrations (data not shown). Selecting the concentration of 5 $\mu$M, thermostability of astaxanthin was examined together with the effect and thermostability of β-carotene. As shown in Table 2, the suppressive effect of astaxanthin and β-carotene on the invasion weakened when they were boiled. However, the loss of anti-invasive activity due to boiling was more prominent in astaxanthin than in β-carotene. Thus, β-carotene seems to be a thermostable component in crab shell extract that suppresses the invasion of AH109A cells. Both carotenoids had little effect on the proliferation of AH109A (data not shown).

From the present results, astaxanthin and β-carotene were found to suppress the *in vitro* invasion of hepatoma cells, although the thermostability seemed to be slightly different between the two carotenoids. This supports our assumption that compounds with antioxidative activity would generally suppress the invasion of cancer cells. Astaxanthin has recently been reported to suppress a stress-promoted metastasis of mastcytoma in mice [11], this finding being well consistent with our present results. The precise mechanism of the inhibitory action of the carotenoids on the AH109A invasion remains to be elucidated.

## 4. Acknowledgement.

This work was supported in part by a grant from the Towa Shokuhin Kenkyu Shinkoukai Foundation, Tokyo, Japan.

### 5. References

1. Liotta, L., Rao, C.N., and Barsky, S.H. (1983) Tumor invasion and the extracellular matrix, *Lab. Invest.*, **49**, 636-649.
2. Miura, Y., Shiomi, H., Sakai, F., and Yagasaki, K. (1997) Assay systems for screening food components that have anti-proliferative and anti-invasive activity to rat ascites hepatoma cells: *In vitro* and *ex vivo* effects of green tea extract, *Cytotechnology*, **23**, 127-132.
3. Furuse, T., Shiomi, H., Miura, Y., and Yagasaki, K. (1998) Coffee's inhibitory action on the invasion of hepatoma cells co-cultured with methothelial cells. *Animal Cell Technology: Basic & Applied Aspects*, **9**, 127-130.
4. Zhang, G., Miura, Y., and Yagasaki, K. (1998) Effects of green, oolong and black teas and related components on proliferation and invasion of hepatoma cells. *Animal Cell Technology: Basic & Applied Aspects*, **10**, in press.
5. Miura, Y., Furuse, T., and Yagasaki, K. (1997) Inhibitory effect of serum from rats administered with coffee on the proliferation and invasion of rat ascites hepatoma cells, *Cytotechnology*, **25**, 219-223.
6. Palozza, P. and Krinsky, N. I. (1992) Astaxanthin and canthaxanthin are potent antioxidants in a membrane model. *Arch. Biochem. Biophys.*, **297**, 291-295.
7. D'Abramo, L. R., Baum, N. A., Bordner, C. E., and Conklin, D. E. (1983) Carotenoids as a source of pigmentation in juvenile lobsters fed a purified diet. *Can. J. Fish. Aquat. Sci.*, **40**, 699-704.
8. Komatsu, W., Yagasaki, K., Miura, Y., and Funabiki, R. (1997) Stimulation of tumor necrosis factor and interleukin-1 productivity by the oral administration of cabbage juice to rats. *Biosci. Biotech.*

*Biochem.*, **61**, 1937-1938.
9. Komatsu, W., Miura, Y., and Yagasaki, K. (1998) Suppression of hypercholesterolemia in hepatoma-bearing rats by cabbage extract and its component, S-methyl-L-cysteine sulfoxide. *Lipids*, **33**, 499-503.
10. Akedo, H., Shinkai, K., Mukai, M., Mori, Y., Tateishi, R., Tanaka, K., Yamamoto, R., and Morishita, T. (1986) Interaction of rat ascites hepatoma cells with cultured mesothelial cell layers: a model for tumor invasion. *Cancer Res.*, **46**, 2418-2422.
11. Yang, Z., Asami, S., Toyoda, Y., Hujii, W., Suwa, Y., and Tanaka, T. (1997) Protective effect of astaxanthin on the promotion of cancer metastases in mice treated with restraint-stress. *J. Jpn. Soc. Nutr. Food Sci.*, **50**, 423-428.

# THE NEURITE-INITIATING EFFECT OF MICROBIAL EXTRACELLULAR GLYCOLIPIDS

HIROKO ISODA[1], HIROSHI SHINMOTO[2], FUJIKO OZAWA[3] and TADAATSU NAKAHARA[4]

[1]*Inst. of Biol. Sci., Univ. of Tsukuba, Tennodai, Tsukuba, Ibaraki, Japan*
[2]*National Food Res. Inst, MAFF, Kannondai, Tsukuba, Ibaraki, Japan*
[3]*Depart. of Radiol. Sci., Ibaraki Pref. Univ. of Health Sci., Ami, Ibaraki, Japan*
[4]*Inst. of Appl. Biochem., Univ. of Tsukuba, Tennodai, Tsukuba, Ibaraki, Japan*

## 1. Introduction

Glycolipid-type biosurfactants are known as microbial extracellular or cell-associated biosurfactants, and it is commonly isolated type of biosurfactant. Mannosylerythritol lipids, MEL-A and MEL-B, were produced by *Candida antarctica* T-34 in soybean oil with a production of 40 g per liter of culture broth . MEL-A and MEL-B were identified as 4-*O*-(di-*O*-acetyl-di-*O*-alkanoyl-β-D-mannopyranosyl)-erythritol and 4-*O*-(mono-*O*-acetyl-di-*O*-alkanoyl-β-D-mannopyranosyl)-erythritol, respectively. Polyol lipid was produced by *Aureobasidium* sp. A-21 with a productivity of about 35 g per liter of culture medium containing no $CaCO_3$ as a neutralizing agent, which is a mixture of fatty acid esters of arabitol and mannitol, and the two main components of t he lipophilic moiety of the lipids proved to be 3,5-dihydroxy-decanoic and 5-hydroxy-2-decenoic acids as identified by their lactones, (+)-3-hydroxydecan-5-olide and (R)-(-)-2-decene-5-olide, that is, R-(-)-massoilactone, respectively. Rhamnolipid was reported as a growth stimulant), having surface activity and emulsifying capability, produced in the culture broth by a hydrocarbon-using bacterium, *Pseudomonas aeruginosa* S7B1, which consisted of rhamnose and β-hydroxydecanoic acid. Sophorose lipid was produced by *Torulopsis bombicola* ATCC 22214 in a mixture of glucose and safflower oil with a production of 70 g per liter, which contains the dimeric sugar sophorose and a long-chain carboxylic acid with a hydroxyl function on the penultimate or terminal carbon. Two succinoyl trehalose lipids, STL-1 (2,3,4,2'-di-*O*-succinoyl-di-*O*-alkanoyl–α–α-trehalose) and STL-3 (2,3,4,2'-2 monosuccinoyl- trialkanoyl-trehalose) were by *Rhodococcus erythropolis* SD-74 and *Rhodococcus* sp. TB-42 with a production of about 30 to 40 g per liter, respectively.

In this study, we investigated whether several microbial extracellular glycolipids including MEL-A, MEL-B, PL, RL, SL, STL-A and STL-B induced a positive neurite initiation of PC12 cells. The PC12 cell line, derived from a rat pheochromocytoma, provides a relatively simple, homogenous system for studying various aspects of neuronal differentiation. PC12 cells can also survive and proliferate in vitro without requiring the presence of neurotrophic factors. Application of NGF causes PC12 cells to differentiate into a neuronal like phenotype characterized by numerous morphological and physiological changes. In PC12 cells, nerve growth factor (NGF) interacts with two dinstinct plasma membrane receptor proteins: $p75^{NGFR}$, a cysteine-rich glycoprotein having a relatively low affinity for NGF, and p140trk, a receptor tyrosine kinase which bound NGF with a high affinity and activated by NGF, and resulting in the rapid tyrosine autophosphorylation of the receptor and activated of signal-transducing proteins.

In this report, we indicate the neurite-initiating effect of microbial extracellular glycolipids in PC12 cells, and discuss on the mechanisms of such function.

## 2. Materials and Methods

### 2.1. Materials

Nerve growth factor (NGF), 2.5S, was purchased from Funakoshi, Tokyo, Japan. The anti low affinity p75NGFR antibody was purchased from Cosmo Bio, Tokyo, Japan.

### 2.2. Production of microbial extracellular glycolipids

Microbial extracellular glycolipids were produced by the methods as described in Isoda *et al*., (1997).

### 2.3. Cells and Cell culture

PC12 cell line was obtained from Riken Cell Bank (Tsukuba, Ibaraki, Japan) and routinely grown in Dulbecco's modified Eagle's medium (DMEM; Nissui Pharmaceutical Co., LTD, Tokyo) supplemented with 5% fetal bovine serum (Sanko Junyaku, Tokyo), 10% horse serum (Sanko Junyaku, Tokyo), 100 μg/ml streptomycin (GIBCO BRL, USA), and 100 units/ml penicillin (GIBCO BRL, USA) in tissue culture flasks at 37 °C humidified 5% $CO_2$ incubator. For the neurite outgrowth assay, the PC12 cells were washed and plated at $2 \times 10^4$ cells /cm$^2$ in serum free RPMI 1640 supplemented 5 μg/ml insulin, 10 μg/ml iron-free human transferrin, 25 μM ethanolamine and 25 nM selenite (RD-1; Kyokuto Pharmaceutical Kogyo Co., Tokyo, Japan) onto fresh collagen-coated dishes (IWAKI GLASS Co., Tokyo, Japan) with various glycolipids and/or 40 ng/ml of NGF.

### 2.4. Determination of neurite differentiation of PC12 cell

We classfied the cells morphologically as to the number of neurites by the methods of Saito *et al*. (1988).

## 3. Results

### 3.1. Effect of glycolipids on neurite outgrowth

We tested several glycolipids including MEL-A, MEL-B, Polyol lipids, Rhamnolipid, SL, STL-1 and STL-3, and found that MEL-A, MEL-B and SL induced the neurite initiation during 48 h treatment with low concentration of glycolipids. The potentiations of MEL-A, MEL-B and SL were with a maximal effect of this induction of MEL-A was observed at the concentrations of 5.0 μM, 5.0 μM and 6.3 μM, respectively. A maximal response was observed at 48 h after treatment with MEL-A, with about 30% of the PC12 cells extended the neurites. We also examined the glycolipids on the neurite outgrowth of PC12 cells in the presence of NGF, and found that MEL-A, MEL-B, PL, SL stimulated the action of NGF to produce the extension of neurite. The costimulation of MEL-A and NGF of PC12 cells resulted the most additional enhancement of the number of cells extended the neurites (Fig. 1).

### 3.2. MEL-A could induce neurite outgrowth after treatment with anti- low affinity $p75^{NGFR}$ antibody

After treatment with anti- low affinity p75NGFR antibody, NGF or MEL-A was treated on PC12 cells. The numbers of cells with neurites were decreased in NGF treated cells (38% to 8%), while slightly decreased in MEL-A treatment (28% to 25%). Treatment with antibody alone, there was hardly effective on neurite outgrowth (Fig. 2). From these results, MEL-A could induced neurite outgrowth after treated PC12 cells with anti- low affinity $p75^{NGFR}$ antibody which obstructed NGF action.

## 4. Discussion

In the previous study, we have found that microbial extracellular glycolipids, Succinoyl trehalose lipid (STL-A) induced monocytoic differentiation while Mannosylerythritol lipid (MEL-A) induced granulocytic differentiation of human promyelocytic leukemia cell line HL60 (Isoda *et al*., 1997), and both STL and MEL at the concentration which induced differentiation of HL60 cells exhibited a significant decrease of protein kinase C activity (Isoda *et al*., 1997). Furthermore, we reported that

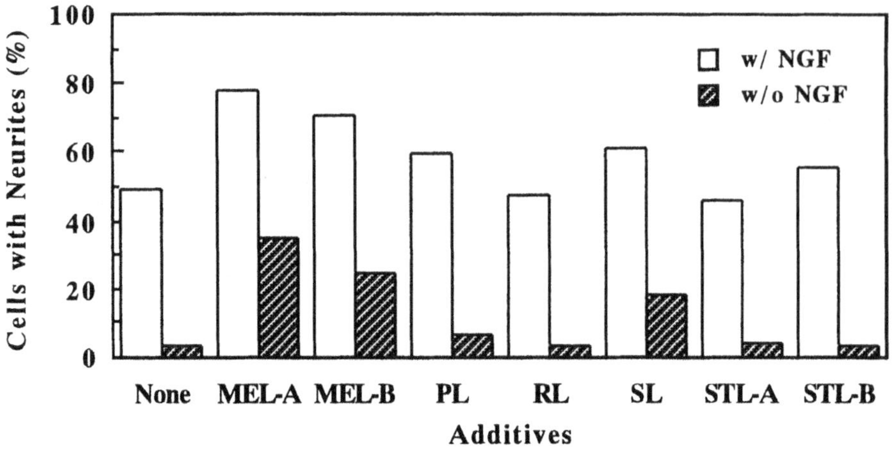


Fig.1. Effect of microbial extracellular glycolipids on neurite outgrowth. PC12 cells were treated for 48 h with 5.0 μM MEL-A, 5.0 μM MEL-B, 12.5 μg/ml PL, 25.0 μM RL, 6.3 μM SL, 15.0μM STL-A, 3.8 μM STL-B in the presence or absence of 40 ng/ml NGF.

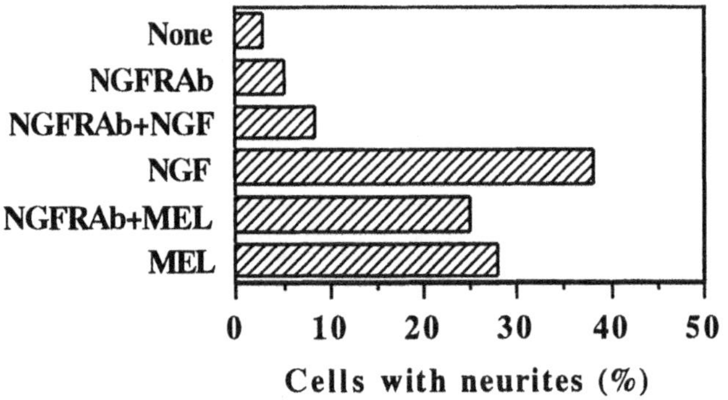


Fig.2. Effect of antibodies against p75NGFR on neurite extension of PC12 cells treated with NGF or MEL-A. Cells were treated with MEL-A or NGF after pretreatment of 10 μg/ml antibody against p75NGFR and scored the percentage of neurite-bearing cells.

Mannosylerythritol lipid induced granulocytic differentiation and inhibits the tyrosine phosphorylation of human myelogenous leukemia cell line K562 (Isoda *et al.*, 1997). In this paper, we showed the neurite-initiating effect of microbial extracellular glycolipids in PC12 cells.

Compared with neurites induced by other known molecules (NGF, bFGF, dBcAMP), MEL-A-induced neurites were unique with respect to morphology, and time course appearance and disappearance. The transitory neurite outgrowth induced by MEL-A suggests that MEL-A may play a role as a neurite initiation factor, whereas NGF is responsible for the stabilization and long-term maintenance of the differentiated state. MEL-A caused the growth of one or two long neurites per cell, and this restricted neurite number could not be caused by NGF, bFGF or dBcAMP. To clarify the physiological significance of fewer neurites, it is necessary to examine in more detail the morphological, biochemical and functional changes caused by MEL-A, and to investiagate the type of intracellular signalling that is involved after MEL-A inhibits the target protease. Since dBcAMP could not mimic the MEL-A response, it is expected that the first intracellular system stimulated by MEL-A does not use cAMP as a messenger. Tsuji *et al*. (1988) reported that addition of ganglioside (GQ1b) to neuroblastoma cells induced one long neurite, and that some protein kinase systems on the plasma membrane (ecto-type) may play an important role in GQ1b-induced neurite outgrowth. That is, there is a possiblity that MEL-A activated mechanisms included an ecto-type kinase and consequently make PC12 cells grow one or two long neuritres.

MEL-A is amphiphilic moiety, therefore MEL-A has both hydrophobic and hydrophilic moieties. It may be taken up by the all kinds of cell components such as cell membrane, cytoplasm, nuclear, *etc*. Further studies on cellular and molecular mechanisms of triggering of microbial extracellular glycolipid signal transduction, and comparison of a role of microbial extracellular glycolipids structure in induction of neuronal differentiation of PC12 cells are needed.

## 5. References

Isoda, H., Kitamoto, D., Shinmoto, H., Matsumura, M., and Nakahara, T., Microbial extracellular glycolipid induction of differentiation and inhibition of the protein kinase C activity of human promyelocytic leukemia cell line HL60. *Biosci. Biotech. Biochem.*, 71, 609-614 (1997).

Isoda, H., Shinmoto, H., Kitamoto, D., Matsumura, M. and Nakahara, T. (1997) Differentiation of Human Promyelocytic Leukemia Cell Line HL60 by Microbial Extracellular Glycolipids., *Lipids,* 32, 263-271.

Isoda H. and Nakahara T., (1997) Mannosylerythritol lipid induced granulocytic differentiation and inhibited the tyrosine phosphorylation of human myelogenous leukemia cell line K562., *Cytotechnology,* 25, 191-195.

Saito, Y. and Kawashima, S. (1989) The neurite-initiating effect of a tripeptide aldehyde protease inhibitor on PC12h cells. J. Biochem., 106, 1035-1040.

Tsuji, S., Yamashita, T., and Nagai, Y. (1988) A novel, Carbohydrate signal-mediated cell surface protein phosphorylation: ganglioside GQ1b stimulates ecto-protein kinase activity on the cell surface of a human neuroblastoma cell line, GOTO. *J. Biochem.*, 104, 498-503.

# INDUCTION OF BASOPHILIC AND EOSINOPHILIC DIFFERENTIATION IN THE HUMAN LEUKEMIC CELL LINE KU812

A. ICHIKAWA, Y. MOCHIZUKI, Y. KATAKURA, K. TERUYA, AND S. SHIRAHATA
*Graduate School of Genetic Resources Technology, Kyushu University, 6-10-1 Hakozaki, Higashi-ku, Fukuoka 812-8581, Japan*

**ABSTRACT.** We had demonstrated that the basophilic leukemia cell line KU812 can be induced to differentiate into basophil-like cells with the expression of high affinity Fcε-receptor (FcεRI) when cultured with hydrocortisone (HC). In this study, we report that sodium nitroprusside (SNP), an intracellular NO donor, also induces the expression of FcεRI on the cell surface. FcεRI expression was detected in about 20 % of KU812 cells 2 weeks after the addition of SNP as well as the cells treated with HC. It was suggested from reverse transcription-PCR (RT-PCR) analysis that the enhancement of FcεRI expression was due to the increase of FcεRI γ chain expression. The SNP treated KU812 cells expressed both eosinophil derived neurotoxin and eosinophil peroxidase. Moreover, it was observed that some important transcription factors, such as AP-1, NF-κB and NF-AT, were all activated in HC treated KU812 cells, but remarkably inactivated in SNP treated cells. These results strongly suggested that KU812 possess bi-directional cell differentiation ability depending upon extracellular factors and that intracellular NO is an important factor to decide the eosinophilic differentiation of KU812 cells. KU812 cells may be a good model for not only basophilic but also eosinophilic differentiation.

## 1. Introduction

Mast cell and basophil play central roles in allergic diseases. Cross-linking of cell surface high-affinity IgE receptors caused by the binding of IgE with polyvalent antigens results in degranulation of mast cells or basophils and the release of various inflammatory mediators that trigger allergic inflammation. Thus, activation of mast cells or basophils through FcεRI is necessary to initiate allergic reactions. However, no human basophil and mast cell line which possess physiological functions *in vitro* exists even today. Human leukemic cell line KU812 is known to be immature basophilic cell. However, to date, there have been no reports on inducibility of differentiation to mature basophils of this cell line. If mature basophils, mast cells or eosinophils are induced from KU812 cells, it will be useful to study allergic reaction and to analyze physiologically functional substances in allergy.

## 2. Materials and Methods

***Cells and cell culture.*** KU812 cells were maintained in RPMI-1640 medium supplemented with 10 % heat-inactivated fetal bovine serum, 2 mM L-glutamine, 10 mM HEPES buffer and antibiotics (100 U/ml penicillin G and 100 mg/ml streptomycin). The cells were passaged every 3-4 days.

***Detection of cell surface and intracellular antigens by flow cytometry.*** To detect cell surface FcεRI, cells (1 x $10^6$ cells) were incubated with mouse monoclonal antibody (mAb) specific for human FcεRI α chain (CRA-1), stained with FITC-conjugated goat anti-mouse immunoglobulins and analyzed by a flow cytometer. Mouse IgG were used as isotype-matched control antibodies. To detect intracellular antigens, we used a FIX & PERM Cell permeabilization kit (Caltag Labpratories, CA, USA). Cells (1 x $10^7$ cells) were incubated with Reagent A ( fixation medium) for 15 minutes at room temperature. After centrifugation and removal of supernatant, cells were incubated with Reagent B ( permeabilization medium) and mouse mAb specific to human FcεRI α chain or human tryptase for 15 minutes at room temperature. After washing with PBS, cells were incubated with FITC-conjugated goat anti-mouse immunoglobulins or FITC-conjugated goat anti-rabbit immunoglobulins for 30 minutes at 4 ℃ and analyzed by a flow cytometer.

***Induction of differentiation of KU812 cells.*** To test for basophilic differentiation, KU812 cells were cultured at 1 x $10^5$ cells/ml in 90 mm dishes. Each dish contained 10 ml of medium supplemented with 100 nM HC as described previously (3). After 7 days of culture, the cells were resuspended and samples were harvested for analysis. To test for eosinophilic differentiation, KU812 cells were cultured with 1 pM to 1 nM SNP. After 14 days of culture, the cells were resuspended and samples were harvested for analysis.

***Detection of mRNA by RT-PCR.*** Total RNAs were isolated from KU812 cells using TRIZOL reagent (GOBCO BRL, MD, USA). cDNA was synthesized from the total RNAs with oligo(dT) primers. Each cDNA was served as template for PCR amplification using specific primers. The primer pairs for FcεRI α chain, γ chain, tryptase, EDN and EPO were as follows: FcεRI α chain (sense) 5'-ATG AAG AAG ATG GCT CCT GC-3', (antisense) 5'-ATT GTG GAA CCA TTT GGT GG-3', FcεRI γ chain (sense) 5'-GAT GAT TCC AGC AGT GGT CTT GCT-3', (antisense) 5'-TAG GGC CAG CTG GTG TTA ATG GCA-3', tryptase (sense) 5'-AGC AAG TGG CCC TGG CAG GTG A-3', (antisense) 5'-AGA GGA AAT GGC GGT GGG AGG C-3', EDN (sense) 5'-CCA GCA CAT CAA TAT GAC CTC C-3', (antisense) 5'-GTG AAC TGG AAC CAC CGG ATA-3', EPO (sense) 5'-CTG CTG GAT GCT GCC TAC AAT T-3', (antisense) 5'CAA GAG GGA GAA AGC CAT T-3'.

***Nuclear extract preparation and electrophoretic mobility shift assays (EMSA).*** Cells cultured with HC or SNP for 4 or 8 days were harvested and pelleted in 15 ml of phosphate-buffered saline by centrifugation for 5 min at 1000 x g. Nuclear extracts were prepared according to Dignam *et al.* (1) with minor modifications. Protein concentrations were measured according to the method described by Bradford (2) with a commercial reagent (Bio-Rad). The double stranded oligonucleotides containing the NF-κB, NF-AT and AP-1 consensus sequences were used in EMSA, and were end-labeled with [γ-$^{32}$P]dATP. Binding reactions (in total 10 μl) were performed by incubating 0.8 μg of the nuclear extract with the reaction buffer containing 10 mM Tris-HCl (pH 7.5), 50 mM KCl, 1 mM DTT, 1 mM EDTA, 12.5 % glycerol, 0.1 % Triton X-100, 250 μg/ml BSA, 50 μg/ml poly(dI-dC)·poly(dI-dC) and [$^{32}$P]-labeled double-stranded oligonucleotide containing the NF-κB, NF-AT or AP-1 consensus sequences in the presence or absence of a competitor for 15 min at room temperature. The resulting DNA protein complexes were analyzed using 5 % native polyacrylamide gel electrophoresis. Bands were visualized by autoradiography.

## 3. Results and Discussion

We examined the expression of FcεRI on untreated KU812 cells by flow cytometry. Cell surface expression of FcεRI was very low (Fig. 1). However, FcεRI accumulated in the cytoplasm of KU812 (data not shown). These results suggested that additional signals require to induce the expression of FcεRI on the cell surface.

In our previous study, we found hydrocortisone (HC), one of the steroid hormones, could induce the expression of FcεRI on the cell surface (3). In this study, we report that sodium nitropruside (SNP), intracellular NO donor, also induced the expression of FcεRI on the cell surface. After several days culture of KU812 cells in the presence of various concentrations of HC or SNP, the cells were analyzed the expression of FcεRI on the cell surface. KU812 cells cultured with 100 nM of HC for 7 days expressed FcεRI on their surface (Fig. 1). Similarly, KU812 cells treated with 10 pM SNP for 14 days also expressed FcεRI on their surface (Fig. 1).

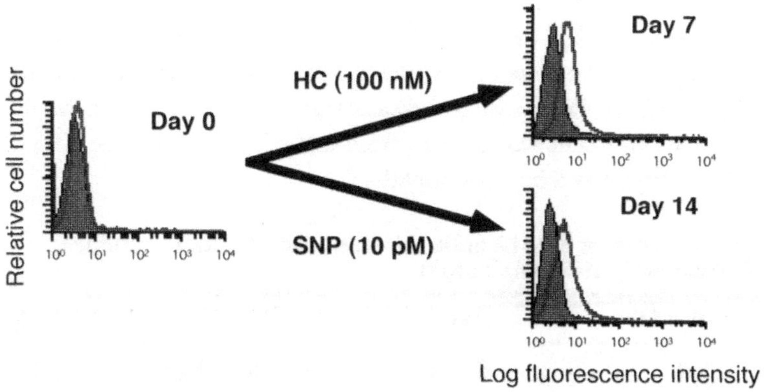


**Fig. 1 Induction of FcεRI expression on KU812 cells by hydrocortisone (HC) and sodium nitropruside (SNP)**

Since the γ and/or β chain is required for the expression of the α chain, it was likely that KU812 treated with HC and SNP expressing FcεRI α chain possess γ and/or β chain. To clarify this point, we measured the mRNA level of α and γ subunit by RT-PCR. α

Chain mRNA was detected in KU812 cells treated with or without HC and SNP. Whereas a significant increase of γ chain mRNA was observed when HC and SNP were added to the culture (Fig. 2). These results demonstrate that the lack of FcεRI expression was due to deficient γ chain transcription and that HC or SNP up-regulated FcεRI γ chain expression at the gene transcription level.

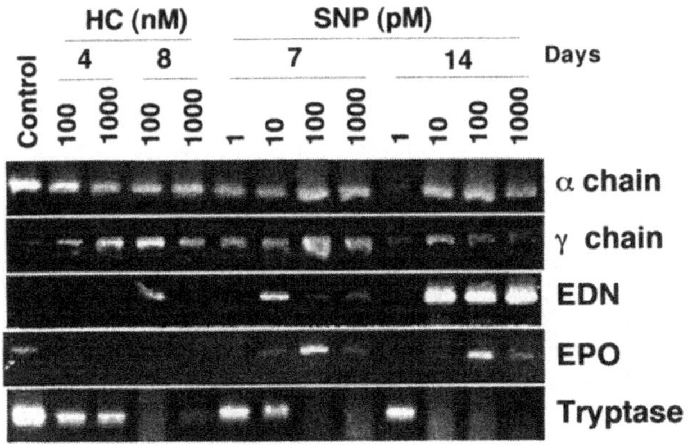


**Fig. 2 RT-PCR analysis for specific gene expression in HC or SNP treated KU812 cells**

Hydrocortisone and SNP have brought about the same change with respect to enhancement of FcεRI expression on the cell surface. However, enhancement of FcεRI expression not always means differentiation into the same cell types. In order to examine the morphological changes of KU812 cells by the treatment with SNP, we performed the Wright-Giemsa staining. KU812 cells cultured with HC showed no apparent changes in morphology. However, SNP induced the differentiation of KU812 cells to eosinophil-like cells in 4 to 9 days (Table 1). These results suggested that the effect of SNP on the induction of differentiation in KU812 cells was different from that of HC and that the KU812 cells cultured with SNP differentiated into eosinophils.

**Table 1 Counts of KU812 cells differentiated to eosinophils by SNP on Wright-Giemsa stain**

| SNP (nM) | Day 4 | Day 9 |
|---|---|---|
| 0.1 | 42.9 % | 44.1 % |
| 1 | 37.8 % | 16.2 % |
| 10 | 13.1 % | 7.9 % |

To further characterize the eosinophilic differentiation on KU812 cells by SNP, we analyzed the expression of eosinophil specific granule proteins, such as eosinophil derived neurotoxin (EDN) and eosinophil peroxidase (EPO). Gene expressions of EDN, EPO and tryptase, which is mast cell marker protein, were detected by RT-PCR using specific primers. Gene expression of tryptase decreased in the KU812 cells treated with SNP and HC depending on the culture periods. Gene expressions of EDN and EPO were detected in the KU812 cells cultured with SNP, but not with HC (Fig. 2). Decrease of tryptase gene expression indirectly suggested that KU812 cells cultured with HC and SNP were induced into basophilic and eosinophilic differentiations, respectively. These results are consistent with the data obtained by Wright-Giemsa staining analysis.

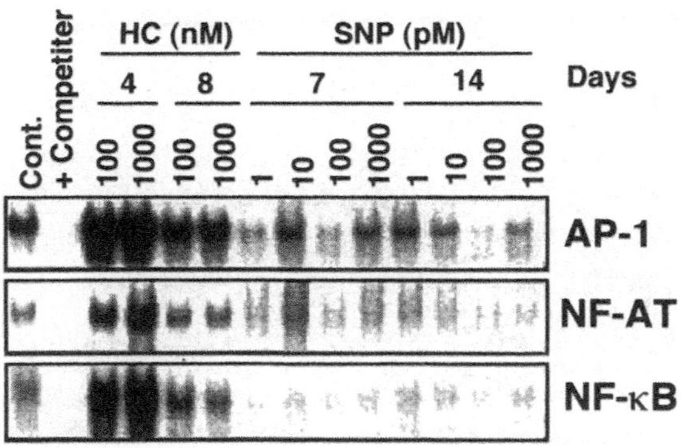


**Fig. 3 Electrophoretic mobility-shift assay for transcription factor-binding activities**

We expected that the bi-directional cell differentiation ability depending on extracellular factors, such as HC and SNP, affected to the intracellular signal transduction pathways. Then, we analyzed the activation of some transcription factors concerning with activation of immune cells by the electrophoretic mobility sift assay. It was observed that some important transcription factors, such as AP-1, NF-AT and NF-κB, were all activated in HC treated KU812 cells, but remarkably inactivated in SNP treated cells (Fig. 3). These results suggested that the different signals introduced into the nuclei caused different genes expression in the KU812 cells cultured with HC and SNP.

## 4. References

1. Dignam, D., Lebovitz, R.M. and Roeder, R.G. (1983) Accurate transcription by RNA polymerase II in a soluble extract from isolated mammalian nuclei, *Nucleic Acids Res.* **11**, 1475-1489.
2. Bradford, M.M. (1976) A rapid and sensitive method for the quantitation of microgram quantities of protein utilizing the principle of protein-dye binding, *Anal. Biochem.* **72**, 248-254.
3. Hara, T., Tachibana, H. and Shirahata, S. (1997) Induction of human leukemia cell line to differentiate into FcεRI expressing cells, *Animal Cell Technology: Basic & Applied Aspects*, **8**, 565-570.

# MANNOSYLERYTHRITOL LIPID INDUCED APOPTOSIS AND DIFFERENTIATION OF MELANOMA B16 CELLS

X ZHAO, T SUDO, Y WAKAMATSU, M SHIBAHARA and T NAKAHARA
*Institute of Applied Biochemistry, University of Tsukuba, Ibaraki 305-0006, Japan*
T MURATA and K YOKOYAMA
*Tsukuba Life Science Center, The Institute of Physical and Chemical Research, Tsukuba, Ibaraki 305-0074, Japan*

**Abstract** We report here that an extracellular glycolipid, mannosylerythritol lipid (MEL), from yeast inhibited the growth of mouse melanoma B16 cells markedly in a dose-dependent manner. Exposure of B16 cells to MEL at 10 μM and higher concentrations caused the condensation of chromatin, a hallmark of cells that are undergoing apoptosis. Moreover, exposure of MEL stimulated the expression of markers of the differentiation of melanoma cells such as tyrosinase activity and the enhanced production of melanin, an indication that MEL triggered both apoptotic and cell-differentiation programs. Forced expression of Bcl-2 protein in stably transformed B16 cells had a dual effect: it interfered with MEL-induced apoptosis but increased both tyrosinase activity and the production of melanin as compared with these phenomena in vector-transfected MEL-treated control B16 cells. These results provide the first evidence that growth arrest, apoptosis and the differentiation of mouse malignant melanoma cells can be induced by a microbial extracellular glycolipid.

## Introduction

Mannosylerythritol lipid (MEL), a microbial extracellular glycolipid, is produced by *Candida antarctica* T-34(1). We have recently reported that the exposure of MEL could induce the differentiation of human leukemia cells HL60 (2). This observation prompts us the possible use of MEL for the therapeutic reagents for not only leukemia but other cancer cells resistant to chemotherapy. Apoptosis has become a focus of attention in studies of the biology of cancer cells and it has been proposed that the progression of a tumor might not only be a function of cell proliferation but might also be a product of the aberrant survival of cells that results from the inappropriate suppression of apoptosis (3). Various compounds can trigger apoptic events. To our knowledge, there are no reports of induction by microbial extracellular glycolipids of an apoptotic response in mammalian cells.

In this present study, we found that the treatment of malignant melanoma B16 cells with MEL resulted a dose-dependent inhibition of growth. MEL was also found to be a potent inducer of both apoptosis and the differentiation of B16 cells.

Overexpression of human Bcl-2 conferred resistance to MEL-induced apoptosis but further enhanced the expression of the differentiation-associated markers. These findings might provide the groundwork for the use of microbial extracellular glycolipids as novel therapeutic reagents in the treatment of melanoma.

## Materials and methods

***Preparation of MEL.*** MEL was prepared and purified essentially as described by Kitamoto *et al.* (1).

***Cell culture.*** Mouse melanoma B16 4A5 cells (referred as B16 cells), obtained from the RIKEN Cell Bank (Tsukuba, Ibaraki, Japan), were maintained in Dulbecco Modified Eagle's Medium (DMEM) supplemented with 10% fetal bovine serum (FBS) at 37 °C. For treatment with MEL, cells were cultured in serum-free DMEM-ITES medium (RD-1; Kyokuto Pharmaceutical Kogyo Co., Tokyo, Japan). MEL was dissolved in distilled water and administered to cells at the indicated concentrations. A cell-counting kit (WST-1; Dojin Laboratories, Kumamoto, Japan) was used to monitor numbers of viable cells.

***Detection of chromatin condensation.*** B16 cells, cultured in the presence or in the absence of MEL, were collected and washed with phosphate-buffered saline without $Mg^{2+}$ and $Ca^{2+}$ ions (PBS). For detection the condensation of chromatin, cells were fixed in formalin, stained with 100 µM Hoechst 33342 (Sigma Chemical Co., St Louis, MO) and examined with a fluorescence microscope (Olympus Inc., Tokyo, Japan).

***Transfection with the bcl-2 gene.*** B16 cells were cotransfected with pCAGGS vector (4) that included the full-length human *bcl-2* gene and pcDNA3, which conferred resistance to geneticin (G418; GIBCO BRL, Rockville, MD). Control cells were cotransfected with pcDNA3 and pCAGGS without the *bcl-2* gene. After selection with 500 µg/ml G418, expression of *bcl-2* was confirmed by the reverse transcription-polymerase chain reaction (RT-PCR; one-step RT-PCR kit; TOYOBO Co., Osaka, Japan) using total RNA as template and the following pair of primers, which are specific for human *bcl-2*: sense primer, 5'-ACGCTGGGAGAACGGGGTAC-3'; and antisense primer, 5'-GCGGCTGTATGGGGCGTGTG-3'.

***Melanin content and tyrosinase activity*** Melanin content of cells was determined by the method of Johnston (5). Tyrosinase activity was analyzed according to the method of Shoji (6) with slight modification. One unit of tyrosinase activity is defined as the activity which could increase 0.001 at $A_{280}$ nm per minute.

## Results

***MEL inhibits the growth of B16 melanoma cells in a dose- and time-dependent manner.*** We examined the effects of MEL on the proliferation and viability of melanoma B16 cells. We incubated B16 cells with MEL of concentrations from 2.5 to 10 µM and monitored changes in the number of viable cells during a 3-day incubation in the absence of serum (Fig. 1). At 2.5 µM, MEL had no significant inhibitory effects on cell proliferation. At 5 µM, MEL suppressed growth

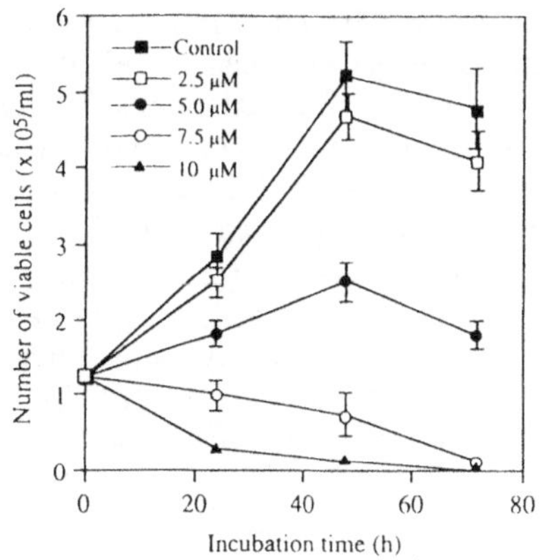


Fig.1 Effects of MEL on growth of B16 cells. The means and standard deviations (bars) of results of five independent tests are shown.

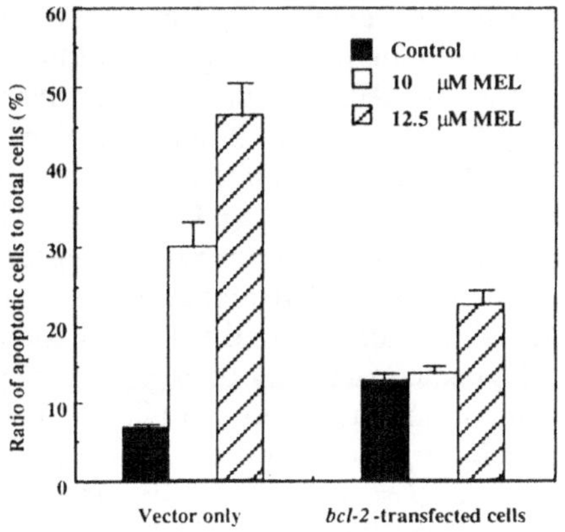


Fig.3 Enhanced expression of Bcl-2 interfered MEL-mediated apoptosis in B16 melanoma cells

significantly during exposure of cells to MEL for 48 to 72 h, with a slight effect on cell viability, as determined by the trypan blue exclusion test. At 10 μM, MEL completely blocked cell proliferation, with cytotoxicity being apparent after exposure to MEL. ***MEL induces apoptosis of B16 melanoma cells.*** We next examined whether the accumulation of dead cells that occurred in response to 10 μM MEL was a result of apoptotic cell death. we analysed the changes in the condensation of chromatin. Staining of cells with Hoechst 33342 revealed nuclear condensation and the fragmentation of cells after treatment with 10 μM MEL for 24 h, while untreated cells appeared normal (Fig. 2). DNA fragmentation and the sub-$G_1$ peak of histogram of flow cytometry, which are other hallmarks of cells that are undergoing apoptosis, were also be detected (data not shown). These results demonstrated that MEL acted as a potent trigger of apoptotic cell death.

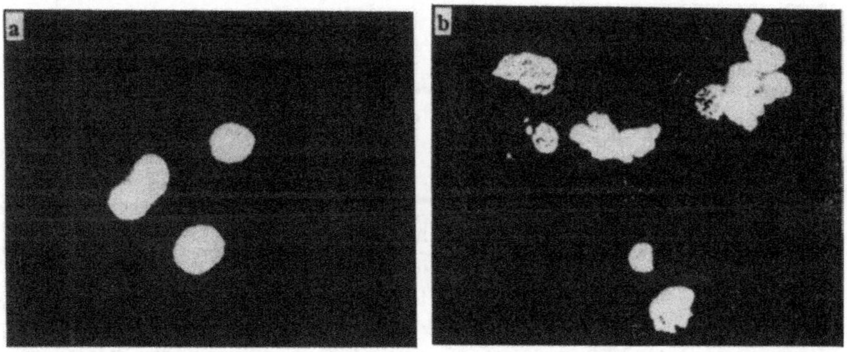


Fig.2 Induction of apoptosis of b16 cells by MEL. Photographsof Hoechst 33342- stained B16 cells after incubation without (a) and with (b) 10 μM MEL for 24 h.

***Ectopic expression of Bcl-2 prevents MEL-mediated apoptosis.*** In an attempt to gain some insight into the molecular pathways that lead to cell death in response to MEL, we induced the overexpression of Bcl-2 in B16 cells by stably transfecting them with full-length cDNA for human Bcl-2 and then challenged the cells with MEL. Ectopic expression of Bcl-2 effectively blocked MEL-induced apoptosis, as assessed the numbers of cells in the sub $G_0/G_1$ peak by flow cytometry (Fig. 3).

***MEL triggers not only apoptosis but also the differentiation of melanoma cells.*** Differentiation of melanoma cells towards melanocytes can be followed by monitoring the accumulation of melanin and increased tyrosinase activity (7). Tyrosinase activity of B16 cells that had been treated with 10 μM MEL for 24 h was about three times that in untreated cells (Fig. 4A). Under the same conditions, we detected a slight but significant increase in melanin content (Fig. 4B). A more profound effect was observed when MEL-induced differentiation was examined in cells that overexpressed Bcl-2 (Fig. 4A, B). Our results suggested that cells had escaped MEL-induced apoptosis as a result of the protective expression of Bcl-2 and that they had undergone further differentiation indicated that MEL can trigger both cell death and differentiation.

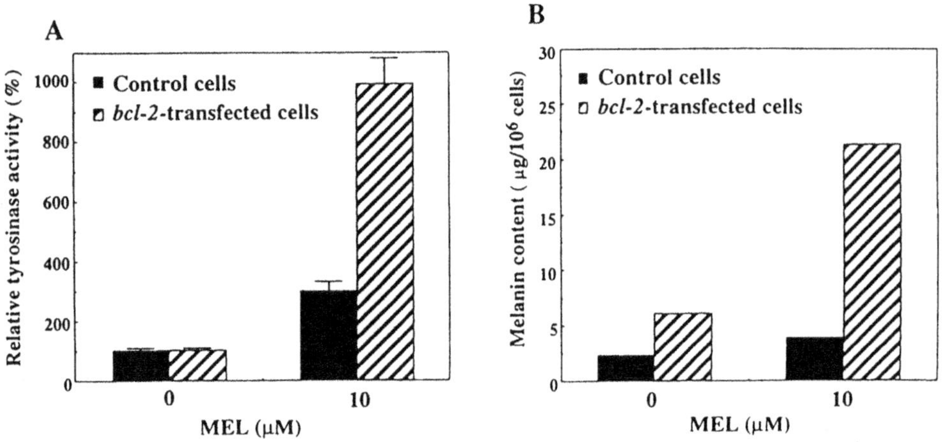


Fig.4 Analysis of Tyrosinase activities (A) and melanin content (B) of B16 cells after treatment with MEL in control (vector-transfected) and *bcl-2*-transfected cells.

## Discussion

In this study, we found that MEL had a dose-dependent anti-proliferative effect on mouse B16 melanoma cells. The growth-suppressive effect of MEL on B16 cells was significantly greater at 5 μM MEL than that at 2.5 μM. The critical micelle concentration (CMC) of MEL in aqueous solution is 2.7 μM (8). Thus, the concentration at which MEL was effective in inhibiting the growth of B16 cells was above the CMC. MEL is a glycolipid-type biosurfactant, the mechanistic links between the formation of micelles of MEL on cell membranes and the observed biological effects are yet to be clarified.

This report is to our knowledge the first that demonstrated the induction of apoptosis in cancer cells by a microbial extracellular glycolipid. MEL at 10 μM stimulated tyrosinase activity and the production of melanin, which are markers of the differentiation of melanoma cells. Thus, MEL had various effects on B16 melanoma cells, inducing apoptosis, as well as differentiation.

By introducing a *bcl-2* expression plasmid into B16 cells, we found that the extent of MEL-induced apoptosis in melanoma cells was diminished by Bcl-2. It has been reported that Bcl-2 protects against multiple signals that lead to cell death suggesting that Bcl-2 regulates a common cell-death pathway and functions at a point where various signals converge (9). The mechanism(s) by which MEL acts to induce apoptosis in B16 cells remain to be elucidated.

In conclusion, microbial glycolipid MEL induced apoptosis in melanoma cells. Enhanced expression of Bcl-2 interfered with MEL-mediated apoptosis and stimulated the MEL-induced differentiation of B16 cells. Further studies are needed to elucidate the precise mechanism(s) of action of MEL in apoptosis and cell differentiation.

## References


1. Kitamoto, D., Akiba,S., Hioki, T ., and Tabuchi,T. (1990) Extracellular accumulation of mannosylerythritol lipids by a strain of Candida antarctica, *Agric. Biol. Chem* **54**, 31-36.
2. Isoda, H., Shinmoto, H., Kitamoto, D., Matsumura, A., and Nakahara, T. (1997) Differentiation of human promyelocytic leukemia cell line by HL60 by microbial extracellular glycolipids, *Lipids* **32**, 263-271.
3. Marx, J. (1993) Cell death studies yield cancer clues. *Science*, **259**, 760-761,
4. Niwa, H., Yamamura, K., and Miyazaki, J. (1991) Efficient selection for high-expression transfectants with a novel eukaryotic vector. *Gene*, **108**, 193-200.
5. Johnston, D., Orlow, S., Levy, E., and Bystryn, J, C. (1992) Induction of B16 melanoma melanogenesis by a serum-free synthetic medium, *Exp. Cell. Res* **201**, 91-98.
6. Shoji, T., Aida, K., Kobri, M., Shimoto, H., and Tsushida, T. (1995) Assays of the inhibitor of melanogenesis, *Tissue Culture* **21**, 293-297.
7. Niles, R. M., and Makarski, J. S. (1978) Control of melanogenesis in mouse melanoma cells of varying metastatic potential. *J. Natl. Cancer Inst.*, **61**, 523-526.
8. Kitamoto, D. Studies on the production of mannosylerythritol lipid as biosurfactants by *Candida antarctica*. Ph.D. thesis of The Univ. of Tsukuba, Japan, 1992.
9. Herrmann, J. L., Bruckheimer, E., and McDonnell, T. J. (1996) Cell death signal transduction and Bcl-2 function. *Biochem. Soc. Trans.*, **24**, 1059-1065.

## A non-radioactive *in vitro* bioassay for recombinant human IL-11

H. YOKOTA, M. KISHIMOTO, H. SAITO, T. SAKAI, S. YOKOTA, S. KOJIMA, Y. TANIGUCHI*, A. MOTOKI*, H. KANIWA, N. SAISHO.

*Yamanouchi Pharmaceutical Co., Ltd., *Toray Research Center, Inc.*

### Abstract

A non-radioactive, cell culture-based *in vitro* bioassay was developed to measure the biological activity of recombinant human interleukin-11 (rhIL-11). The bioassay measures induced proliferation of T-10 cells using the WST-1 colorimetric detection system. The precision of the bioassay was improved by randomization of the sample and standard positions in 96-well microplates. This randomization was accomplished by making two mirror-image microplates in which serially diluted sample and standard lines were interleaved. This method without using radioactive reagents such as $^{3}$H-thymidine is simple, sensitive and quantitative. Although the precision of a bioassay is greatly inferior to that of a physicochemical analytical method in general, the coefficient of variation for this bioassay was less than 10%. Consequently, the stability of rhIL-11 can be evaluated accurately with this bioassay method.

### 1. Introduction

IL-11 is a multifunctional cytokine which affects the proliferation, differentiation and maturation of various types of hematopoietic cells in synergy with other endogenous growth factors [1, 2]. It also has been shown to exert effects on osteoclastogenesis and neurogenesis. Recombinant rhIL-11 (*des*-Pro hIL-11 [3]) has been developed as a biotechnological protein drug for the prevention of thrombocytopenia. To evaluate quality of a biotechnological drug, a bioassay is the most critical method among many analytical methods available. However, a major concern with using a bioassay is its precision. Improvement of bioassay precision would be of great benefit to accurately evaluate quality and stability of biotechnological drugs. In this study, a simple and quantitative bioassay for rhIL-11 was developed and the precision of this bioassay was improved by randomization of the sample and standard positions in 96-well microplates.

## 2. Experimental

### 2. 1. Cells

The T-10 cell line was obtained from the Genetics Institute, which is subline of the IL-6-dependent murine plasmacytoma T-1165 cell line [4].

### 2. 2. Bioassay

Sample and reference materials were diluted to a concentration of 50 ng/mL and then serially diluted two-fold with the RPMI 1640 medium, supplemented with 10% fetal bovine serum, 1% L-glutamine, 1% penicillin, 1% streptomycin, and 0.1% $\beta$-mercaptoethanol, on a 96-well microplate. Following addition of the T-10 cell suspension containing $3.5 \times 10^6$ cells/mL, the microplate was incubated for 2 to 3 days at 37 ℃ in a humidified 5% $CO_2$ atmosphere. The WST-1 solution was then added, and incubation continued for an additional day. The optical density was measured using a microplate reader at a wavelength of 450 nm with a reference wavelength of 650 nm. The data were fit to a sigmoidal dose-response curve using a four parameter logistic function. The biological activity of a sample was then determined by comparing the mid-points of the dose-response curves of the sample and the standard.

## 3. Results and discussion

### 3. 1. Dose-response relationship

The proliferation of T-10 cells was dependent on the concentration of rhIL-11. A typical dose-response curve with this method is shown in Fig. 1.

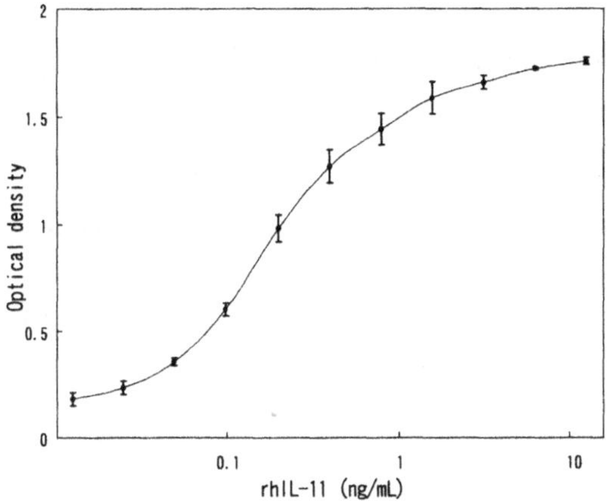


*Fig. 1.* Dose-response curve of T-10 cells for rhIL-11.

### 3. 2. Plate format

Two kinds of sample layout formats, shown in Fig. 2, were assessed. Two independent plates were used in each format. In format A, sample and standard lines were interleaved on 96-well microplates, and inverted in the two plates. In format B, the samples were allocated to the top half of each microplate, and standards were allocated to the bottom half.

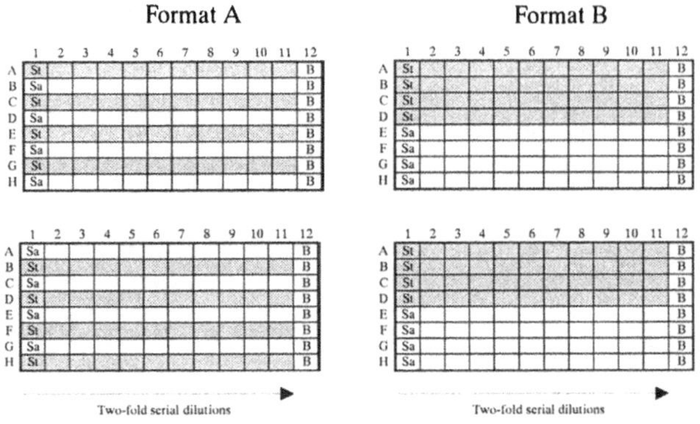


*Fig. 2.* Positions of samples and standards on 96-well microplates.
Abbreviations: Sa, samples; St, standards; B, blank.

## 3. 3. Assay precision

Analyst-to-analyst intermediate precision was examined using these two plate formats. The results are shown in Table 1.

Table 1 Analyst-to-analyst intermediate precision with Formats A and B

| Analyst | Format A ($\times 10^8$ units/mL) | Format B ($\times 10^8$ units/mL) |
|---|---|---|
| | 1.99 | 2.08 |
| A | 2.15 | 1.63 |
| | 1.81 | 2.20 |
| | 1.72 | 2.10 |
| B | 1.75 | 1.84 |
| | 1.69 | 1.74 |
| | 1.83 | 1.75 |
| C | 1.67 | 1.80 |
| | 1.68 | 1.51 |
| | 1.84 | 1.67 |
| D | 1.86 | 1.53 |
| | 1.79 | 1.61 |
| Average | 1.81 | 1.79 |
| RSD (%) | 7.7 | 12.8 |

The coefficients of variation were 7.7% in case of format A, but 12.8% in case of the format B. These results show that the positions of samples and standards on 96-well microplates affect the precision of this bioassay and that the precision with format A is superior to that with format B.

### 3. 4. Stability studies

To confirm this effect, both formats were used to evaluate the stability of rhIL-11. The results are shown in Fig. 3. These results show that format A is superior to format B in practice.

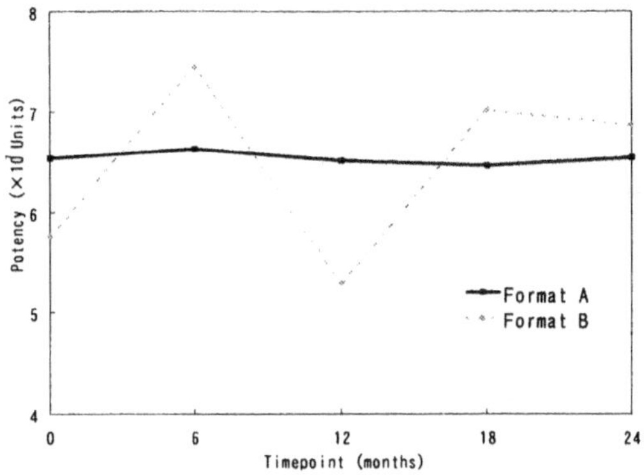


*Fig. 3.* Results of rhIL-11 stability studies with Formats A and B.

#### References

[1] Paul, S. R., Bennett, F., Calvetti, J. A., Kelleher, K., Wood, C. R., O'hara, R. M, Leary, A. C., Sibley, B., Clark, S. C., Williams, D. A., and Yang, Y. C. (1990) Molecular cloning of a cDNA encoding interleukin 11, a stromal cell-derived lymphopoietic and hematopoietic cytokine, *Proc. Natl. Acad. Sci.* **87**, 7512-7516.

[2] Du, X., and Williams D. A. (1997) Interleukin-11: Review of molecular, cell biology, and clinical use, *Blood* **89**, 3897-3908.

[3] Czupryn, M. J., McCoy, J. M., and Scoble, H. A. (1995) Structure-function relationships in human interleukin 11, *J. Biol. Chem.* **270**, 978-985.

[4] Nordan R. P., and Potter, M. (1986) A macrophage-derived factor required by plasmacytoma for survival and proliferation in vitro, *Science* **233**, 566-569.

# ESTABLISHMENT OF ASSAY SYSTEM FOR IMMUNOREGULATORY FACTORS USING WHOLE CELL CULTURE OF MOUSE SPLENO-CYTES

Mikako TAKASUGI, Yuki TAMURA*, Kyoko YAMADA*, Hirofumi TACHIBANA*, Michihiro SUGANO** and Koji YAMADA*
*Department of Human Nutrition, Faculty of Human Life Science, Yamaguchi Prefectural University, 3-2-1 Sakurabatake, Yamaguchi 753-8502 JAPAN*
**Laboratory of Food Science, Department of Food Science and Technology, Faculty of Agriculture, Kyushu University, 6-10-1 Hakozaki, Higashi-ku, Fukuoka 812-8581 JAPAN*
***Faculty of Human Life Science, Prefectural University of Kumamoto, 3-1-100 Tsukide, Kumamoto 862-8502 JAPAN*

**Key Words** Immunoglobulin Adhesive cells Immunoregulatory factor Milk protein mitogen

Abstract. We established the assay system for immunoregulatory factors using whole cell culture of mouse splenocytes to screen immunoregulatory factors and to clarify the mechanisms by which they regulate immunoglobulin (Ig) production. Adhesive cells such as macrophages and antigen-presenting cells enhanced Ig production of B lymphocytes in the presence of lipopolysaccharides, but the susceptibility to adhesive cells was different with Ig classes. In the absence of adhesive cells, mitogens, such as lipopolysaccharides, pokeweed mitogen and phytohemagglutinin stimulated Ig production, and adhesive cells enhanced the stimulatory effect of mitogens. The effect to mitogens and adhesive cells was different with Ig classes. Milk proteins, such as lactoferrin, β-lactoglobulin, α-casein and β-casein stimulated IgA production, and adhesive cells modified their IgA production stimulating activities. These results indicate that the assay system is useful for screening of immunoregulatory factors and clarification of the mechanisms by which they regulate Ig production of mouse splenocytes.

## 1. Introduction

Humoral immune responses is mediated by Ig produced by B lymphocytes and prevent from pathogen infections. *In vivo*, B lymphocytes interact with T lymphocytes, macrophages or other immunocompetent cells. Ig production by B lymphocytes is class specifically regulated by cytokines, such as IL-4, IL-5 and IFN-γ [1-3]. It was reported that mitogens, some bioproducts and food components regulate Ig production of rat lymphocytes [4, 5]. Such modification in Ig productivity of B lymphocytes may be affected by adhesive cells, such as macrophages or antigen-presenting cells, as well as T lymphocytes. To clarify the mechanisms by which they regulate Ig production, mouse or human cell culture systems are useful, because of the abundance of immunological reagents. To examine not only direct effect on B lymphocytes but also indirect effect on interaction among immunocompetent cells, we established the assay system for immunoregulatory factors using whole cell culture of mouse splenocytes.

## 2. Materials and Method

### 2.1. MATERIALS

Concanavalin A (Con A) and pokeweed mitogen (PWM) were obtained from EY Lab. (San mateo, CA), lipopolysaccharides (LPS) from Difco Lab. (Detroit, MI), phytohemagglutinin (PHA) from Vector Lab. (Burlingame, CA). β-Lactoglobulin (β-LG), α-, β-, κ-casein (CA) and transferrin (TF) were purchased from Sigma Chemical Co. (St. Louis, MO) and lactoferrin (LF) from Wako Pure Chemicals (Osaka, Japan). These compounds were dissolved in a phosphate buffered saline (pH 7.4) and used for cell culture experiments. All other reagents used were analytical grade.

### 2.2 CELLS AND CELL CULTURE

Mouse spleen was excised from male 8-9 weeks old BALB/c mice under diethyl ether anesthesia and splenocytes were squeezed out into RPMI1640 medium. The cells were incubated at 37°C for 0 to 6 hr to remove adherent cells and then the cells suspension was recovered. The non-adhesive cells were rinsed three times with medium and cultured in the RPMI1640 medium supplemented with 10% fetal bovine serum (FBS) in humidified atmosphere containing 10% $CO_2$ at 37°C. After cultivating the cells in the presence or absence of various mitogens or milk proteins, culture supernatants were isolated by a centrifugation at 400 x *g* for 5 min and Ig content in the solution was determined by enzyme-linked immunosorbent assay.

## 3. Results and Discussion

### 3.1 EFFECT OF ADHESIVE CELLS ON IMMUNOGLOBULIN PRODUCTION BY MOUSE SPLENOCYTES

Mouse splenocytes were cultured for various periods with 25 μg/ml of LPS with or without 30 min incubation to remove adhesive cells (Fig.1). In the absence of adhesive cells, IgA and IgM concentrations in the culture medium of splenocytes were very low and slightly increased during the 10-days cultivation. On the other hand, IgG concentration was increased markedly at day 7 after a 5 days lag period and reached plateau level. Adhesive cells enhanced the production of these Igs, though the susceptibility was different with Ig classes.

To examine the effect of adhesive cells on Ig production, splenocytes were incubated at 37°C for various periods to remove adhesive cells, and cultured for 7 days in the presence of

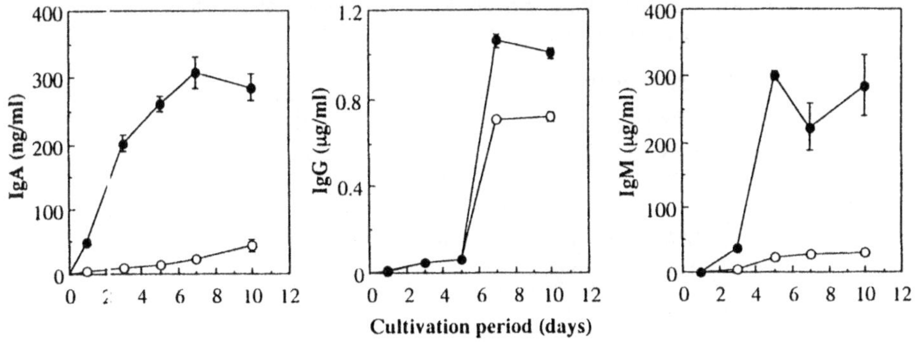


**Fig.1. Effect of adhesive cells on immunoglobulin production by mouse splenocytes**
Open circle; without adhesive cells, closed circle; with adhesive cells.
Values are means ± SE (n=3 or 4).

25 μg/ml of LPS. As shown in Table 1, IgA, IgG and IgM productivity decreased sharply during the first 2 hr adhesion period and was lost almost completely after 6 hr adhesion. Adhesion period to give half Ig productivity in the presence of adhesive cells was 1.23 hr in IgA, 0.58 hr in IgG and 0.35 hr in IgM, respectively. This suggests that the effect of adhesive cells is different with Ig classes. This may be due to the difference in the sensitivity of B cells to cytokines produced by adhesive cells, or the difference of sensitivity to signals from T cells which stimulated by adhesive cells. We used the 6-hr adhesion period for complete removal of adhesive cells, thereafter.

TABLE 1. Effect of adhesion period on immunoglobulin production by mouse splenocytes stimulated with lipopolysaccharides

| Ig | Adhesion period (hr) | | | | | |
|---|---|---|---|---|---|---|
| | 0 | 0.5 | 1 | 2 | 4 | 6 |
| IgA | 100 ± 9.5 | 80.8 ± 6.5 | 57.3 ± 6.4 | 22.1 ± 2.4 | 11.3 ± 2.2 | 4.8 ± 0.5 |
| IgG | 100 ± 3.9 | 54.3 ± 3.4 | 16.4 ± 1.2 | 8.6 ± 4.3 | 3.6 ± 0.3 | 0.1 ± 0.0 |
| IgM | 100 ± 10.0 | 30.3 ± 0.8 | 19.3 ± 0.1 | 11.6 ± 1.0 | 5.6 ± 0.4 | 1.4 ± 0.2 |

Cells were cultured for 7 days in the presence of 25 μg/ml of LPS.
Values are means ± SE (n=3 or 4).

## 3.2 EFFECT OF MITOGENS, MILK PROTEINS AND ADHESIVE CELLS ON IMMUNOGLOBULIN PRODUCTION BY MOUSE SPLENOCYTES

### 3.2.1. *Effect of Mitogens and Adhesive Cells on Immunoglobulin Production by Mouse Splenocytes*

To examine the effect of mitogens and adhesive cells, splenocytes were cultured in the presence or absence of 25 μg/ml of mitogens and adhesive cells for 7 days. Data are expressed as relative values to the level in the absence of both adhesive cells and mitogens (Table 2). In the absence of adhesive cells and mitogens, Ig concentration was 1.1 ng/ml in IgA, 5.0 ng/ml, in IgG, 460 ng/ml in IgM. LPS, PWM and PHA stimulated Ig production in the absence of adhesive cells, and adhesive cells enhanced the stimulatory effect of

TABLE 2. Effect of mitogens and adhesive cells on immunoglobulin production by mouse splenocytes

Without Adhesive Cells

| Mitogens | IgA | IgG | IgM |
|---|---|---|---|
| None | 1.0 ± 0.5 | 1.0 ± 0.2 | 1.0 ± 0.1 |
| LPS | 25 ± 4 | 460 ± 20 | 340 ± 2 |
| Con A | not detected | 0.2 ± 0.0 | 1.2 ± 0.0 |
| PWM | 16 ± 7 | 190 ± 36 | 97 ± 3 |
| PHA | 39 ± 3 | 50 ± 6 | 25 ± 2 |

With Adhesive Cells

| Mitogens | IgA | IgG | IgM |
|---|---|---|---|
| None | 5.5 ± 0.3 | 2.0 ± 0.0 | 5.0 ± 0.6 |
| LPS | 40 ± 16 | 1800 ± 130 | 890 ± 31 |
| Con A | 2.8 ± 0.5 | 0.8 ± 0.0 | 2.0 ± 0.1 |
| PWM | 43 ± 3 | 650 ± 13 | 560 ± 16 |
| PHA | 50 ± 13 | 58 ± 1 | 58 ± 1 |

Data are means ± SE (n=3 or 4). LPS; lipopolysaccharides, Con A; concanavalin A, PWM; pokeweed mitogen, PHA; phytohemagglutinin.

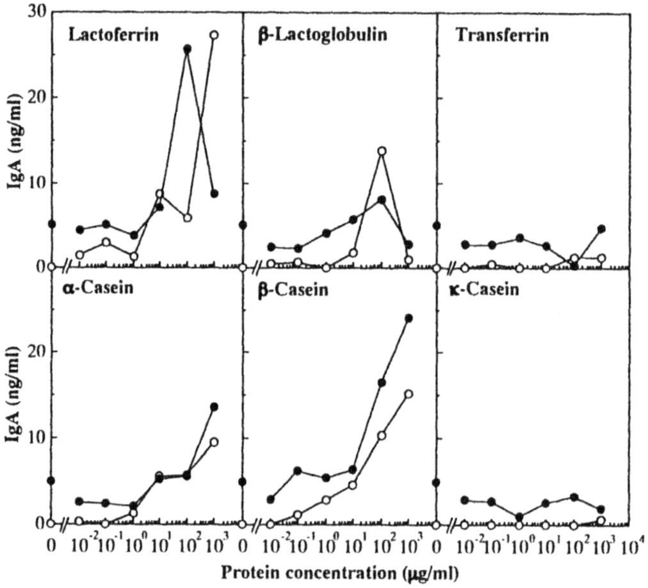


**Fig.2. Effect of milk proteins on IgA production by mouse splenocytes**
Cells were cultured for 7 days in the presence of various concentrations of milk proteins. Open circle; without adhesive cells, closed circle; with adhesive cells.

mitogens. But the susceptibility to mitogens and adhesive cells was different with Ig classes.

3.2.2 *Effect of Milk Proteins on IgA Production by Mouse Splenocytes*

Using the assay system we examined IgA production stimulating activity of milk proteins and TF, an iron-binding protein like LF. Splenocytes were cultured in the presence of adhesive cells, and various concentrations of milk proteins were added to the medium. As shown in Fig. 2, 100 μg/ml of LF, β-LG, α-CA and β-CA stimulated IgA production and adhesive cells modified their IgA stimulating activity. TF and κ-CA had little effect on the production of IgA. Yamada *et al.* found that IgM production by human-human hybridoma was stimulated by LF and CA but not by β-LG [7]. β-LG has been reported to stimulate specific Ig production of B lymphocytes *in vitro* and *vivo* [8, 9], but β-LG-specific IgA was not detected in this study. These results suggest that these milk proteins enhance IgA production with a non-specific manner. Such enhancement of IgA production may lead to the activation of the gut immune system, the first defensive line against the invasion of allergens or pathogens.

In conclusion the assay system is useful for screening of immunoregulatory factors, and for clarification of the mechanism which regulate Ig production in mouse splenocytes.

## 4. References


1. Ochel, M., Vohr, H.W., Pfeiffer, C., and Gleichmann, E. (1991) IL-4 is required for the IgE and IgG1 increase and IgG1 autoantibody formation in mice treated with mercuric chloride, *J. Immunol.* **146**, 3006-3011.
2. Pene, J., Rousset, F., Briere, F., Chretien, I., Wideman, J., Bonnefoy, J.Y., and De, Vries, J.E. (1988) Interleukin 5 enhances interleukin 4-induced IgE production by normal human B cells. The role of soluble CD23 antigen, *Eur. J. Immunol.* **18**, 929-935.
3. Rousset, F., Robert, J., Andary, M., Bonnin, J.P., Souillet, G., Chretien, I., Briere, F., Pene, J., and De, Vries, J.E. (1991) Shifts in interleukin-4 and interferon-γ production by T cells of patients with elevated serum IgE levels and the modulatory effects of these lymphokines on spontaneous IgE synthesis, *J. Allergy Clin. Immunol.* **87**, 58-69.
4. Kaku, S., Yamada, K., Hassan, N., Watanabe, T., and Sugano, M. (1997) Effect of vegetable extracts on immunoglobulin production by mesenteric lymph node lymphocytes of Sprague-Dawley rats, *Biosci. Biotech. Biochem.* **61**, 558-560.
5. Lim, B.O., Yamada, K., and Sugano, M. (1994) Effects of bile acids and lectins on immunoglobulin production in rat mesenteric lymph node lymphocytes, *In Vitro Cell. Dev. Biol.* **30A**, 407-413.
6. Yamada, K., Ikeda, I., Nakajima, H., Shirahata, S., and Murakami, H. (1991) Stimulation of proliferation and immunoglobulin production of human-human hybridoma by various types of caseins and their protease digests, *Cytotechnology* **5**, 279-285.
7. Fritsche, R., Pahud, J.J., Pecquet, S., and Pfeifer, A. (1997) Induction of systemic immunologic tolerance to β-lactoglobulin by oral administration of a whey protein hydrolysate, *J. Allergy Clin. Immunol.* **100**, 266-273.
8. Piastra, M., Stabile, A., Fioravanti, G., Castagnola, M., Pani, G., and Ria, F. (1994) Cord blood mononuclear cell responsiveness to β-lactoglobulin: T-cell activity in 'atopy-prone' and 'non-atopy-prone' newborns. *Int. Arch. Allergy Immunol.* **104**, 358-365.

# Author Index

# Subject Index

GPSR Compliance

*The European Union's (EU) General Product Safety Regulation (GPSR) is a set of rules that requires consumer products to be safe and our obligations to ensure this.*

*If you have any concerns about our products, you can contact us on ProductSafety@springernature.com*

In case Publisher is established outside the EU, the EU authorized representative is:

Springer Nature Customer Service Center GmbH
Europaplatz 3
69115 Heidelberg, Germany

**Batch number: 10809439**

Printed by Printforce, the Netherlands